SECOND THOUGHTS

FIFTH EDITION

To our mothers . . .
Mary A. Ruane, a much treasured and loved source of second thoughts and valuable knowledge
Lina T. Cerulo, who convinced the men in the family that girls deserved a college education too

SECOND THOUGHTS

Sociology Challenges Conventional Wisdom

FIFTH EDITION

Janet M. Ruane | Karen A. Cerulo
Montclair State University *Rutgers University*

Los Angeles | London | New Delhi
Singapore | Washington DC

Los Angeles | London | New Delhi
Singapore | Washington DC

FOR INFORMATION:

SAGE Publications, Inc.
2455 Teller Road
Thousand Oaks, California 91320
E-mail: order@sagepub.com

SAGE Publications Ltd.
1 Oliver's Yard
55 City Road
London EC1Y 1SP
United Kingdom

SAGE Publications India Pvt. Ltd.
B 1/I 1 Mohan Cooperative Industrial Area
Mathura Road, New Delhi 110 044
India

SAGE Publications Asia-Pacific Pte. Ltd.
33 Pekin Street #02-01
Far East Square
Singapore 048763

Acquisitions Editor: David Repetto
Editorial Assistant: Maggie Stanley
Production Editor: Eric Garner
Copy Editor: Kim Husband
Typesetter: C&M Digitals (P) Ltd.
Proofreader: Theresa Kay
Indexer: Molly Hall
Cover Designer: Anupama Krishnan
Marketing Manager: Erica DeLuca
Permissions Editor: Karen Ehrmann

Printed in the United States of America

Library of Congress Cataloging-in-Publication Data

Ruane, Janet M.

Second thoughts : sociology challenges conventional wisdom / Janet M. Ruane, Karen A. Cerulo. — 5th ed.

p. cm.
Includes bibliographical references and index.

ISBN 978-1-4129-8809-4 (pbk.)

1. Sociology. 2. Sociology—Quotations, maxims, etc.
3. Maxims. I. Cerulo, Karen A. II. Title.

HM585.R867 2012
301—dc22
2011011718

This book is printed on acid-free paper.

11 12 13 14 15 10 9 8 7 6 5 4 3 2

Contents

Preface

It is not uncommon for those assigned to teach entry-level sociology courses to experience some trepidation, even dread, about the teaching task ahead. In many ways, intro to sociology is a tough sell. Some students perceive the discipline as being nothing more than a rehash of the obvious; it speaks to everyday life, something many students believe they already know and understand. Other students confuse sociology courses with disciplines such as psychology or social work; they take our courses hoping to figure out the opposite sex, learn to better work the system, or overcome their personal problems with regard to deviant behavior or family relations. To muddy the waters even more, many intro students are likely to be in our courses not because of some desire to learn sociology but because the courses satisfy a general education requirement. Taking all of these factors into account, sociology instructors can face substantial resistance. Getting students to adjust their visions of the world so as to incorporate the sociological eye is no small feat.

Despite these challenges, it remains essential to achieve success in entry-level sociology courses. From an instrumental point of view, the discipline recruits future sociologists from these courses, thus mandating a sound foundation. Furthermore, departments may gain significant institutional resources by keeping intro course enrollments up. Intellectual concerns also contribute to the importance of entry-level courses. Many sociology instructors believe that intro courses offer a guaranteed dividend for the student: The sociological vision represents an essential tool for understanding and surviving our increasingly complex social world. Thus, intro courses provide instructors with a valuable opportunity to plant and nurture the sociological imagination in each new cohort of college students. Thought of in this way, failing the intro student can carry long-term social costs.

Second Thoughts offers a tried and true approach to successfully nurturing sociological thinking in the newcomer. The book provides a vehicle with which to initiate dialogue; it allows instructors to meet their students on common ground. Each chapter in this book begins with a shared idea—a conventional wisdom that both instructors and students have encountered by virtue of being consumers of popular culture. Once this common footing is established, *Second Thoughts* introduces relevant sociological concepts and theories that mesh with each conventional wisdom. Sociological ideas and perspective are used to explain, qualify, and sometimes debunk conventional wisdom.

Throughout the chapters, we provide tools for discussion in the form of "Your Thoughts" and "Second Thoughts About the News" boxes. At the conclusion of each chapter, we provide a vehicle by which students can apply their new sociological knowledge beyond the classroom. We have incorporated a set of exercises—many Internet based—linked to the subject matter covered in the chapter. The exercises, too, are grounded in the familiar. We encourage students to turn to everyday, common resources for some firsthand learning experiences.

Our own classroom experiences prove that the familiar is a user-friendly place to jump-start discussion, thus laying the foundation for critical thinking and informed analysis. In the classroom, we also have found the familiar to be a useful tool with which to delineate the sociological vision. This book attempts to pass along some of the fruits of our own learning. In pushing beyond the familiar, *Second Thoughts* also exposes students to the sociological advantage. At minimum, readers will accrue the benefits that come from taking time to give conventional ideas some important second thoughts.

Acknowledgments

There are, of course, a number of people who have contributed to this book. First and foremost, we would both like to acknowledge the students encountered in the many sociology courses taught at SUNY Stony Brook, Rutgers University, and Montclair State University. These students challenged us to make the sociological imagination a meaningful and desirable option on students' learning agendas. Thanks also go to our wonderful research assistant, Susan Kremmel, an extremely talented graduate student at Rutgers. Her inquisitiveness, tenacity, and professionalism brought great improvements to this edition of *Second Thoughts.* Thanks also go to our wonderfully supportive editor, Dave Repetto, and to the solid support staff at Pine Forge, including Maggie Stanley, Eric Garner, and Kim Husband. We would also like to thank several family members (Mary Agnes, Anne, and Mary Jane) for consistently asking about the book and thus encouraging us to stick with the program and to stay on schedule.

Introduction

The Sociological Perspective

In this introduction, we discuss the roots of conventional wisdom. We also contrast such knowledge with that acquired via the sociological perspective. In this way, we introduce students to a sociological mode of thinking.

Conventional wisdom is a part of our everyday lives. We are exposed to its lessons from early childhood, and we encounter its teachings until the day we die. Who among us was not taught, for example, to "be fearful of strangers" or that "beauty is only skin deep"? Similarly, we have all learned that "stress is bad for our well-being" and that "adult life is simply incomplete without children."

Conventional wisdom comes to us in many forms. We encounter it via folk adages, old wives' tales, traditions, and political or religious rhetoric. We find it in advice columns, cultural truisms, and the tenets of common sense. **Conventional wisdom** refers to that body of assertions and beliefs that is generally recognized as part of a culture's common knowledge. These cultural lessons are many, and they cannot be taken lightly. They are central to American society, and they are frequently the source of our beliefs, attitudes, and behaviors.

Conventional wisdom
Body of assertions and beliefs that is generally recognized as part of a culture's common knowledge

To be sure, conventional wisdom often contains elements of truth. As such, it constitutes a starting point for knowledge (Mathisen 1989; Ruane 2005). Consider, for example, the well-known truism "Actions speak louder than words." Many laboratory studies indeed have shown that those assessing an individual who says one thing but does another are influenced more strongly by the individual's actions (Amabile and Kabat 1982; Bryan and Walbek 1970; Ekman and Frank 1993; Ekman et al. 1991; Van Overwalle 1997). Similarly, a long line of research supports the adage that warns, "Marry in haste, repent at leisure." When we define *haste* as "marrying too young or marrying too quickly," we find that those who "marry in haste" report less satisfaction over the course of the marriage and experience higher divorce rates than those who make a later or a slower decision (Furstenburg 1976; Glenn and Supancic 1984; Grover, Russell, and Schumm 1985; Kitson et al. 1985; Martin and Bumpass 1989; National Marriage Project 2009).

Complete faith in conventional wisdom, however, can be risky. Social patterns and behaviors frequently contradict the various wisdoms we embrace. Many studies show, for instance, that adages encouraging the fear of strangers often are misguided; most crimes of personal violence are perpetrated by those we know (see Essay 13). Similarly, research documents that beauty may be merely "skin deep," but its importance cannot be underestimated. Physically attractive individuals fare better than those of more average appearance in almost all areas of social interaction (see Essay 8). Many studies suggest that stress is not always bad for one's well-being; it can sometimes be productive for human beings (see Essay 5). And despite all of the accolades to the presence of children in our lives, research shows that many adults report their highest levels of lifetime happiness take place *before* they have children or *after* their children leave home (see Essay 3).

Second Thoughts: Seeing Conventional Wisdom Through the Sociological Eye addresses the gaps that exist between conventional wisdom and social life. The book reviews several popular conventional wisdoms, noting the instances in which such adages cannot be taken at face value. Each of the following essays uses social research to expose the gray area that is too often ignored by the bottom-line nature of conventional wisdom. In so doing, *Second Thoughts* demonstrates that social reality is generally much more involved and complex than these cultural truisms imply. The book suggests that reviewing conventional wisdom with a sociological eye can lead to a more complete, detailed understanding of social life.

When Conventional Wisdom Isn't Enough

Although there may well be a kernel of truth to much of conventional wisdom, too often these adages present an incomplete picture. Why is this the case? The answer stems, in part, from the source of most conventional wisdom.

Much of the conventional wisdom we embrace originates from a particular individual's personal experiences, observations, or reflections. Often, such adages emerge from a highly specific circumstance; they are designed to address a particular need or event as experienced by a certain social group at a specific place or historical moment. For example, consider this well-known adage: "There's a sucker born every minute!" P. T. Barnum coined this now-familiar phrase. But recall Barnum's personal circumstance—he was one of the most famous circus masters in history. When one considers Barnum's unique history, both the source and the limits of his wisdom become clear.

Now consider this maxim: "Don't switch horses in midstream." Abraham Lincoln originated this quote. (His actual words were: "It is not best to swap horses when crossing streams.") But note that Lincoln's frequently cited advice actually represents the political rhetoric of a historical moment. Lincoln coined the phrase as a kind of campaign slogan when seeking reelection to the United States presidency.

Finally, consider the famous quotation "Good fences make good neighbors." Robert Frost forwarded this thought in his 1914 poem *Mending Wall.* Contrary to popular belief, however, Frost never intended to promote social separatism—quite the opposite. In *Mending Wall,* Frost criticized the character who uttered the adage, writing, "He will not go behind his father's saying"—in other words, the character will not break with tradition. In so doing, Frost suggested that the wisdom linking good fences to good neighbors was that of *another* generation in a *former* time; it was not wisdom for all time.

Each of these examples shares a common thread. In each case, conventional wisdom was born of a particular experience or a specific social situation. The wisdom took root and grew as it resonated with other people who faced similar events and circumstances. Yet each of these examples also illustrates an inherent weakness of conventional wisdom. The "truth" revealed by such wisdom is tied to the particular circumstances of every maxim's origin. This characteristic can make conventional wisdom a precarious source of generalized knowledge. Because such wisdom is individualistic or situation-specific information, it may not carry the general applications that most people assume of it.

For the sociologist, reliable knowledge mandates that we move beyond individualistic or circumstantial information. Sociologists contend that there is more to the story than any one person's life or the lives of one's associates can reveal.

Can one safely conclude that my experiences with an aging parent or my neighbors' experiences in raising their 4-year-old will provide others with sufficient knowledge for handling the events of their lives? It is difficult to say. If these experiences are atypical, the wisdom they provide will offer little in the way of general conclusions regarding the treatment of elderly parents or 4-year-olds. The wisdom will fail to transcend one individual's personal world. Similarly, wisdom born of experience may or may not transcend various social contexts. The maxim "Delay is the best remedy for anger" may prove fruitful in a variety of social sites: romance, work, friendship, parenting. Yet the adage that instructs you to "keep your cards close to the vest" may lead to success on the job but spell failure for a personal relationship.

Although your life may convince you that "birds of a feather flock together," my experience may reveal that "opposites attract." One situation may convince you that "haste makes waste," although another may convince you to "strike while the iron is hot." To be sure, experientially based or situation-specific information offers us knowledge, but that knowledge presents a fragmented, and thus incomplete, picture of the broader social world.

Relying on individualistic, circumstantial information can prove especially problematic when pursuing information regarding broad social patterns. Consider for a moment the ways in which one's geographic location might influence a person's estimate of general population patterns. The life experiences of Maine residents might lead them to conclude that 96% of the U.S. population is white. Such an estimate would greatly exaggerate the racial homogeneity of the nation. In contrast, the experiences of those living in our nation's capitol (Washington, D.C.) might lead them to estimate that only 40% of the U.S. population is white, a vast underestimation of population homogeneity. (The U.S. average for percentage of white Americans is 79.8.2%.) Based on experience, Californians would have no trouble believing that Hispanics are now the largest minority in the United States (Hispanics constitute 36.6% of California's population). Yet the experiences of those in Mississippi, Alabama, or Louisiana would identify blacks, not Hispanics, as the largest minority group. (Blacks constitute 37.2%, 26.4%, and 32% of Mississippi's, Alabama's, and Louisiana's populations while Hispanics constitute 2.2%, 2.9%, and 3.4% of Mississippi's, Alabama's, and Louisiana's populations, respectively.) On the basis of experience, residents of Alaska or New Jersey would never guess that the United States averages 80 inhabitants per square mile. Alaska averages one inhabitant per square mile, while New Jersey averages 1,134! And personal experience might leave residents of Kentucky or Mississippi baffled by Californians' or New Yorkers' concerns over the number of foreign-born individuals entering the nation and settling in their states. Only 1.4% of Mississippi's population and 2% of Kentucky's population are foreign born, whereas 26% of California's population and 20% of New York's population hail from other nations (U.S. Census Bureau 2010a).

The point we are trying to make here is really quite simple. Accurate knowledge about society requires us to move beyond the limitations of experientially based conventional wisdom. That leap represents one of the most compelling features of sociology. Sociologists are interested in social patterns. **Social patterns** are general trends that can be seen only when we force ourselves to stand back and look beyond any one, two, or three cases. In essence, sociologists search for the big picture: the view that emerges when many individual stories are aggregated into a whole.

Social patterns
General trends that can be seen when we look beyond any one or two cases

The sociologist's emphasis on patterns does not necessarily mean that she or he is never interested in personal stories and experiences. Rather, sociology's strength lies in its ability to place or situate individual stories in a social context. **Social context** refers to the broad social and historical circumstances surrounding an act or an event. Once the sociologist discovers the general trends within a particular group or society, she or he is in a better position to assess the relative meaning of any one individual's personal experiences. General patterns must be documented before we can assess one's personal experiences as typical or as exceptional.

Social context
The broad social and historical circumstances surrounding an act or an event

Obstacles to the Sociological Vision

Approached in this way, the task of sociology sounds straightforward and even appealing. The discipline encourages us to move beyond the personal and adopt a broader social vision—a vision that promises to improve the accuracy of our knowledge. With such gains at stake, why do so many approach the sociological vision with skepticism or confusion?

Certain obstacles can make it difficult to adopt a sociological view of the world. For example, the sociological vision contrasts with Americans' long-standing cultural value of individualism. A **cultural value** is a general sentiment regarding what is good or bad, right or wrong, desirable or undesirable. In the United States, we like to think of ourselves as special and unique individuals. We view ourselves as masters of our own fates. Thus, the notion that our behaviors follow patterns or are the product of social forces is at odds with an individualistic mentality. To illustrate this point, consider our typical reactions to a serious and growing problem in the United States today: **obesity**.

Cultural value
A general sentiment regarding what is good or bad, right or wrong

Obesity
A condition of having a body mass index of 30 or higher or being about 30 pounds overweight

Our individualistic mentality encourages us to see obesity as a problem of the person, an issue involving one's self-control or self-restraint: Thus, individuals are overweight because of their personal eating habits, their lack of exercise, their laziness, or their emotional baggage. Contrast such thinking with the sociological perspective. A sociological view of obesity encourages us to push beyond the individual; it forces us to look at obesity in light of broader social patterns and contexts. In analyzing obesity, for example, a sociologist would consider the following facts: 67% of adult Americans aged 20 and over are overweight or obese (about one-third of adults are overweight while a little more than one-third are obese—i.e., have a body mass index of 30 or higher; Centers for Disease Control and Prevention 2009a; Centers for Disease Control and Prevention 2010a, 2010b). The **poor** and **working poor** are more likely to be overweight than other economic groups. The prevalence of obesity is higher for blacks and for Hispanics (51% and 21% higher) than it is for whites (Centers for Disease Control and Prevention 2009b). Children are particularly vulnerable to the ills of excessive weight: Obese children are more likely to become obese adults (Centers for Disease Control and Prevention 2010b). And obesity is a common underlying factor in school bullying—it increases the risk of being the target of bullies by 63% (Lumeng et al. 2010). The

Poor
Those with incomes below the minimal level needed to meet basic needs

Working poor
Individuals who have spent at least 27 weeks in the labor force but have incomes below the poverty level

long-term costs of obesity are severe: As weight increases, so too does the risk of many health problems: heart disease, type 2 diabetes, cancers, stroke, and so on. There are economic costs associated with obesity: lost income due to decreased productivity and premature death. Finally, our obesity epidemic is a relatively recent phenomenon, largely a product of the last 40 years. The sociological perspective urges us to connect the dots between these facts. In doing so, we discover some important social patterns and social structural sources of the American obesity problem. For instance, when one adopts the sociological perspective, it becomes easier to see that changes in farm, trade, and economic policies during the past few decades have contributed to our national bulk problem. During the 1970s, secretary of agriculture Earl Butz pushed for significant changes in the production, pricing, importing, and exporting of grains and oils, which helped bring down the cost of food for consumers. In the same period, discoveries and innovations in food science and manufacturing led to major reductions in food production costs. But as we are now learning, these changes came on the back of high-fructose sweeteners with questionable health properties and unexpected metabolic costs. Further, some researchers argue that an expanding economy naturally produces expanding waistlines. Changes in our work and lifestyles make it easier and cheaper to be fat than thin. The economy also thrives on our obsession with weight as we spend billions of dollars each year trying to shed our extra pounds (Finkelstein and Zuckerman 2008). Adopting a sociological perspective also helps us discover that Americans' attitudes toward food have changed considerably. During the past three decades, our growing enamorment with snack foods, fast foods, and super-sized portions has put more and more between-meals and away-from-home calories into consumers' stomachs. And school budget cuts, changes in physical education curricula, and increasing hours of television viewing all mean that more and more of the calories we consume stay with us (Critser 2003). With the sociological vision, we can appreciate that our obesity problem is not just one of individual control. Rather, obesity is the product of larger developments in politics, food production, marketing, and lifestyles. Indeed, the Centers for Disease Control and Prevention refer to American society as "obesogenic."

Cultural values, such as those that champion individual control, are not the only obstacle to the sociological perspective. Adopting the sociological vision also can be hindered by our general preference for certain rather than probable answers. The study of large-scale patterns commits sociologists to predictions that are based on odds or probabilities. In other words, sociologists can identify outcomes that individuals from particular groups and places or in particular circumstances are *likely* to face. However, they cannot predict the definitive outcome for *any one* individual. In a culture that favors definitive answers for specific cases (usually, our own), this feature ensures a certain amount of resistance to the sociological approach. Indeed, sociology instructors often note a familiar complaint among newcomers to the field: If sociology can't predict what will happen to me, then what good is it?

Developing a sociological vision also can be undermined by the dynamic quality of social existence. Social reality is not static—it changes constantly. So just when we think we know the patterns, new patterns may be emerging. In addition, the very act of examining social phenomena can inevitably influence the entity we are studying. (If you need a concrete example of this, think about how sensitive the stock market is to people's ideas about the economy.) Such dynamics mean that sociologists' work, in a sense, is never done. Furthermore, the conclusions they reach must often remain tentative and open to change. Unlike a physics formula or a mathematical proof, sociological knowledge is rarely final. That dynamic quality often leaves the onlooker questioning its legitimacy.

Another obstacle facing the sociological viewpoint is doubt about the value of *socially* informed knowledge. As conventional wisdom indicates, many people trust only their own personal experiences to teach them about the world, arguing that such knowledge works for them. And, in a certain sense, sociologists must concede the point. Often, it does *appear* as if personal experience is more relevant, or truer, than sociological knowledge. Consider the fact that as thinking human beings, we have some capacity to create our own social reality. If we think people are not trustworthy, for example, we won't trust them, and we certainly won't give them the chance to prove us wrong. This course of action, no matter how ill conceived, serves to substantiate our own life experiences and to validate our own personally informed knowledge. In clinging to such a stance, we create a self-fulfilling prophecy. A **self-fulfilling prophecy** is a phenomenon whereby that which we believe to be true, in some sense, becomes true for us. In this way, self-fulfilling prophecies make personal experience seem like the clear victor over social knowledge.

Self-fulfilling prophecy
A phenomenon whereby that which we believe to be true, in some sense, becomes true for us

Your Thoughts . . .

In what areas of your own life are you more inclined to trust knowledge derived from your personal experiences over more socially informed knowledge? Why?

Finally, the sociological viewpoint often is ignored by those who believe they already possess sociological expertise. One of the earliest figures in American sociology, William Graham Sumner (1963), noted the tendency of people to think they know sociology by virtue of living in societies. "Being there" affords the opportunity to make social observations and, arguably, social observations are the ingredients of which sociology is made. Thus, "being there" mistakenly is deemed by many as sufficient for generating social knowledge and sociological insights. As our previous discussion indicates, however, personal experience is not the same as the sociological perspective.

If you consider all of these obstacles, you will better understand why the sociological vision is not more readily pursued or adopted by all. It requires effort to move beyond our personal views or experiences and develop what C. Wright Mills (1959) called the sociological imagination. **Sociological imagination** refers to the ability to see and evaluate the personal realm in light of the broader social/cultural and historical arenas.

Sociological imagination
The ability to see and evaluate the personal realm in light of the broader social/cultural and historical arenas

Why Read This Book?

By introducing this broader picture of reality, *Second Thoughts* encourages readers to step back and sharpen their analytic focus on the familiar. The essays that follow highlight the complex reality of modern-day society—a complexity often missed when we restrict our knowledge to personal experience and the common knowledge or popular assumptions born of those experiences.

Second Thoughts also introduces readers to many concepts central to sociology. In this way, the book can serve as an initiation into the ways in which sociologists frame the world around them. For those who find their sociological eye activated, we provide some of the tools needed for additional research. Each essay concludes with several suggested readings that elaborate on key concepts and ideas introduced in the essay. Furthermore, each essay includes several reliable sources from which facts and figures were derived.

Readers may also find that some of the information presented here moves them beyond curiosity and toward action. To assist such individuals, we close many chapters with the names and URLs of organizations where individuals might further pursue their interests. These listings are not meant as publicity for anybody or any cause. Rather, we offer them as preliminary leads, starting blocks for those who feel directed toward change.

In moving through the text, it will become clear that we have organized *Second Thoughts* according to topics typically covered in introductory-level courses. Those who wish to consider broader applications of this material should consult the "concepts covered" sections in the table of contents. These lists suggest a variety of issues for which one might use a specific conventional wisdom to jump-start critical thinking and discussion.

In Closing

When we open our eyes and carefully examine the world around us, we must concede that the realities of social life often run contrary to our stock of common knowledge. In the pages that follow, we aim to highlight some of these contradictions and, in so doing, to demonstrate that reviewing conventional wisdom with a sociological eye can provide a valuable correction to our vision of the world around us.

LEARNING MORE ABOUT IT

To learn more about developing a sociological vision, see Peter Berger's classic book *Invitation to Sociology* (New York: Anchor, 1963). C. Wright Mills also provides a brilliant theoretical treatise on this subject in *The Sociological Imagination* (London: Oxford, 1959). A more recent and very readable treatment of these issues is offered by Earl Babbie in *What Is Society? Reflections on Freedom, Order, and Change* (Thousand Oaks, CA: Pine Forge, 1994).

A compelling and humanistic introduction to sociology and its core concepts is offered in Lewis Coser's classic work *Sociology Through Literature: An Introductory Reader* (Englewood Cliffs, NJ: Prentice Hall, 1963).

In *The Fattening of America: How The Economy Makes Us Fat, If It Matters and What To Do About It* (New York: John Wiley & Sons, 2008), Erik Finkelstein and Laurie Zuckerman investigate the economic and technological underpinnings of America's current obesity problem.

GOING ONLINE FOR SOME SECOND THOUGHTS

Public Agenda Online is a nonpartisan organization that offers users a chance to access public opinion studies on major social issues. The site also provides educational materials on various policy issues. You will find *Public Agenda Online* at http://www.publicagenda.org/.

NOW IT'S YOUR TURN

1. Think about the social arrangements of your life—that is, your family relations, your community, your school, and work experiences. If your knowledge of the world were restricted to just these arenas, identify five important national facts that you would fail to know. (Entering your zip code on the "American FactFinder" page may prove helpful to you in completing this task: http://factfinder.census.gov/home/saff/main.html?_lang=en.)

2. The media give us one view of our social world. Select one week's worth of prime-time TV programs and use them to learn about U.S. society. Put together coding sheets that will allow you to collect basic data on all the program characters you encounter; that is, record each character's age, education, ethnicity, family status, family size, gender, occupational level, race, residence patterns, and so on. Tabulate summary statistics from your data. For example, determine the percentage of characters that are male and female, the average education level, and so on. Consult the "American Community Survey" link on the Census "American FactFinder" page (http://factfinder.census.gov/home/saff/main.html?_lang=en) and compare the data you obtain via TV with comparable real-life demographics for the U.S. population. What can you conclude about the media's picture of American society? How did your particular selections of prime-time programming bias or influence your data?

3. Visit the Census Bureau's State and County Quick Facts page at http://quickfacts.census.gov/qfd/states/23000.html and compile a list of additional facts that clearly show how state (or county) context affects our social vision.

4. Visit the Public Agenda Online site: http://www.publicagenda.org/. Spend some time reviewing the various issues of the day. Select a topic that needs some second thoughts and, as the semester progresses, prepare a brief presentation of facts that challenges our conventional thinking about your selected topic.

METHODS

ESSAY 1

Conventional Wisdom Tells Us . . . Numbers Don't Lie

Americans like to "run the numbers." No matter what the realm—sports, business, politics, or entertainment—numbers often provide the bottom line. Why is our faith in numbers so strong? Conventional wisdom tells us that numbers don't lie. Is such wisdom accurate? Can we be confident in the "realities" claimed by national polls, social scientific surveys, and other quantitative studies? In this essay, we note several important elements to consider in establishing the "truth" of numbers.

Numbers are everywhere. Indeed, in today's world, it is virtually impossible to avoid statistical calculations, data sets, measurements, and projections. We judge the quality of our athletes by their "numbers"—their hits, their home runs, their stolen bases. We use numbers to gauge our financial futures. The Dow is up; the NASDAQ falls. Such numbers drive our investment decisions. The percentages gathered in public opinion polls guide both policymakers and citizen voters in making critical political decisions, and numbers form the cornerstone of research models in fields as diverse as education, medicine, the social sciences, and quantum physics.

Numbers, data, and statistics: Why is our commitment to these entities so strong? Many believe that numbers provide us with precision and objectivity. With numbers, we count rather than guesstimate; we measure rather than suppose; we capture reality rather than assume it. Armed with an empirical blueprint of the day's occurrences, many believe that we move one step closer to seeing the way things *really* are. In essence, many believe that numbers provide us with truth (Babbie 2009; Best 2001, 2004, 2008).

Is this conventional wisdom about numbers accurate? Is it true that numbers don't lie? Can we be confident in the "realities" presented by national polls, social scientific surveys, and other quantitative studies?

To be sure, some of the numbers we see and read may indeed provide us with accurate pictures of the world. Yet the "truth" of numbers cannot be taken for granted. Before one can feel comfortable with the conclusions drawn from any body of data, one must consider several research-related factors. As we find ourselves increasingly bombarded by more and more numbers, we would do well to take a little extra time to review certain aspects of every study's design. By posing certain key questions in evaluating numerical findings, one can better gauge the veracity of any statistical claim.

In assessing the truth of numbers, one must first raise this question: *Exactly who do the numbers in question represent?* This concern focuses our attention on issues of sampling.

Social researchers generally wish to draw conclusions about groups of people or things. For example, a researcher may wish to learn something about all women in the United States or about first graders enrolled in New Jersey's public schools or about Asian Americans living in Los Angeles. Similarly, a researcher may be interested in the content of *Time* magazine covers published from 1980 to 2010; he or she may wish to study the characters appearing in second-grade readers published in 1950 versus those published in 2000. Such groups of interest constitute what researchers call a **population**. A population is a collection of individuals, institutions, events, or objects about which one wishes to generalize or describe.

Population
The totality of a collection of individuals, institutions, events, or objects about which one wishes to generalize

Sample
A portion or subset of the population of interest

Representative sample
Refers to a group or subset of elements that mirrors the characteristics of the larger population of interest

While social scientists are interested in populations, constraints such as time and money make it difficult, if not impossible, to study an entire population. Consequently, researchers often work with a **sample**, that is, a portion of the population. When selecting a sample, ideally researchers strive to select a **representative sample**. A representative sample refers to a group that mirrors the characteristics of the larger population of interest. With a representative sample, a researcher can make accurate inferences about a large population while working with a small, manageable group.

Representative samples are critical to the truthfulness of numbers. If researchers use nonrepresentative samples to make inferences or generalizations about a population, their conclusions may be misleading or erroneous. The following example helps illustrate the problem. Suppose that you wish to study dating habits among the students at your college. In particular, you are interested in the ways in which students meet potential partners. To research the issue, you choose a campus dorm and interview the first 20 people who leave the dorm one Thursday evening. Based on the answers offered by your 20 subjects, can you draw conclusions about the college population at large? Unfortunately, the answer is no. Because you took no steps to ensure that your sample was representative, your 20 subjects will provide only a limited picture of overall college dating practices. For example, if your subjects were members of a coed dorm, then their experiences could differ significantly from those of noncoed dorm residents. The fact that you are interviewing people *leaving* the dorm on a Thursday night may also prove significant. Those residents who choose to stay in for the night may exhibit very different dating patterns from those who leave the dorm. Similarly, the dating experiences of your 20 dorm residents may differ quite dramatically from the experiences of students commuting to your college from home or from off-campus residences. By drawing a sample of college students based on convenience, you failed to represent the general character of the entire student body systematically. Therefore, the data you collect will tell us something about the 20 individuals who live in a particular dorm. However, the numbers generated from these 20 interviews cannot provide a useful picture of overall student dating practices.

Attention to sampling represents a first step in determining the "truthfulness" of numbers.

However, if we wish to confirm the veracity of data, we must ask other questions as well. For example, *How were a researcher's numbers collected?* A researcher's data collection methods and measurement instruments will tell us much about the ultimate value of her or his research conclusions.

Consider, for example, the area of public opinion research. In today's world, there is scarcely a day when some dimension of public opinion is absent from the news. Public opinion reports typically result from a data collection method known as survey research. **Survey research** involves the administration of a carefully designed set of questions; the questions are posed during a face-to-face interview, during a telephone interview, or via a written or an online questionnaire.

Survey research
Data collection technique using a carefully designed set of questions

If survey researchers construct their questions thoughtfully and subjects answer those questions honestly, then surveys should provide us with an accurate picture of the world, right? Maybe, but maybe not. Questionnaire design presents social scientists with one of the most difficult and challenging tasks of research. Because of this, even the most experienced researchers can fall prey to design problems that may inadvertently influence the accuracy of their numbers.

For example, recent research convincingly shows that the ordering of survey questions can dramatically affect respondents' answers to the questions they are asked. In other words, exposure to one question on an interview schedule can influence a respondent's interpretations and responses to subsequent survey questions. Keeping this phenomenon in mind, imagine a survey designed to solicit likeability ratings for various public figures. In such a survey, a subject's rating for one individual—say, Hillary Clinton—will vary significantly depending on where and when Mrs. Clinton is presented for evaluation. Hillary Clinton may prove quite likable if she is rated immediately following other cabinet members such as Timothy Geithner or Eric Holder. However, she may not fare as well if rated immediately following other Secretaries of State such as Thomas Jefferson, John Foster Dulles, or Henry Kissinger. In essence, the "numbers" on Hillary Clinton are strongly tied to the instrument by which her likeability is gauged. In survey research, the series of individuals considered by a respondent can make an impact on a respondent's perception of any one person (Krosnick and Presser 2010; Meyers and Crull 1994; Tanur 1992; Wansink, Seymour, and Bradburn 2004).

Your Thoughts . . .

If you were designing a survey, what could you do to overcome the problems presented by the ordering of questions?

The ordering of survey questions can have other important effects as well. Methodology experts such as Earl Babbie note that the ordering of survey questions can sometimes alter a respondent's perception of current events. Such effects must be considered in evaluating the "truth" of numbers. Suppose, for example, that a survey researcher questions a respondent regarding the dangers of violent crime. After several questions on this topic, imagine that the researcher then asks the respondent a seemingly unrelated question: "What do you believe to be the single greatest threat to public stability?" Under these conditions, it is highly likely that the respondent will nominate violent crime more often than any other social problem. Why? The researcher's initial questions on violent crime can unintentionally focus the respondent on a specific set of concerns. By directing the respondent's attentions toward one particular subject, the researcher can inadvertently blind the respondent to other areas of consideration (Babbie 2009).

The ordering of questions represents one important element of data collection, but in judging the veracity of survey numbers, it is also critical to consider the specific questions that generated the numbers. All too often, critical information is lost in the reporting of survey results. For example, responses can be generalized in ways that misrepresent the questions posed in a survey. Similarly, answers to different questions may be reported and used to suggest changes in public attitudes. Thus, to correctly interpret the actual information that survey numbers provide, we must carefully trace survey data to their original source.

The following example helps illustrate the importance of tracing survey data to their source. In the early 1990s, the American Jewish Committee commissioned the Roper organization to survey Americans' attitudes on the Holocaust. One of the questions posed in the Roper survey read as follows: "Does it seem possible or does it seem impossible to you that the Nazi extermination of the Jews never happened?" Twenty-two percent of Roper's respondents answered that it was possible that the extermination never happened. (Roper was working with a representative sample of American adults.) Many Jewish leaders were stunned by the survey's results. The numbers suggested that approximately one in five Americans were terribly misinformed about the Holocaust. Researchers were surprised by these results as well. Hence, they decided to redo the survey. Again, researchers questioned a representative sample of American adults. This time, however, they posed the Holocaust question in a slightly different way. Respondents were asked, "Do you believe that the Holocaust: (a) definitely happened, (b) probably happened, (c) may have happened, (d) probably did not happen, (e) definitely did not happen?" In the second survey, only 2.9% of those questioned said that the Holocaust "definitely" or "probably" did not happen (Kifner 1994; Ruane 2005).

What happened here? Did Americans dramatically change their position on the Holocaust between the first and second surveys? On the surface, it may seem that way, but by tracing the data to their source, we can comfortably eliminate that possibility. The shift in Americans' attitudes toward the Holocaust is connected to the change in the researchers' survey questions. In the first survey, researchers posed a confusing question to respondents: "Does it seem possible or does it seem impossible to you that the Nazi extermination of the Jews never happened?" Consider the poor wording of this question. Choosing the seemingly positive response "possible" resulted in respondents expressing a negative view on the authenticity of the Holocaust. In essence, a poorly designed question resulted in a flawed measurement of attitudes. In the second survey, the Holocaust question was worded more clearly. Thus, the second survey produced very different results. Moreover, the clarity of the second survey question suggests that its results are a more accurate representation of American attitudes.

The Holocaust example highlights some other important benefits of tracing data to their source. In checking the source, we verify the quality of the data collection process. For example, examining a survey's original question allows us to assess the validity of a researcher's measurement tools. A **valid measure** is one that accurately captures or measures the concept or property of interest to the researcher. Examining a survey's original question also provides some sense of a measure's reliability. A **reliable measure** proves consistent and stable from one use

Valid measure
One that accurately captures or measures the concept or property of interest to the researcher

Reliable measure
A measure that yields consistent or stable results

to the next. The greater the validity and reliability of a researcher's measures, the greater the confidence we can have in the data generated.

Tracing data to their source represents an important step in determining the veracity of numbers. However, in reviewing numbers, it is equally important to consider the issues behind the questions. When we review a single study or when we compare the findings from two or more studies, it is important to ask, *What did the researcher hope to measure, and how did she or he operationalize that concept?* **Operationalization** refers to the way in which a researcher defines and measures the concept or variable of interest. Without a full understanding of a researcher's operationalizations, those reviewing a study's findings may misinterpret the researcher's intentions. Under such circumstances, a researcher's conclusions can be inadvertently applied to issues beyond the scope of the study. Similarly, when a researcher's operationalizations are misunderstood, projects addressing different concepts may be mistakenly compared, creating conflict and confusion.

Operationalization
The way in which a researcher defines and measures the concept or variable of interest

For example, in Essay 13, you will read about two important studies of criminal activity: the *FBI Uniform Crime Reports* and the *National Crime Victimization Survey.* Both studies represent highly reputable data analysis projects. Both studies address the annual incidence of crime in the United States. Yet each study presents completely different estimates of crime. For example, while the *Uniform Crime Reports* estimate approximately 454.5 violent crimes per every 100,000 Americans (U.S. Department of Justice 2008a), the *National Crime Victimization Survey* estimates 1,930 violent crimes per every 100,000 Americans—nearly four times the number recorded by the FBI (U.S. Department of Justice 2008b)!

Which report presents the "true" number of crimes? If we failed to note the ways in which each study operationalizes crime, we would probably conclude that one set of numbers is in error. But by examining each study's operationalizations, we learn that the numbers in both reports are credible. The *Uniform Crime Reports* are based on crimes reported to the police. Thus, to be counted in the FBI's statistics, a crime must be known to and officially recorded by some police agency. In contrast, estimates forwarded in the *National Crime Victimization Survey* are based on the self-reports of a nationally representative sample. Note that crimes self-reported by victims may be completely unknown to the police. Furthermore, a violent crime's classification in the *National Crime Victimization Survey* remains solely in the hands of the "victim." Thus, in some cases, a victim's report of a "crime" may not meet the normal standards of the law.

When we consider the very different ways in which these two studies measure violent crime, it becomes easy to understand the discrepancy between the two data sources. The FBI statistics capture "official crime" as reported by the police. In contrast, the *National Crime Victimization Survey* offers "unofficial crime" as reported by ordinary citizens. In essence, each study presents a different operationalization of crime. Each study provides us with different dimensions of violent crime in the United States.

Whom do research numbers represent? How did the researcher collect his or her numbers? How were concepts operationalized? All of these questions should be posed by those assessing the veracity of data. But a complete assessment of data requires one additional question as well: *Who is conducting the research?* Most researchers strive to remain value-free when conducting their work. A **value-free researcher** is one who keeps personal values and beliefs out of the collection and interpretation of data. In

Value-free researcher
One who keeps personal values and beliefs out of the collection and interpretation of data

some cases, certain ideologies or self-interests can color the nature of a project. For example, should one believe numbers that document cigarettes' effects on health if one learns that the study generating such numbers is funded by a major tobacco company? Similarly, can one feel comfortable with data that suggest racial differences in IQ if one knows that the researcher presenting such conclusions is an avowed White Supremacist? The motives and interests of those executing or sponsoring a study must be carefully considered before one can determine the "truth" behind the numbers.

It is important to note that even the best of researchers—scholars trying to maintain a value-free stance—can unintentionally allow certain biases to color their interpretations of data. Researchers are, after all, social beings. They carry with them certain cultural assumptions and understandings. When executing research, these assumptions and understandings can unknowingly influence that which falls within a researcher's "viewfinder." Anthropologists have been especially effective at uncovering situations in which a researcher's vision is unintentionally distorted. How does such a thing happen?

Imagine a researcher who is interested in studying modes of interpersonal, nonverbal communication. She or he observes and records such exchanges among both American and non-American **dyads** (a dyad is a group of two). Now suppose that in observing interactions among non-American dyads, the researcher notes several instances in which one member of the dyad sticks out her or his tongue at the other member. Within American culture, such a gesture generally suggests teasing and mockery. Some researchers would allow this American standard to guide their interpretation of these observations of non-Americans. But were the researcher to impose the "American" meaning on interpretations of *all* tongue-sticking incidents, she or he would forward data that were biased by ethnocentrism. **Ethnocentrism** is a tendency to view one's own cultural experience as a universal standard. When ethnocentrism intrudes on research, one's data can present a completely false picture. Consider the fact that sticking out one's tongue in South China, for example, is a sign of deep embarrassment. Among inhabitants of the Caroline Islands, the gesture is used to frighten demons away. In New Caledonia, sticking out one's tongue at another carries a wish for wisdom and vigor. And in India, the gesture is a sign of incredible rage.

Dyad
A group of two

Ethnocentrism
Tendency to view one's own cultural experience as a universal standard

As data consumers, we should make a reasonable effort to learn the cultural background of the researcher. We may also wish to explore the researchers' efforts to overcome their own biases, for if we remain unaware of their cultural "blinders," we can fall prey to the same misinterpretations made by the professional observer.

Of course, even if one covers all of the bases mentioned here—attending to sampling rules, questionnaire design, validity, reliability, and so forth—a survey may still produce flawed results. The reason for this may rest with the respondents themselves. Depending on the topic about which people are questioned, respondents may feel reluctant to answer questions candidly. They may give survey researchers the answer they believe the questioner wants to hear rather than the answer that reflects respondents' true beliefs. This phenomenon is especially common with regard to race. Dubbed the **Bradley Effect**, it was first systematically studied after the 1982 California gubernatorial election. At the time, polls showed that black candidate Tom Bradley held a double-digit lead over his opponent,

Bradley Effect
Discrepancies in poll results and actual outcomes, usually linked to respondents' deception with reference to sensitive topics

Republican George Deukmejian. Most expected that Bradley would win the election and become the nation's first black governor. But in spite of the numbers, Bradley was defeated by a thin margin. Apparently, many voters felt negatively toward Bradley because of his race but were uncomfortable sharing that information with pollsters. As a result, the poll numbers were flawed. One might think of the Bradley Effect as a case of respondent deception. And research shows that such deception is not confined to race. The Bradley effect regularly occurs with regard to issues surrounding gender, religion, and sexuality as well (Stout and Kline 2008).

Your Thoughts . . .

Suppose you were doing a survey on some sensitive issues—that is, sexual habits, voting preferences, and the like. What are some things you could do to overcome the Bradley Effect?

Having reviewed the "red flags" discussed in this essay, you may now feel completely doubtful of the conventional wisdom on numbers. How can numbers ever be trusted? But keep in mind that as a research project unfolds, a careful researcher is asking the very same questions posed in this essay. A careful scholar is attending to the veracity of data even as they are produced. When one couples a careful researcher with an astute data consumer, the product can be a set of numbers that sheds much light on aspects of the social world.

LEARNING MORE ABOUT IT

Janet Ruane offers an engaging and very readable discussion of good methodological technique in *Essentials of Research Methods* (Malden, MA: Blackwell, 2005).

Earl Babbie ponders the problems of doing research in *Observing Ourselves: Essays in Social Research* (Prospect Heights, IL: Waveland Press, 1998).

The Gallup organization offers some firsthand insight into selecting a representative sample, formulating questions, and so on. Visit its website at http://www.gallup.com. Looking under the Resources and then the FAQs link will lead to several helpful links about the polling process.

Jon A. Krosnick and Stanley Presser instruct us on good questionnaire design in "Question and Questionnaire Design," pages in P. Marsden and J. Wright (eds.), *Handbook of Survey Research* (San Diego, CA: Elsevier, 2010).

Joel Best provides a practical "manual" to help us identify faulty survey data in *Stat-Spotting: A Field Guide to Identifying Dubious Data* (Berkeley: University of California Press, 2008).

Jane Miller discusses the ins and outs of reporting statistics in *Writing About Numbers: Effective Presentation of Quantitative Information* (Chicago: University of Chicago Press, 2004).

GOING ONLINE FOR SOME SECOND THOUGHTS

The General Social Survey provides regularly collected data on the structure and development of American society. Because data are collected at regular intervals, this resource allows us to both monitor social change within the United States and compare U.S. society and culture to that of other nations. You can find the General Social Survey at http://www.norc.org/GSS+Website/.

NOW IT'S YOUR TURN

1. Earlier in this essay, we described a study exploring dating patterns of college students. The essay presented the wrong way to draw a sample for such a project. Now it's your turn. Review the material on sampling and write a brief description of one possible method for drawing a better, more representative sample of a college student community.
2. Consult a sociology dictionary for a definition of *alienation.* Then visit the website for the General Social Survey (GSS; http://www.norc.org/GSS+Website/) and find the questions used on the GSS to measure alienation. (Once you've accessed the GSS home page, click on Browse GSS Variables, then Subject Index, then A, and then Alienation.) Now it's your turn to assess the adequacy of the GSS questions relative to your working definition of alienation.
3. Visit the website PollingReport.com (http://www.pollingreport.com/). Find several surveys that address the same topic, but ask slightly different questions. Now it's your turn to decide. Do the differences in polling numbers offer any insight into the meaning of statistical data's validity?

CULTURE

ESSAY 2

Conventional Wisdom Tells Us . . . Winning Is Everything

Conventional wisdom suggests that competition and achievement go hand in hand. In this essay, however, we highlight the many studies that show the benefits of cooperation over competition. In so doing, we review American cultural values, strategies of action, and the connection of these elements to both positive and negative outcomes.

Think back to the last Little League, soccer, or professional hockey game you attended. Or take note of the stores and businesses that celebrate the "salesperson of the week." Now consider the megadollar spending and the hardball tactics embraced by contenders for national political office. And who can forget the thrill of victory and the agony of defeat as read on the faces of the most recent World Series, Super Bowl, or Olympic contenders?

These snapshots of American life remind us that competition is central to our culture. As children, we are taught to play hard and fight to win. As adults, we learn to value winning. We equate winning with the most talented or the "best man," and we regularly remind ourselves that "nice guys finish last." In the United States (as well as in most capitalist societies), the emphasis is on beating one's opponent. (And as recent scandals in professional baseball and football reveal, some feel that steroids or other forms of cheating are legitimate strategies by which to accomplish that goal.) We want to be the one on top, the king of the hill, the one left standing after a fair fight. Al Davis, managing partner of the Oakland Raiders NFL team, captures the sentiment: "Just win, baby!" (as quoted in Wetzel 2007).

Cultural value
A general sentiment regarding what is good or bad, right or wrong

The conventional wisdom on competition represents a **cultural value**. A cultural value is a shared sentiment regarding what is good or bad, right or wrong, desirable or undesirable. In the United States, competition is a positive cultural value (Aronson 1980; Cavanagh 2005; Hunt 2000; Stinchcombe 1997; Williams 1970). Yet despite our commitment to healthy competition, research shows that the practice may not always be in our best interest. A growing literature suggests that in many areas of social life, cooperation leads to more profitable outcomes than does competition.

Social psychologists David and Roger Johnson reviewed nearly 200 studies on human performance. The results of their research indicate that cooperation promotes higher individual achievement, higher group productivity, better problem solving, and more effective learning than do competitive strategies of interaction. These same studies show that the more cooperative people are, the better their performance. Thus, when group members periodically take the time to review their efforts while executing a task or attempting to solve a problem—that is, when group members reflect on their actions, ensure the equal distribution of responsibility, and protect open communication channels—the benefits afforded by cooperative strategies often increase (Johnson and Johnson 1989, 2002; also see Alper, Tjosvold, and Law 2000; Balliet 2010; Bendor and Swistak 1997; De Dreu, Weingart, and Kwon 2000; Jensen, Johnson, and Johnson 2002; Johnson and Johnson 2005; Kohn 1986; Madrid, Canas, and Ortega-Medina 2007; Muthusamy, Wheeler, and Simmons 2005; Quiggin 2006; and Wilkinson and Young 2002).

The success of cooperation stems from the strategies of action that it stimulates. **Strategies of action** are the means and methods social actors use to achieve goals and fulfill needs. Research indicates that the cooperative stance allows individuals to engage in more sophisticated and advanced thinking and reasoning than that which typically occurs in competitive environments. Social psychologists refer to these sophisticated thinking strategies as higher-level reasoning and metacognitive strategies.

Strategies of action
Means and methods social actors use to achieve goals and fulfill needs

Why do cooperative environments enable sophisticated thinking? Research suggests that interaction within cooperative settings typically evolves according to a process that sociologists refer to as a dialectic. A **dialectic** is a process by which contradictions and their solutions lead participants to more advanced thought. The dialectic process consists of three steps. In Step 1, thesis, the group experiences conflict. Here, members propose different ideas, opinions, theories, and information regarding the task or problem at hand. This conflict or disequilibrium sparks Step 2, antithesis. In antithesis, members actively search for more information and additional views, thus maximizing their knowledge about the task or problem they face. When the search is complete, the group begins Step 3, synthesis. Synthesis is a period in which group members reorganize and reconceptualize their conclusions in a way that merges the best thinking of all members (Cerulo 2006; De Dreu et al. 2000; Fuchs 2010; Glassman 2000; Johnson and Johnson 1989, 2002; Wichman 1970).

Dialectic
A process by which contradictions and their solutions lead participants to more advanced thought

In addition to bettering group and individual performance, cooperation—according to more than 180 studies—enhances the quality of interpersonal relationships. Friends, workers, and intimates who cooperate with one another rather than competing report feeling greater levels of acceptance from their colleagues and partners. As a result, cooperators become more caring and committed to their relationships. Furthermore, those involved in cooperative relationships report higher self-esteem and less psychological illness than those who compete with their friends, colleagues, and partners. Indeed, competitiveness has been repeatedly linked to psychological pathology

(Combs 1992; Finlinson, Austin, and Pfister 2000; Johnson and Johnson 1989, 2002, 2005; Jurd 2005; Kohn 1986; Muthusamy et al. 2005; Tjosvold 2008; Wilson 2000).

Some studies also link cooperation to diminished feelings of prejudice. **Prejudice** refers to the prejudgment of individuals on the basis of their group memberships. For example, individuals who cooperate with those whom they previously had stigmatized or negatively stereotyped report an increased liking toward such individuals. In contrast, individuals placed in competitive situations with members of a previously stigmatized group report a greater dislike for their competitors. On the basis of such studies, many researchers suggest that interracial, interethnic, and intergender cooperative tasks should become a regular orientation strategy in workplaces, schools, civic groups, and neighborhood organizations. Many believe that such cooperation exercises could help reduce prejudice and bigotry in the sites of our daily interactions (Aronson and Cope 1968; Aronson and Thibodeau 1992; Bukowski et al. 2009; Holtz and Miller 2001; Jehn and Shah 1997; Johnson and Johnson 2000; Lu, Tjosvold, and Shi 2010; McConahay 1981; Peterson 2007; Rabois and Haaga 2002; Slavin and Madden 1979; Stapel and Koomen 2005; Vanman, Paul, and Ito 1997). Indeed, several researchers have documented such outcomes as they pertain to periods of economic development, civil rights movements, and political revolutions (Deutscher and Messner 2005; Mihailescu 2005; Miller 2006; United Nations 2005).

Prejudice
A prejudgment directed toward members of certain social groups

If cooperation leads to so many benefits, why do social actors continue to choose the competitive stance? The persistence of competition is a good example of the power of cultural values and the ways in which cultural values can promote a phenomenon that sociologists refer to as **culture against people**. Culture works against people when the beliefs, values, or norms of a society lead to destructive or harmful patterns of behavior. When it comes to the value of competition, the culture-against-people phenomenon couldn't be any stronger. Indeed, research conducted over the past several decades suggests that, even when individuals are made fully aware that they have more to gain by cooperating with others than by competing with them, they often continue to adopt competitive strategies of action (Deutsch and Krauss 1960; Hargreaves-Heap and Varoufakis 2005; Houston et al. 2000; Kelley and Stahelski 1970; Miceli 1992; Minas et al. 1960; Schopler et al. 1993; Schultz and Pruitt 1978; Van Avermaet et al. 1999).

Culture against people
When beliefs, values, or norms of a society lead to destructive or harmful patterns of behavior

In one experiment, for example, a college professor told his students that they were participating in an investment research project for the *Wall Street Journal.* The project consisted of several simple exercises. In each exercise, students in the class would be asked to write a number, either 1 or 0, on a slip of paper. Before casting their votes, students were informed that the number of "1" votes cast by class members would determine a financial payoff for each student. The professor explained that each exercise was designed such that a unanimous class vote of 1 would maximize every class member's payment as well as the total class pot. The class was also instructed that a single 0 vote could increase one voter's payoff, but such split votes would always result in a smaller payment for the remaining class members, as well as a smaller payment for the class overall.

Here is a concrete example of the payoff schedule. If 30 people wrote the number 1, a total of $36.00 would be evenly divided by the class—$1.30 for each class member. But if 29 people chose 1 and one person chose 0, only

$35.30 would be paid to the class. The one individual who voted 0 would be paid $1.66; the 29 students who voted 1 would each receive only $1.16. Now, consider a situation in which 10 students choose the number 0 and 20 choose the number 1. Here, the class would divide only $29.00. Students choosing 0 would each receive $1.30, but those choosing 1 would receive only $.80. (The professor provided students with a breakdown of all possible payoffs before the class voting began.)

The professor took several votes in his class, allowing class members to debate strategies before each vote. Yet even when the entire class recognized and agreed that a unanimous 1 vote would be the fairest and most lucrative strategy overall, several class members continued to vote 0 in an effort to maximize their own individual gain (Bishop 1986).

Your Thoughts . . .

Suppose you were a member of the class involved in the *Wall Street Journal* experiment. What arguments would you present to your classmates to encourage cooperation over competition?

Similar results have emerged from experiments conducted in laboratory settings. The Prisoner's Dilemma, for example, is an experimental game designed to test an individual's preference for cooperative versus competitive strategies. The game is based on a hypothetical problem faced by two suspects and a district attorney, all gathered in a police station. The district attorney believes that both suspects have committed a crime but has no proof connecting the suspects to the crime. Thus, he separates the two, telling each prisoner that he has two alternatives: confess to committing the crime in question or not confess. Prisoners are told that if neither confesses, both will be convicted of only a minor offense and receive only a minor punishment (1 year in prison). If both confess, each will be convicted of a major crime and face fairly severe penalties (10 years in prison). If only one of the prisoners confesses, the confessor will receive full immunity, while the nonconfessor will receive the maximum penalty allowed by law (15 years in prison).

The Prisoner's Dilemma is designed to encourage cooperation. Clearly, silence on the part of both prisoners maximizes the favorable chances of each one. Yet in experimental trials, prisoners repeatedly favor competition. When faced with both the cooperative and competitive options, players consciously choose confession—the strategy that offers them the potential to maximize their own gain. In other words, players choose to compete even when such a strategy proves riskier than cooperation (Hargreaves-Heap and Varoufakis 2005; Houston et al. 2000; Kollock 1998; Rapoport 1960; Schopler et al. 1993; Van Avermaet et al. 1999).

The *Wall Street Journal* experiment (devised by three economists, Charles Plott, Mark Isaac, and James Walker), the Prisoner's Dilemma, and other similar games and experiments all illustrate the way in which our cultural value of competition can work against people. These examples suggest that, even when a goal can be realized best via a common effort, significant numbers of individuals will espouse cooperation but act in a competitive way. Significant numbers of individuals will act to maximize their own gain rather than acting in the best interest of the group as a whole—even if that maximization proves risky and the payoff uncertain.

BOX 2.1 SECOND THOUGHTS ABOUT THE NEWS: NO TIME OFF FROM COMPETITION

The "staycation" made the news with the arrival of our latest recession. As gas prices, food costs, and unemployment rates were going up, more and more Americans were staying home for their vacations. Perhaps more interesting than this financially driven staycation, however, is the fact that Americans are really out of step with the rest of the world with regard to the *need* for vacations.

Vacations are often justified as essential for recharging our batteries and making us more productive workers. And many nations strongly commit to this idea: Of the richest nations of the world, only the United States fails to *require* paid vacation time for workers. Workers in nations enjoying the same standard of living as in the United States typically receive 20 to 25 days of paid vacation a year. (France requires 30 days of paid vacation for its workers!)

Even as European nations are looking for ways to fight their own debt problems, tinkering with vacation time is not seen as a viable option. Since the United States does not mandate paid vacations, it should come as no surprise that Americans average just 9 vacation days a year—only 10% take 2 weeks off, and one in four Americans don't take any vacations from work at all!

Economist John Schmidt says Americans have a "tortured" relationship with vacationing, especially during recessions. Skipping vacations, while consistent with our Puritan work ethic, is also seen as a way to stay one step ahead of workplace competition (CBS News, 2010). In 2010, freshman Congressman Grayson of Orlando, Florida (a major vacation spot), strongly advocated for legislation requiring a week of paid vacation for workers at companies with 100 or more employees. Unfortunately, in the 2010 midterm elections, the voters decided to send Representative Grayson on at least a 2-year vacation—he lost his bid for re-election. It appears that the United States is not yet ready to be a vacation nation.

Can we overcome the allure of competition? Are there any circumstances that motivate individuals to choose the cooperative path? Studies show that certain factors can enhance the chances of cooperation. For example, one study suggests that the body may play a role. When experimental subjects were asked to synchronize their bodily movements (i.e., rocking in synchrony or clapping their hands in synchrony), subjects became more aware of those with whom they were interacting, and they were more likely to cooperate with them in the solution of problems and the completion of tasks (Valdesolo, Ouyang, and DeSteno 2010). Another study showed that cooperation begets cooperation. In experimental settings, those persuaded to contribute cooperatively to a task become more and more likely to repeat the behavior—doing so more quickly with each subsequent trial (Fowler and Christakis 2010). Still another line of experimental work suggests that those faced with some type of external threat will abandon their competitive stance and band together with others in cooperative strategies of defense or protection. Within organizations, for example, pitting teams or departments against one another can enhance cooperation *within* groups as people join together in the service of winning. The same is true in sports, in academic settings, and, some argue, in the international political realm

(Blake and Moulton 1979; Burton-Chellew, Ross-Gillespie, and West 2010; Deutsch and Krauss 1960; Dion 1979; Lanzetta 1955; Puurtinen and Mappes 2009; Sherif 1966; Sherif et al. 1961; Wilder and Shapiro 1984). Recently, these findings have been replicated in real-world settings, particularly in studies devoted to the aftermath of natural disasters or terrorist attacks (Cerulo 2006; Halbert 2002; Heeren 1999; Press 2007; Turkel 2002). Equally exciting is a growing line of research that examines the role of cooperation in capitalism. For centuries, economists have argued that competition is essential to a sustainable, dynamic economy. However, a number of current studies suggest that locally, nationally, or globally, the healthiest economies are those that find a balance between competition and cooperation within institutional structures (Amodeo 2001; Brandenburger and Nalebuff 1996; Deutsch 2000; Deutscher and Messner 2005; Glaser 2010; Moore 1997; Royo 2006; Wisman 2000). Thus, in the final analysis, working together may be the most profitable strategy of all.

LEARNING MORE ABOUT IT

Those interested in a thorough examination of the conditions that foster cooperation and in learning how to promote cooperation should consult Robert Axelrod's *The Evolution of Cooperation* (New York: Penguin, 2006) or a collection of articles edited by Brandon A. Sullivan, John L. Sullivan, and Mark Snyder entitled *Cooperation: The Political Psychology of Effective Human Interaction* (Malden, MA: John Wiley and Sons, 2008).

The notion of "culture against people" stems from a line of work that includes Philip Slater's *The Pursuit of Loneliness: American Culture at the Breaking Point* (Boston: Beacon Press, 1970); Richard Sennett's *The Fall of Public Man* (New York: Knopf, 1977); and Robert Bellah and colleagues' *Habits of the Heart: Individualism and Commitment in American Life* (Berkeley: University of California Press, 1985).

For an engaging look at Americans' fixation on competition, see Francesco Duina's *Winning: Reflections on an American Obsession* (Princeton: Princeton University Press, 2011).

GOING ONLINE FOR SOME SECOND THOUGHTS

The following organizations host websites that can help you learn more about cooperation:

Cooperative Learning Center 60 Peik Hall, University of Minnesota, Minneapolis, MN 55455, http://www.co-operation.org

Future Problem Solving Program (Fosters creative thinking in problem-solving efforts) http://www.fpspi.org/

Grace Contrino Abrams Peace Education Foundation http://www.peaceeducation.com

Institute on Global Conflict and Cooperation http://igcc.ucsd.edu/

Up With People (Builds understanding and cooperation among those of different cultures through a special educational program) http://www.upwithpeople.org/

NOW IT'S YOUR TURN

1. To determine the cultural importance of competition, try the following experiment. Solicit several friends or relatives to join you in some traditionally competitive activity, such as basketball, bowling, cards, or tennis. Vary the conditions of play with each individual you choose. For example, use normal game rules with some of your subjects; tell others that you don't want to play for points or keep score. Now it's your turn to record the different reactions, if any, to the different conditions of play. Does playing without keeping score affect the willingness of some to participate? Affect the quality of interaction? Do you note any effects of gender or age on the reactions you observe?
2. It's your turn to explore the significance of competition for college grading. Prepare a serious proposal that testing be conducted under conditions of cooperation rather than competition. (Cooperative conditions might include group testing, open discussion during the exam, or an adjustment in grade calculation methods such that each student's grade is an average of her or his actual performance as well as the performance of the best and worst students in the class.) Conduct a limited survey of your colleagues. Do they support or reject such a proposal? What reasons do they offer for their position? Is their reasoning consistent with American values on competition? On cooperation?
3. It's your turn to access 20 newspaper stories reporting your state's annual budget negotiations for this year. Read these stories carefully and try to summarize the give and take that led to a final budget. Based on your review, would you say that the workings of the governor and state legislature are more reflective of competition or cooperation? What is your evidence?

ESSAY 3

Conventional Wisdom Tells Us . . . Children Are Our Most Precious Commodity

We frequently hear it said: Children are our future. They are our most valuable resource. Here, we present research suggesting otherwise. Children may be the most overlooked, the most neglected segment of the population despite current talk of family values and the future of American youth.

Children—who doesn't love them? In the United States, we refer to children as "our nation's future." Our conventional wisdom celebrates them as society's most precious commodity. And national opinion polls repeatedly document that Americans consider children one of life's true rewards. Witness how our popular icons—Madonna or Angelina Jolie—can cause a media frenzy over their efforts to adopt and garner that reward. Indeed, couples who want children but cannot have them are ensured the sympathies and support of their fellow members of society. In contrast, couples who don't want children have in the past found themselves objects of contempt or suspicion.

Much of today's political rhetoric is fueled by the cultural value America places on children. Citizens and elected officials are urged to "act now" in the long-term interest of the nation's youth. Politicians vow to cut today's spending and spare our children and grandchildren a troubled and debt-laden tomorrow. Many advocate curbing Social Security and Medicare costs so as to protect benefits for future generations. Global warming is frequently discussed with an eye to the environmental legacy we will leave our heirs. Such prescriptions underscore the importance of children in our youth-oriented society.

Threats to our children can mobilize American sentiments in a way that few other issues can. Consider some landmark moments of the past few decades that joined American citizens together in public outrage: the bombing of a federal building in Oklahoma City; school shootings at Columbine High School, "Virginia Tech," or that one-room schoolhouse in Pennsylvania's Amish country; the Amber Hagerman kidnapping, which inspired the creation of the Amber Alert system; the murder and rape of Megan Kanka, which resulted in Megan's Law; or the series of brutal killings of schoolchildren in China in the spring of 2010. In such instances, public outrage was fueled in large measure by the fact that these acts threatened and/or took the lives of so many children.

Photo 3.1 Copyright Anne Glassman. Reprinted by permission of Anne Glassman.

Can we take the prochild rhetoric of conventional wisdom at face value? How do our prochild sentiments compare with the realities of American children's lives? A review of worldwide infant mortality rates offers one perspective on the matter. **Infant mortality rates** gauge the number of deaths per 1,000 live births for children under 1 year of age. Such statistics represent a commonly consulted measure, or social indicator, of a society's behavior toward its children. **Social indicators** are quantitative measures or indices of social phenomena.

Infant mortality rates
The number of deaths per 1,000 live births for children under 1 year of age

Social indicators
Quantitative measures or indices of social phenomena

Despite the prochild sentiments of American culture, the United States trails many other nations in the fight against infant mortality. To be sure, infant deaths typically are highest within less developed nations of the world community, such as Afghanistan, Liberia, Angola, or the Congo. Yet the U.S. infant mortality rate—7 deaths for every 1,000 live births—is comparable to that found in Latvia, Serbia, and Slovakia—nations with far less wealth or international power than the United States (see Figure 3.1). A variety of other major (and minor) industrial nations, such as Poland, Australia, Canada, Croatia, Cuba, Germany, France, Japan, Italy, and Iceland, have been more effective than the United States in fighting infant deaths. Indeed, the U.S. infant mortality rate is higher than the summary rate (5) for all industrialized countries (UNICEF 2009a).

Photo 3.2 Copyright Karen A. Cerulo. Reprinted by permission of Karen A. Cerulo.

Child inoculation rates are another informative measure of a society's behavior toward its children. Ten years into the new millennium, 25% of 2-year-olds in the United States do not

Figure 3.1 Infant Mortality Rate for the U.S. and Selected Countries of the World (2007)

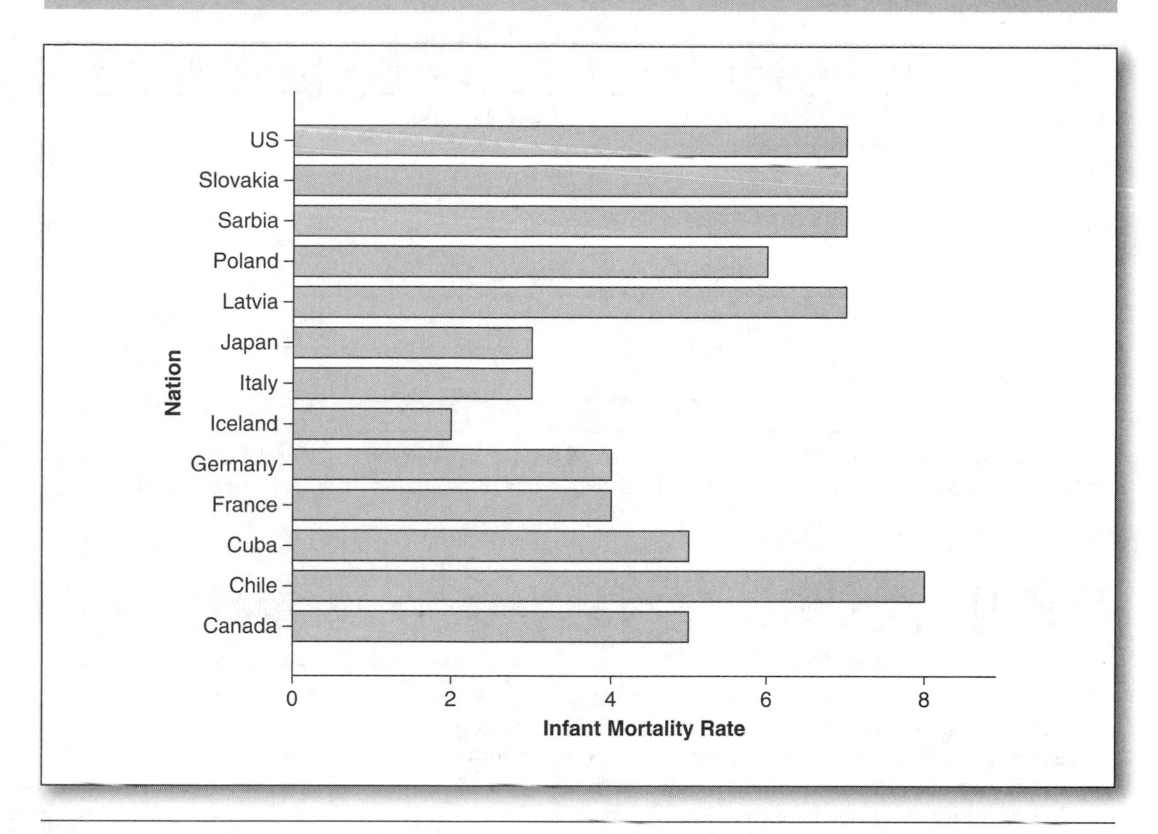

Source: UNICEF The State of the World's Children 2009 http:www.unicef.org/sowc09/index.php

Note: Infant mortality rate refers to the number of deaths of children under one year of age per 1,000 live births.

have the full series of childhood vaccinations. (Children's Defense Fund 2010). And while recent UNICEF data indicate that 93% of 1-year-olds in the United States are fully immunized against polio, that figure lags behind rates found in far less developed nations, such as Albania (99%), Cyprus (97%), Cuba (99%), Kazakhstan (99%), and Sri Lanka (98%) and Tonga (99%; UNICEF 2009).

The wealthiest nation in the world doesn't guarantee a healthy start to all of its newborns. Eight percent of U.S. babies are born at a low birth weight, a number that has been *increasing* in recent years (Martin et al. 2006). This 8% rate is higher than that found in many other countries of the world: Tonga (3%); Belarus, China, Iceland, and Republic of Korea (4%); Croatia, Cuba, Denmark, and Serbia (5%); Ireland, Spain, and Switzerland (6%); and Germany, France, and Kuwait (7%; UNICEF 2010). And while progress is being made with regard to health insurance for children via the Children's Health Insurance Program (CHIP), approximately 10% of U.S. children still lack such insurance. This is a condition with serious consequences. Uninsured children are nearly twice as likely as the insured to be in fair or poor health and they are twice as likely as their insured peers to not have seen a dentist in more than 2 years (Children's Defense Fund 2010). (See Figure 3.2 for the states doing the worst and best by children in terms of health insurance coverage.)

To be sure, it is not easy nor is it safe to be young in America (see Figure 3.3). Consider, for

Figure 3.2 States with the Highest and Lowest Percentages of Uninsured Children

Highest % of Uninsured		Lowest % of Uninsured	
Texas	20.8%	Massachusetts	4.5%
Florida	18.7%	Iowa	5.5%
Nevada	17.8%	Wisconsin	5.5%
New Mexico	16.5%	Hawaii	5.6%
Mississippi	15.3%	Michigan	5.6%
National average: 10.3%			

Source: Children's Defense Fund. 2010. The State of America's Children 2010 Report: Child Health Section. http://www.childrensdefense.org/child-research-data-publications/data/the-state-of-americas-children-2010-report-health.pdf.

Figure 3.3 "Select" Moments in the Lives of U.S. Children (2009)

- **Every 32* seconds** a baby is born into poverty.
- **Every 41 seconds** a child is confirmed as abused or neglected.
- **Every 42* seconds** a baby is born without health insurance.
- **Every minute** a baby is born to a teen mother.
- **Every minute*** a baby is born at low birth weight.
- **Every 18* minutes** a baby dies before his first birthday.
- **Every 3 hours** a child or teen is killed by a firearm.
- **Every 5 hours** a child or teen commits suicide.
- **Every 6 hours** a child is killed by abuse or neglect.
- **Every 15* hours** a mother dies from childbirth complications.

Source: Children's Defense Fund: Moments in America for Children (2009). http:www.childrensdefense.org/child-research-data-publications/moments-in-america-for-children.html

*Indicates worsening condition from previous years

example, that in 2005, more preschoolers were killed by firearms than were law enforcement officers in the line of duty: 69 versus 53. Indeed, in the United States, gunfire kills a child or teen every 3 hours (Children's Defense Fund 2009). Homicide is the second leading cause of death for those aged 10 to 24. Note too that in 2006, more than 720,000 of those aged 10 to 24 received emergency room treatment for injuries from violence. And America's love of automobiles can be lethal for children. Injuries from car accidents are the leading cause of death for U.S. children. To make matters worse, these fatalities are often the result of children riding with drunken adults and/or failure by adults to use proper child car seats or restraints (Centers for Disease Control and Prevention 2009c, 2009d).

These facts and figures on the physical well-being of U.S. children unveil glaring discrepancies between what we say and what we do with regard to children. However, the discrepancies extend well beyond the realm of health and mortality. Although conventional wisdom celebrates the child, the reality is that millions of American children confront poverty and its many ills throughout their childhoods. Currently, 19% of those under 18 years old are living in poverty and 41% live in **"low-income" families**—that is, families with income above the federal **poverty** level but below the level deemed necessary for meeting basic needs (National Center for Children in Poverty 2009). For those 6 years of age or younger, the rate climbs to 22%. (Contrast these figures with 12% of the 18 to 64 population and 10% of the 65+ population living below the poverty level; see Wight and Chau 2009). After a decade of declining poverty rates in the 1990s, rates for children living in poverty have increased by 21% since 2000 (Wight, Chau, and Aratani 2010). If we compare the United States to a select group of our "global peers"—that is, countries with comparable economic opportunities and that belong to the Organization for Economic Cooperation and Development (OECD)—our record on fighting childhood poverty is truly dismal. Among the OECD nations, the United States currently ranks second highest in per-capita income (Norway is first). Yet the United States has the highest childhood poverty rate (21.9% as measured by international standards). Among the "rich" OECD nations, the United States also has the distinction of having the lowest social expenditures for fighting poverty (Economic Policy Institute 2010).

Low-income families
Families with income above the federal poverty level but below the level deemed necessary for meeting basic needs

Poverty
An economic state or condition in which one's annual income is below that judged necessary to support a predetermined minimal standard of living

It should also be noted that for many poverty scholars, official poverty rates fail to capture economic hardship in the United States. This is because the poverty level is calculated using what many consider an outdated metric. (That metric is based on 1950s data and assumes that the average family spends one-third of its income on food.) Using the official equation, the poverty level for a family of four is approximately $22,000. But for many, this figure is unrealistically low. It fails to give sufficient weight to the growing and disproportional impact of nonfood family expenses such as health care, child care, housing costs, transportation expenses, and so on. Critics argue that a family of four needs about twice the official poverty level to meet its basic needs (or approximately $44,000 a year). Families with this level of income, however, are classified by the U.S. Census Bureau as "low income" rather than as impoverished. If we were to revise our poverty statistics to utilize these new income cutoffs, we would find roughly 41% of U.S. children living in impoverished conditions (Wight, Chau, and Aratani 2010).

To fully understand the plight of our nation's children, consider these images. In a "group portrait" of the U.S. poor, more than one-third of the faces would belong to children (U.S. Conference of Catholic Bishops 2009). About 8 million children—1 in 10—have no health insurance (even though nearly all of these children have at least one employed parent). One in six children are born to mothers who did not receive any prenatal care in the first trimester (Children's Defense Fund 2007, 2008, 2010).

Living conditions play a role in this group portrait of the poor. Financially speaking, two-parent families represent the most favorable condition for children. But the percentage of children under 18 living in such households has fallen from 77% in 1980 to 67% in 2008. Today, 23% of children under the age of 18 live in mother-only families. Such arrangements can exact a heavy cost for children: As of 2007, 43% of the children in mother-only families live in poverty. Contrast this with the fact that only 9% of children in married-couple families are poor (Federal Interagency Forum on Child and

Family Statistics 2009). The high poverty rate for children in single-mother families is in part due to "deadbeat dads." In divorced families, child support constitutes about one-quarter of total family income. It is responsible for lifting about half a million children out of poverty each year (Center for Law and Social Policy 2007). In 2007, there were 13.7 million custodial parents who were due $34.1 billion in child support. Just about 47% of these families received the full amount of support payments that were due (Grall 2009). One quarter of these custodial parents (83% of whom were mothers) had incomes below the poverty level. For these families, child support represents nearly 50% of the family income. Yet in 2007, only 40% of impoverished custodial parents received all the child support payments they were due; 31% received no payment at all (Grall 2009).

Poverty undermines the education of our children as well. Before entering kindergarten, children in the highest socioeconomic groups have cognitive scores that are 60% above the average scores for children in the lowest socioeconomic group. As children move through the system, the effects of poverty persist. Low-income third graders with undereducated parents have vocabularies of about 4,000 words. In contrast, their peers from middle-income families with well-educated parents have vocabularies of 12,000 words (Klein and Knitzer 2007). Note too that poor children are more likely to be retained in a grade—in 2007, for K–8th-grade students, 27% of those from poor families had been retained compared to 5% from nonpoor families (Planty et al. 2009). The poor also have high school dropout rates that are more than five times higher than their wealthy peers (U.S. Department of Education 2008). Fifty-five percent of poor high school graduates immediately enroll in college compared to 78.2% of high-income graduates (Planty et al. 2009)

For many children, poverty goes hand in hand with homelessness; nationally, about 40% of the homeless population consists of children under 18 years of age. Further, families with children are among the fastest-growing segments of the homeless population (National Coalition for the Homeless 2009). These conditions have prompted at least one writer to observe that child poverty is one of our country's most stunning failures (Madrick 2002).

But even among children who have a roof over their heads, life is not always easy. National surveys suggest rates of physical and sexual child abuse that are completely inconsistent with America's prochild rhetoric. The U.S. Department of Health and Human Services estimates that approximately 772,000 children were victims of abuse and neglect in 2008. Seventy-one percent of these children experienced neglect, 16.1% were physically abused, and 9% were sexually abused. Approximately 1,750 children died from abuse and neglect in 2008 (U.S. Department of Health and Human Services 2010).

BOX 3.1 SECOND THOUGHTS ABOUT THE NEWS: FLIGHT ATTENDANT TO THE RESCUE?

On a recent airline flight, I found myself sitting in front of a mother with two very loud, misbehaving children. It was late evening and many passengers were trying to catch some shuteye. I had no such luck. As the unbuckled kids kept the havoc going, my mind wandered to another flight story that was just in the news.

A Southwest Airline attendant witnessed a mother slap her crying and kicking 13-month-old child and proceeded to take the child away from the mother for the rest of the flight. The incident sparked a debate about the appropriateness of the attendant's actions. Some were calling her heroic while others empathized a bit more with the mother, admitting to being in a similar place with their own kids and wondering if the attendant overreacted to parental discipline. In fact, flight attendants are empowered by federal law to take action to ensure the safety of passengers.

Southwest called ahead and the plane was met by authorities and paramedics. After the parents were questioned extensively and the child examined, the family was permitted to go on with their travels. No charges were filed.

While it may be hard to fathom mistreating helpless babies and infants, note that it is the *youngest* children who are the most vulnerable to abuse and neglect. Infant boys (under 1) had the highest rates of fatalities: 19.3 deaths per 100,000. Infant girls had a death rate of 17.22 per 100,000. In 2008, nearly 80% of children who died from abuse and neglect were under 4 years of age (U.S. Department of Health and Human Services 2010).

As children make their way into the school system, danger persists. Student days are certainly not carefree days for children. Thirty percent of sixth to tenth graders have reported being the targets or the perpetrators of bullying. At the high school level, more than 12% have reported being in a physical fight *on school property.* Slightly more than 27% report having their property stolen or deliberately damaged *while in school* (Centers for Disease Control 2009c, 2009e). In the United States, school grounds simply don't offer students a sufficiently safe haven.

Alcohol and drugs play a role in the lives of our schoolchildren as well. According to the latest Youth Risk Behavior Survey, 45% of high school students reported drinking alcohol on one or more days in the preceding month. Twenty-six percent have engaged in heavy **binge drinking** (five or more drinks in a row). Eleven percent reported driving after drinking (Centers for Disease Control and Prevention 2008a). With regard to drug use, 38% of high school students have used marijuana, 13% have used inhalants, nearly 8% have used hallucinogenics, nearly 6% have used ecstasy, and just over 7% have used cocaine. Sex is a factor to be reckoned with as well. Nearly 48% of high school students have already had sexual intercourse, and 35% are currently sexually active (Centers for Disease Control 2008b).

Binge drinking
The consumption of five or more drinks in a row

Our schools also appear to fail our children with regard to academic achievement. Since the early days of the American space program, many have voiced concern regarding the performance levels of American schoolchildren. These concerns are well grounded. While U.S. fourth graders earn above-average scores in reading literacy, they are nonetheless below fourth graders of 10 other nations. When it comes to math literacy, fourth graders in eight countries score higher than their U.S. peers. Furthermore, if we look at the math and science literacy of 15-year-olds, we find that U.S. students place in the bottom quartile and bottom third, respectively, of participating OECD nations (Provasnik, Gonzales, and Miller 2009). Eleven percent of 16- to 18-year-olds are functioning at a *below* basic literacy rate (Kutner et al. 2007). Across the nation, the average high school graduation rate is 73.4% (Planty et al. 2009).

We may idealize childhood (especially in our memories). But these everyday patterns suggest that the reality for today's U.S. children is often anything but ideal. Indeed, despite our prochild stance, many children find childhood too difficult to endure. The suicide rate among American youth increased by 240% from the early 1950s to the late 1970s. In 2007, 14.5% of high school students reported that they seriously considered suicide in the previous year, and just under 7% reported actually attempting to take their own lives. It is no wonder, then, that suicide is the third leading cause of death for 15- to 24-year-olds (Centers for Disease Control and Prevention 2009f). On an average day in the United States, five children or teens will take their own lives (Children's Defense Fund 2009).

Your Thoughts . . .

In 2008, Nebraska enacted a "safe haven" law designed to enable troubled mothers to relinquish their newborns, no questions asked. But in the first 4 months of the program, no newborns were given up. Instead, 19 children ranging in age from 1 to 17 were dropped off by parents. The state quickly amended the law to restrict the program to babies up to 3 days old. Should "safe haven" laws be enacted for older children as well? What do you think?

Ideal culture
Values, beliefs, and norms each society claims as central to its modus operandi; the aspirations, ends, or goals of our behaviors

Real culture
Values, beliefs, and norms actually executed or practiced; behaviors or the means actually used to pursue a society's ends

What meaning can we draw from the discrepancies between conventional wisdom's view of children and the way children actually are treated? Are we simply a nation of hypocrites? We gain some perspective on the issue when we consider a distinction sociologists draw between ideal culture and real culture. **Ideal culture** comprises the values, beliefs, and norms each society claims as central to its modus operandi. In other words, ideal culture has to do with aspirations, the ends or goals of our behaviors. In contrast, **real culture** refers to those values, beliefs, and norms we actually execute or practice. Thus, real culture has to do with behaviors or the means to a society's ends.

Your own life experiences have surely taught you that humans have a remarkable capacity to be inconsistent: We can say one thing and do another. In fact, Americans have a cultural prescription reflecting this capacity: "Do as I say, not as I do." When sociologists examine the fit between ideal and real culture, they are exploring the "Say one thing, do another" phenomenon as it occurs at the social level.

Cultural inconsistency
A situation in which actual behaviors contradict cultural goals; an imbalance between ideal and real culture

For a society to achieve perfect agreement between its ideal and real cultures, it must achieve both consensus on goals and agreement regarding the appropriate methods for achieving those goals. That is, ideal and real cultures are in balance only when a society is free of contradiction between what it says and what it does. If a society cannot synchronize its goals and behaviors, then it experiences a condition sociologists refer to as cultural inconsistency. **Cultural inconsistency** depicts a situation in which actual behaviors contradict cultural goals. Cultural inconsistency indicates an imbalance between ideal and real cultures.

Why do cultural inconsistencies emerge in a society? Conflict theorists offer one possible answer.

Conflict theorists
Sociologists who analyze social organization and social interactions by attending to the differential resources controlled by different sectors of a society

Power
The ability of groups and/or individuals to get what they want even in the face of resistance

Social policies
Officially adopted plans of action

Conflict theorists analyze social organization and social interactions by attending to the differential resources controlled by different sectors of a society. These theorists suggest that the inability to balance ideal and real cultures has much to do with the broader issues of power and social policy. **Power** is the ability of groups and/or individuals to get what they want even in the face of resistance. **Social policies** are officially adopted plans of action.

In American society, social policies often guide social behaviors, or what we are referring to as real culture. Yet social policies rarely emerge from a consensus of the general population; they are rarely directed toward ideal culture. Rather, such prescriptions inevitably are influenced by the actions and relative power of various sectors of the population: special interest groups, political action committees, lobbyists, and so on.

By definition, special interest groups promote or advance the cause of certain segments of the population, such as the New Christian Right, senior citizens, tobacco manufacturers, or trial lawyers. Thus, these groups are unduly responsive to the interests of the few—and necessarily ignore the broader interests of the larger population. Special interest groups vie to prescribe social policy. Ultimately, then, social policy generally reflects the particularized goals of groups sufficiently powerful to influence it.

Lacking control of economic resources or access to the political ballot denies children, as a collective, the typical tools of power. Furthermore, age works against the self-serving collective actions of children. Children are dependents; they must rely on adults to act as their advocates. As a result, the interests and rights of children always will be weighed against those of parents, families, and society at large. The child's voice always will be rendered via an intermediary's perspective.

The drawbacks of the child's indirect political presence are aptly illustrated when we review efforts to combat child abuse in the United States. History reveals a parade of policies consistent with child advocates' views and beliefs regarding the best interests of children. For example, child advocates of the early 1800s believed that abused and neglected children were at risk of delinquency; such advocates saw abused children as threats to society. As a result, social and reform policies of the period demanded the *institutionalization* of abused children. Protecting society was deemed "action in the best interest of the child." In the early 1900s, the newly emerging professions of social work and clinical psychology argued that promoting and protecting intact families would best serve the interests of children. Such policy recommendations remanded abused children to the very sites of their mistreatment (Pfohl 1977).

In the current era, many child advocates continue to cling to the family protection theme. The Family Research Council, for example, is an advisory group that has worked for the past 25 years to promote traditional family values. This group rejects any efforts to view children's rights as an issue separate from the context of the family; it is dedicated to the primacy of paternal authority (Gusdek 1998). The Family Research Council's position, as well as those of other groups, makes the political plight of children clear. Without the ability to organize and lobby solely on their own behalf, children will always be one critical step removed from the social policy process—and the gap between the ideal and real cultures that frame childhood in America will continue to exist.

Given the cultural inconsistency that exists in American society's stance toward children, isn't it hypocritical to continue to espouse the ideals we hold? Should an honorable society promote ends it cannot meet?

Sociologically speaking, there are several important reasons to maintain the concept of an ideal culture even when society fails to practice what it preaches. First, a gap between goals and behaviors—that is, a gap between ideal and real cultures—does not diminish the value of a society's ideals. We can honestly place a high value on children even though our behaviors may fall short of the ideal. Ideals, goals, and values are aspirations, and as such, they are frequently not achieved.

Second, changing ideal culture to fit a society's actual practices might indeed bring an end to cultural inconsistency. However, such a change would not alter an important fact: Children literally are the future of any group or society. For a society to survive, individuals must be persuaded to reproduce themselves. In preindustrial days, economic necessity was an attractive incentive for reproduction. Children furnished valuable labor power to colonial families. Children were valuable sources of family income in the early days of industrialization as well. In the industrialization era, children regularly took their places alongside older workers in factories and sweatshops (LeVine and White 1992; Zelizer 1985).

Today, the economic incentives attached to childbearing have changed dramatically. Children are no longer regarded as valuable labor power or income sources for families. Furthermore, the cost of having and raising children in our society has risen dramatically over the years. The U.S. Department of Agriculture estimates that the expense of raising one child born in 2008 (through to the age of 17) would be just over $210,000 for the lowest-income families, close to $292,000 for middle-income families, and just over $484,000 for the highest-income families! These lifetime expense estimates do not include the cost of a college education. Parents who agree to assist with college costs must be prepared to spend tens of thousands of dollars more to cover college tuition and room and board. A two-parent, one-child family can expect 25% of total household expenditures to be attributable to that child. When the number of children increases to two and three, the percentage of child-related expenditures increases to 40% and 47%, respectively (Lino and Carlson 2009).

Changes in lifestyles and priorities, education, and career commitments render the decision to have children more problematic today than at earlier points in America's history. Individuals no longer automatically equate children with personal fulfillment and/or marital success. Surveys reveal that many married adults report that their highest levels of marital satisfaction occurred before they had children or after their children left home. Indeed, some researchers assert that for a growing number of adults, especially those who seek "soul mate" marriages, having children is no longer a defining or essential life event (Whitehead and Popenoe 2008). Today, fully one in five women complete their childbearing years without having children (Livingston and Cohn 2010).

The economic and personal costs of child rearing may account for the steady decrease in average family size during the past century. The latest census figures put the average family size at 3.2 (U.S. Census Bureau 2008a). Furthermore, surveys indicate that Americans no longer fantasize about "having a house full of children." A recent Gallup poll indicates that 56% of Americans say the ideal family size is zero, one, or two children. When Gallup first started tracking this measure in 1936, Americans expressed a desire for larger families—that is, three or more children. But the trend reversed itself in the mid- to late 1970s, and since the late 1990s, roughly half of all Americans prefer smaller families (Gallup Poll 2007a).

Delayed childbearing also accounts for part of the decline in average family size. Women in the United States are waiting longer to marry (Whitehead and Popenoe 2008) and to have children, and such delays ultimately translate into fewer total births (Martin et al. 2006). In addition, increases in education levels, career aspirations, satisfaction with present life situations, and more reliable birth control all appear to be making it easier for some couples to remain childless (Bennett, Bloom, and Craig 1992; Dalphonse 1997; Wu and MacNeill 2002). Census data indicate that 44.6% of women of childbearing age are childless (Dye 2005). About 20% of women in their early 40s are childless, a twofold increase over the percentage childless in 1976 (Whitehead and Popenoe 2008). A growing trend toward singlehood among the young and middle-aged further complicates overall U.S. fertility patterns. Since 1970, there has been a steady increase in the percentage of males and females who have never married: In 2008, 35% of males and 28% of females 15 years and older fell into the "never-married" category. When we consider all of these changes and trends, it isn't too surprising to see that at present, only 21% of U.S. households embody the traditional nuclear family, that is, married couples with children under age 18 (U.S. Census Bureau 2008a).

Any society interested in its own survival must keep a watchful eye on such developments. In 2002, the U.S. birth rate reached its lowest level since birth records were first kept in the early 1900s: 13.9 births per 1,000 persons, a decline of 17% since 1990 (Zitner 2003)! With this drop, the United States teetered on joining the growing ranks of countries with below-replacement-level fertility: Austria, Ireland, Japan, Lithuania, Portugal, South Korea, and Taiwan. (**Replacement-level fertility** is the level needed for a population to continually renew itself without increasing.) Currently, replacement-level fertility is about 2.1 children per woman. In 2007, the U.S. total fertility rate was 2.12 (National Marriage Project 2009).

Replacement-level fertility
The fertility level needed for a population to continually renew itself without growing

Many population experts regard lower birth rates as foreshadowing dangerous population decline. In general, low fertility produces shrinking labor forces. Declining birth rates also mean that the burden of supporting the social programs for a nation's aging and retiring population will fall on fewer and fewer shoulders. And the longer low fertility rates exist, the harder it becomes to reverse population declines (McDonald 2001; Zitner 2003). Some suspect that the sluggish U.S. economy will further reduce our birth rate and bring us closer to the "zero population growth" situation of other industrialized countries of the world; others see our fertility rates settling into a stabilizing trend (Haub 2003, 2007).

Somewhat ironically, then, nations with increasingly geriatric populations can ill afford record low birth rates—the older a population, the greater its stake in achieving higher birth rates. Viewed in this light, maintaining an ideal culture that values children makes good sense, even when we sometimes fail to practice what we preach. Placing a high premium on children is one important way to ensure that individuals continue to make a financial investment in children. Many industrialized countries of Europe are now promoting pronatalist policies, and they seem to be having some success (Haub 2008). With prochild policies, these countries are promoting the critical raw material for societal survival. While once a rather academic subject matter, discussions of world fertility issues are now finding their way into the mainstream press. The *Population Reference Bureau* has a website that allows users to monitor fertility conditions of low-fertility countries of the world. The U.S. fertility rate, while holding somewhat steady in the last few years, deserves our attention (Haub 2007).

American conventional wisdom on children seems at odds with our behaviors and actions. But this cultural inconsistency is not likely to disappear soon. Indeed, even if the costs of having and raising children continue to increase and the structures of our families continue to change, prochild rhetoric may grow even stronger in the United States in the days ahead.

LEARNING MORE ABOUT IT

For a very readable review of sociological perspectives on childhood, see William Cosaro's book *The Sociology of Childhood* (Thousand Oaks: Sage, 2005).

D. Russell Crane and Tim B. Heaton provide a collection of articles addressing family and poverty. See the *Handbook of Families and Poverty* (Thousand Oaks, CA: Sage, 2007).

For an interesting perspective on historical responses to child abuse in the United States, see Stephen J. Pfohl's "The Discovery of Child Abuse" (*Social Problems* 24 (3): 310–323, 1977). Murray Straus and Denise Donnelly offer a thorough review of corporal punishment and its effects on children in *Beating the Devil Out of Them: Corporal Punishment in American Families and Its Effects on Children* (New Brunswick, NJ: Transaction Publishers, 2001). And Sylvia I. Mignon, Calvin J. Larson, and William M. Holmes provide a comprehensive review of family violence—one that offers an integrated theoretical explanation (sociology, psychology, and biology). See *Family Abuse: Consequences, Theories, and Responses* (Boston: Allyn & Bacon, 2002).

Viviana Zelizer offers a fascinating historical review on the social value of children in *Pricing the Priceless Child* (New York: Basic Books, 1985).

A review of the issue of low fertility and its global spread can be found in S. Philip Morgan and Miles Taylor's article, "Low Fertility at the Turn of the Twenty-First Century" (*Annual Review of Sociology,* 32: 375–399, 2006).

For a discussion on the discrepancy between goals and the paths we choose to achieve them, see Robert Merton's classic 1938 work, "Social Structure and Anomie" (*American Sociological Review* 3: 672–682).

GOING ONLINE FOR SOME SECOND THOUGHTS

The Child Trends Data Bank offers a collection of pertinent indicators for monitoring trends in child and family welfare: http://www.childtrendsdatabank.org/.

The Children's Welfare League of America offers online national and state fact sheets on the status of America's children at http://www.cwla.org/advocacy. Scroll down the page and follow the link for the "CWLA 2010 National & State Factsheets."

The National Center for Children in Poverty offers a link to demographic profiles of children in each of the 50 states: http://www.nccp.org/profiles/.

The following organizations and sites can help you learn more about children in society:

The Administration for Children and Families: http://www.acf.hhs.gov/

Children's Defense Fund: http://www.childrensdefense.org

Children's Rights Council: http://www.crckids.org/

NOW IT'S YOUR TURN

1. Select three family TV programs whose cast of characters includes children. Monitor the content of the programs for a 2- to 3-week period, noting program incidents and themes. On the basis of your observations, do the programs emphasize ideal or real culture in their portrayal of children in our society? What factors can you suggest that might account for your findings?

2. Visit and view the state fact sheets about children posted on the Child Welfare League of America website: http://www.cwla.org/advocacy/statefactsheets/statefactsheets10.htm

   - What are the states (if any) where the poverty rate for children under 18 is lower than the overall poverty rate for the entire nation? What were your most surprising findings? Least surprising? Which states appear to be making the most progress against childhood poverty? What's your evidence? Speculate about the factors that seem to account for progress.

3. Sample some TV commercials from early-morning children's programming. Use the information to assemble a social profile of U.S. children. How does this profile (i.e., race, economic status, living arrangements, etc.) compare with the picture provided by the National Center for Children in Poverty (http://www.nccp.org/profiles/)?

4. Access the National Center for Children in Poverty's "50-State Demographics Wizard" page: http://www.nccp.org/tools/demographics/.

   - Choose "all states" and then select some characteristic/area of interest to you. Create the table that uses the "characteristic as denominator." How do select states compare to the national data for the chosen characteristics?

SOCIAL STRUCTURE

ESSAY 4

Conventional Wisdom Tells Us . . . Love Knows No Reason

In this essay, we explore various social statuses–age, education, gender, income, race, religion–noting the ways in which these factors can guide something as seemingly individualistic as Cupid's arrow.

Love: that invigorating, addictive sensation, the emotional roller-coaster ride that signals the wonder of being human. Conventional wisdom locates love in the world of passion; adages describe it as an experience that can make us irrational, impetuous, and often oblivious to the real-life events that surround us. There is nothing logical about love. Love bows to emotion, not reason.

The conventional wisdom on love is not easily shaken. One might even say it stands shoulder to shoulder with the laws of the land! For example, in 1974, the National Science Foundation awarded an $84,000 grant for a social scientific investigation into the experience of love. But upon hearing of the grant, then-Senator William Proxmire blasted the investment from the floor of the U.S. Senate, saying,

> I'm against this, not only because no one—not even the National Science Foundation—can argue that falling in love is a science; not only because I'm sure that even if they spend 84 million or 84 billion they wouldn't get an answer that anyone would believe. *I'm also against it because I don't want the answer!* (quoted in Harris 1978, our emphasis)

Proxmire embraced the conventional wisdom on love—for him, love knows no reason. Is his assumption supported? Is the fall into love strictly an emotional journey? Does Cupid's arrow defy reason or direction and truly strike us senseless?

Despite romantic visions, research indicates that the experience of falling in love is much more logical than we might like to admit. Although Cupid's arrow may contain the magic that joins two hearts, Cupid's aim appears to be highly selective and heavily influenced by the social status of his targets. **Social status** refers to the position or location of an individual with reference to characteristics such as age, education, gender, income, race, and religion. Consider the love-and-marriage game. Our choices for the "love of our lives" generally occur along highly predictable lines (Arum, Roksa, and Budig 2008; Blackwell and Lichter 2004; Brynin, Oerez and Longhi 2009; Buss et al. 2001; Gardyn 2002; Kalmijn 1998; McPherson, Smith-Lovin, and Cook 2001; Ono 2006; Schwartz and Scott 2007; U.S. Census Bureau 2009a). Indeed, these choices are guided by "rules" that sociologists refer to as norms of homogamy. **Norms of homogamy** encourage interaction between individuals occupying similar social statuses.

Social status
The position or location of an individual with reference to characteristics such as age, education, gender, income, race, and religion

Norms of homogamy
Social rules that encourage interaction between individuals occupying similar social statuses

Research on both opposite-sex and same-sex couples shows that the large majority of Americans falls in love and marry mates within three years of their own age (Max Planck Institute for Demographic Research 2010). Furthermore, the look of love seems routinely reserved for those of like races, religions, and classes. Fewer than 5% of all U.S. marriages are interracial (U.S. Census Bureau 2009a), fewer than 15% are interethnic (Passel, Wang, and Taylor 2010), and only about a third of American marriages occur between people of different religions (The Pew Forum on Religious and Public Life 2008; Rostosky et al. 2008). Further, the large majority of all American marriages unite people of the same educational or occupational status (Blossfeld 2009; Gardyn 2002; Holman, Larson, and Olsen 2001; Hou and Myles 2008; Kalmijn and Flap 2001). Cupid's arrow also is typically reserved for those with similar intelligence levels (Blossfield 2009; Sprecher and Regan 2002) and similar physical appearances (Gardyn 2002; Montoya 2008; Mulford et al. 1998). And for centuries, Cupid's flights were seemingly amazingly short. Research documented a 50/50 chance that individuals would marry people who live within walking distance of their homes or their jobs. This fact still holds true for those living in large urban areas. With the dawn of Internet dating, however, Cupid sometimes take longer trips. Indeed, those who meet a mate via a dating site or virtual community are far more likely to partner with someone outside their home state (Barak 2008).

When it comes to love and marriage, research suggests that "birds of a feather seem to flock together," while "opposites rarely attract." Star-crossed lovers may exist in principle, but in practice, the stars of love are likely to be carefully charted by social forces—and thus glitter in the eyes of a social peer. Such patterns lead sociologists to characterize both romantic love and marriage as endogamous phenomena. **Endogamy** is a practice restricted by shared group membership; it is an "in-group" phenomenon. What explains the endogamous nature of love? Endogamy generally results from a societal wish to maintain group and class boundaries. In some cases, these boundaries are formally stated. Certain religious groups, for example, strongly urge their members toward intrafaith marriages. Similarly, traditional caste systems generally prohibit marriage between individuals of different castes. In other settings, however, boundary maintenance occurs via more informal channels, where certain key social players may quietly enforce endogamy in love and marriage.

Endogamy
Marriage that is restricted by shared group membership ("in-group" phenomenon)

In U.S. society, parents and peers often play the role of enforcer. It is not uncommon for family and friends to threaten, cajole, wheedle, and even bribe individuals in the direction of a "suitable" mate. Suitability generally translates into a partnership based on similar social profiles. Thus, the "American way" of marriage places the most personal decisions of one's life—falling in love and getting married—under "group control." This subtle control serves as a vehicle by which endogamy can be maintained.

Like romantic love, the development and expression of platonic love or friendship are also heavily influenced by the social location of its participants. Research shows that friendship formation is largely a product of one's daily interaction patterns rather than of chance or "good chemistry."

Friendships tend to develop with the people we see most often—those with whom we work, those enrolled in the same classes, or members of our churches, health clubs, and so on. Similarly, friendships tend to form among those who are in close geographic proximity to us, as opposed to those who are farther away—that is, the people in the same apartment complex or those who live in the same neighborhood as we do (Carley and Krackhardt 1996; Festinger, Schachter, and Back 1950; Giordano 2003; McPherson et al. 2001; Pahl and Pevalin 2005; Spencer and Pahl 2006). One study shows that even when people are randomly assigned to seats in a classroom, people are most likely to become friends with those they sit nearest to—even if they have had prior contact with other people in the room (Mitja, Schmukle, and Egloff 2008).

Like romantic love, the platonic love of friendship grows best out of similarity. We tend to build the strongest friendships with those who hold attitudes similar to our own. We also tend to connect with those who share our physical and social characteristics—appearance, income, education level, race, and so on (Adams and Allan 1998; Becker et al. 2009; Cadell 2010; Espenshade 2009; Giordano 2003; McCall et al. 2010; McPherson et al. 2001; Pahl and Pevalin 2005; Spencer and Pahl 2006). These patterns hold true in friendships built in physical or virtual space (Thelwall 2009). Thus, in the case of friendship, familiarity breeds attraction rather than contempt.

Interestingly, our feelings about friendship—the way we define and express it—differ in systematic ways on the basis of our socioeconomic status. **Socioeconomic status** refers to a particular social location—one defined with reference to education, occupation, and financial resources. Studies show, for example, that working-class Americans conceive of friendship as an exchange of goods and services. Gifts and favors come to indicate the strength of a friendship bond. In contrast, material exchange is absent from middle-class definitions of friendship. Middle-class individuals frequently view friendship as an emotional or intellectual exchange; they may also conceive of friendship simply as the sharing of leisure activities.

Socioeconomic status
A particular social location defined with reference to education, occupation, and financial resources; *see also* **Poverty and wealth**

The "faces" of our friends also differ by social class. Thus, if you are a member of America's working class, your friends are highly likely to be relatives—siblings, cousins, parents, and so on. In contrast, middle-class individuals prefer non-blood relations for friends. Furthermore, among working-class people, friendships are overwhelmingly same sex, whereas middle-class people are more open to cross-gender friendships. Finally, if you are from America's working class, your friendships are likely to be local. Thus, working-class friends interact, on average, once a week or more. In contrast, middle-class individuals have as many long-distance friendships as they do local ones. Because middle-class life in America often involves high levels of geographic mobility, members of the middle class are more likely than their working-class counterparts to maintain friendships after individuals move out of the immediate geographic area. This distance factor carries a downside, however. Middle-class friends generally report less-frequent contact than their working-class counterparts (Adams and Allan 1998; Elles 1993; Fischer 1982; Giordano 2003; Gouldner and Strong 1987; Liebler and Sandefur 2002; McCall et al. 2010; Rawlins 1992; Spencer and Pahl 2006; Walker 1995).

Your Thoughts . . .

Actor Peter Ustinov once remarked: "I do not believe that friends are necessarily the people you like best, they are merely the people who got there first." What do you think?

The patterned nature of love also emerges in matters of self-love, or what social scientists refer to as self-esteem. **Self-esteem** refers to the personal judgments individuals make regarding their own self-worth. Like romantic and platonic love, love of self appears quite systematically tied to an individual's social situation.

> **Self-esteem**
> Personal judgments individuals make regarding their own self-worth

On the level of experience, for example, many studies show that self-esteem is directly tied to the love expressed toward that individual by her or his significant others. Not surprisingly, when significant others give positive feedback, self-esteem increases. Conversely, consistently negative feedback from significant others lowers self-esteem (Amato and Fowler 2002; Hattie and Timperley 2007; Jaret, Reitzes, and Shapkina 2005; Mruk 2006; Voss, Markiewicz, and Doyle 1999). Similarly, the character of one's social or work environment clearly influences self-esteem. Those who are situated among optimistic people in positive, upbeat environments have been shown repeatedly to enjoy better self-esteem than those who find themselves in negative environments with depressed or disgruntled colleagues (Cross and Vick 2001; Kacmar et al. 2009; Mruk 2006; Ross and Broh 2000). Finally, various social attributes can influence levels of self-love or self-esteem, with members of upper classes and racial majority groups routinely faring better than the poor or those in racial minorities (Coopersmith 1967; Felson and Reed 1986, 1987; Gergen 1971; McLeod and Owens 2004; Mruk 2006; Shaw, Liang, and Krause 2010).

Romantic love, platonic love, self-love—when findings from these areas are considered together, we must concede that love, in its various forms, is a highly structured phenomenon. There is much rhyme and reason regarding how we find it, define it, experience it, and express it. And that logic is tied to aspects of our social backgrounds and our social locations. Knowing this, we might do better to trade our notions of the irrational heart for knowledge of the social organization of the heart. Indeed, the study of love reminds us that even the most personal of experiences can succumb to the systematic influence of the social.

LEARNING MORE ABOUT IT

For informative and very readable studies on romantic coupling in America, see *The Mating Game: A Primer on Love, Sex, and Marriage,* 2nd ed. by Pamela C. Regan (Thousand Oaks: Sage Publications, 2008), *Seven Stories of Love* by Marcia Millman (New York: HarperCollins, 2002), or *Talk of Love: How Culture Matters* by Ann Swidler (Chicago: University of Chicago Press, 2003).

Amy Baker offers a fascinating study on Internet dating in "Down the Rabbit Hole: The Role of Place in the Initiation and Development of Online Relationships." See pages 163–184 in A. Barak (ed.),

Psychological Aspects of Cyberspace: Theory, Research, Applications (New York: Cambridge University Press, 2008).

For recent reviews of literature on the social similarities of lovers and friends, see "Birds of a Feather," a 2001 article by Miller McPherson, Lynn Smith-Lovin, and James M. Cook (*Annual Review of Sociology* 27: 415–444, 2001), Peggy Giordano's 2003 piece "Relationships in Adolescence" (*Annual Review of Sociology* 29: 257–281, 2003), or Nijole Benokraitis's *Marriages and Families: Changes, Choices, and Constraints* (Upper Saddle River, NJ: Prentice Hall, 2010).

For an interesting and still timely look at the social aspects of friendship, see Lillian Rubin's *Just Friends: The Role of Friendship in Our Lives* (New York: HarperCollins, 1993). A more recent treatment of the social patterns of friendship can be found in George McCall and colleagues' new book *Friendship as a Social Institution* (New York: Aldine/Transaction, 2010).

Most sociological work on self-esteem is steeped in the writings of Charles Horton Cooley. His classic works include *Human Nature and Social Order* (New York: Scribner, 1902) and *Social Organization* (New York: Charles Scribner, 1909). A very readable review of recent research on self-esteem can be found in Christopher Mruk's book *Self-Esteem: Research, Theory, and Practice*, 3rd edition (New York: Springer, 2006).

GOING ONLINE FOR SOME SECOND THOUGHTS

The Matchmaking Institute provides online comprehensive home study kits designed to teach you the art of professional matchmaking. Information on the effort to professionalize this occupation can be found at http://www.matchmakinginstitute.com/home-study-kit/.

NOW IT'S YOUR TURN

1. It's your turn to collect some data. Identify your three closest friends. Then create a social profile of them—their ages, education levels, ethnicities, genders, geographic locations, occupations, races, religion, and so forth. How many of these characteristics are similar to those in your own social profile? Now approach four acquaintances or colleagues and ask them to do the exercise. Review your data. Do these data support or refute the notion of friendship and norms of homogamy?

2. Now collect some data on marriage patterns. Using yourself (if appropriate) and your married friends as case studies, discuss how the rules of homogamy either apply or do not apply in these individuals' selections of marriage partners. Collect similar information about your parents, aunts, and uncles, and compare it with what you found about yourself and your friends. Do you see any important generational changes in the rules for marital homogamy? Are there rule variations that can be linked to class or educational factors?

ESSAY 5

Conventional Wisdom Tells Us . . . Stress Is Bad for Your Well-Being

Or is it? This essay reviews the conditions under which stress can prove beneficial in one's everyday activities. In so doing, we highlight the importance of considering social context in assessing social behaviors.

Stress has become a regular feature of modern-day existence. Finding a parking space at the mall, hooking up a new home entertainment system, navigating the university's new automated registration system, correcting an error on your credit card bill—in today's fast-paced, high-tech environment, stress can weave its way into even the most routine tasks.

Modernization and technological advancement have stress-related costs. To be sure, these phenomena make possible amazing strides, including increased life spans, greater geographic mobility, and heightened industrial and agricultural productivity. Yet these changes also actively alter a society's social structure. **Social structure** refers to the organization of a society—the ways in which social statuses, resources, power, and mechanisms of control combine to form a framework of operations.

Social structure
The organization of a society; the ways in which social statuses, resources, power, and mechanisms of control combine to form a framework of operations

Thus, for many, a society's "amazing strides" may translate into commuter marriages, single parenthood, long widowhoods, "downsized" work environments, unmanageable traffic, pollution—conditions often associated with increased stress. In addition, such advances may expand the ranks of the poor, trigger rapid population growth, and increase competition for resources. Such structural changes can increase the day-to-day stress experienced by those in certain social locations. **Social location** is an individual's total collection of social statuses; it pinpoints an individual's social position by simultaneously considering age, education, gender, income, race, and so on.

Social location
One's social position as indicated by one's total collection of social statuses—that is, age, education, gender, income, race, and so forth

The pervasiveness of stress makes it important

to weigh conventional wisdom's dire warnings on the subject. Will the benefits of modernization ultimately cost us our physical, mental, or emotional health? Just what toll does stress take on our overall well-being?

The links between stress and well-being are complex because the effects of stress vary by social context (Cockerham 2009; Jacobson 1989; Lennon 1989; Pearlin 1989; Pearlin et al. 2005). **Social context** refers to the broad social and historical circumstances surrounding an act or an event. For example, consider the links between stress and health. Many studies link stress to serious physical problems: cancer, heart disease, mental illness, and emotional depression. Research also suggests that stress can trigger increases in smoking, drinking, drug use, and other hazardous behaviors (Centers for Disease Control 2009g; Lantz et al. 2005; Ross and Huber 1985; Sinha, 2008; Stenson 2003; Wheaton 1983). However, these negative effects are largely confined to contexts in which stress is chronic. **Chronic stress** refers to the relatively enduring problems, conflicts, and threats that individuals face on a daily basis. Most researchers agree that chronic stress contexts, such as persistent financial woes, a bad marriage, and constant exposure to crime, violence, overcrowding, or even noise, are harmful to our well-being. In contrast, sporadic, short-term stress generally proves less detrimental to well-being (Aneshensel 1992; Hormone Foundation 2008; House et al. 1986; Kristenson et al. 2004; Pearlin 1989; Sinha, 2008).

Social context
The broad social and historical circumstances surrounding an act or an event

Chronic stress
The relatively enduring problems, conflicts, and threats that individuals face on a daily basis

Now consider the stress generated by certain life events—retirement, children leaving home, the death of a spouse, major changes in the workplace, and so on. Conventional wisdom suggests that such events can be the most stressful experiences of our lives. Again, however, research reveals that the stress associated with these life events varies with the social context in which the event occurs. For example, retirement actually has been shown to alleviate stress if one is leaving an unpleasant or difficult job. Similarly, when a child leaves home, stress actually decreases for those parents who perceived their family relationships to be troubled or strained. And major changes in the workplace can prove beneficial if those changes challenge and tap workers' knowledge and skills rather than threatening their security. To determine the level of stress associated with any life event, one must explore the circumstances and activities that precede and/or accompany the event; one must assess the life event within its proper social context (Aneshensel 1992; Cockerham 2009; Jacobson 1989; Lennon 1989; Martin and Svebak 2001; Mirowsky and Ross 1989; Pearlin 1989; Simon 1997; Simon and Marcussen 1999; Thoits 1983; Verhaeghe et al. 2006; Wheaton 1982).

In assessing the conventional wisdom on stress, it also is important to note that the negative effects of stress are not inevitable. Research documents several coping mechanisms that can temper or even cancel the negative impact of stress on one's well-being. For example, individuals who enjoy strong social support networks often are protected from the harmful consequences of stress. A **social support network** consists of family, friends, agencies, and resources—entities that actively assist individuals in coping with adverse or unexpected events. Studies document, for instance, that widows or widowers who have close friends or confidants report much less stress from the death of a spouse than do individuals who lack such support. Elderly people with strong friendship networks suffer

Social support network
A network that consists of family, friends, agencies, and resources that actively assist individuals in coping with adverse or unexpected events

fewer of aging's stressful effects than do those who lack support networks. Similarly, the stress of divorce appears greatly diminished for those with close friends, confidants, or new romantic interests. And research on children shows that increased parental contact can mitigate the stress experienced by children when families are forced to geographically relocate. For children and adolescents, strong friendship networks can help decrease the stress associated with domestic or school victimization (Boardman 2004; Cockerham 2009; Cohen 2004; DuBois et al. 2008; Haden and Scarpa 2008; Hagan, MacMillan, and Wheaton 1996; House et al. 1986; Hutchinson et al. 2008; Kessler, Price, and Wortman 1985; Lin, Ye, and Ensel 1999; Mui 2001; Roussi, Vagia, and Koutri 2007; Thoits 1995; Treharne, Lyons, and Tupling 2001; Wheaton 1982, 1990; Zimmer-Gembeck and Locke 2007).

Certain resources also can influence the experience of stress. Research indicates that relaxation techniques can buffer individuals from the negative impact of stress. Similarly, problem-solving strategies or strategies that can physically or mentally distance one from the site of stress can help mitigate its harmful effects. Research also indicates significant stress reduction among optimists, those who learn to deemphasize the negative aspects of a stress-producing role, or among those with strong belief systems. And those who perceive some level of personal control over their environment experience less stress than those who feel little or no control (Affleck, Tennen, and Apter 2000; Caplan and Schooler 2007; Chan 2002; Frazier et al. 2000; Nes and Segerstrom 2006; Pham, Taylor, and Seeman 2001; Simon 1997; Thoits 1994, 1995, 2006; Von Ah, Kang, and Carpenter 2007; Wilkinson and Coleman 2010).

Coping mechanisms can offer protection from stress. However, some contend that modern lifestyles may make it difficult for individuals to put these "safeguards" into effect. For example, the geographic mobility that characterizes modern society may place friends and family out of one's immediate reach. Similarly, increased access to information may create a mental overload that eats away at one's relaxation time, and technological advancements that allow one to merge work and home sites may make mental distancing strategies difficult to execute (Hochschild 1997/2001, 2003a; Nippert-Eng 1996; Pearson 1993; Philipson 2002; Robinson 2003).

Your Thoughts . . .

What role do you think social location plays in relation to the availability of coping strategies?

The successful enactment of coping mechanisms may be largely related to the kinds of social relationships that characterize one's social environment. Ferdinand Tonnies (1855–1936) analyzed such relationships using two distinct categories: Gemeinschaft and Gesellschaft. **Gemeinschaft** refers to an environment in which social relationships are based on ties of friendship and kinship. **Gesellschaft** refers to an environment in which social relationships are formal, impersonal, and often initiated for specialized or instrumental purposes. Modern social environments, with their emphasis on privacy and individuality, reflect the Gesellschaft environment.

Gemeinschaft
An environment in which social relationships are based on ties of friendship and kinship

Gesellschaft
An environment in which social relationships are formal, impersonal, and often initiated for specialized or instrumental purposes

As such, the social resources from which coping mechanisms develop may not be readily available to modern women and men.

Are there contexts in which stress positively influences our well-being? Social psychologists have demonstrated that task-oriented stress often can lead to visibly productive consequences. **Task-oriented stress** refers to short-term stress that accompanies particular assignments or settings. Individuals who report feeling completely comfortable or relaxed during the execution of certain mental tasks remember and absorb less information than those who experience moderate levels of task-oriented stress. Indeed, moderate levels of task-oriented stress have been linked to enhanced memory of facts and skills and increased learning ability—important attributes in our postindustrial, knowledge-based society. This suggests that the nervous tingles you experience in studying for the law boards, your driving test, or a public speaking engagement may serve you better than a lackadaisical stance (Driskell, Johnston, and Salas 2001; Ellis 1972; LeBlanc and Bandiera 2007; Mughal, Walsh, and Wilding 1996; Nelson and Simmons 2005; Quick et al. 2003). Social psychologists have also demonstrated that stress can promote self-improvement. This is because stressful events encourage (or force) individuals to acquire new skills, reevaluate their positions, and reconsider their priorities. In this way, stress may serve as a catalyst for personal growth (Calhoun and Tedeschi 2001; Gibson and Quick 2008; Quick et al. 2003; Tennen and Affleck 1999; Van den Broeck 2010; Verhaeghe et al. 2006).

Task-oriented stress
Short-term stress that accompanies particular assignments or settings

Studies also show that stress can work through other physiological or psychological states to produce quite unexpected, and quite positive, behavioral outcomes. For example, when stress leads to a state of emotional arousal such as anxiety or fear, stressed individuals are more likely to befriend or bond with others. (This may explain why your student colleagues always seem more approachable on the day of a big exam.) Furthermore, when stress leads to anxiety or fear, stressed individuals demonstrate a greater tendency to like and interact with people whom they usually dislike or around whom they typically feel uncomfortable. This "benefit" extends to people who differ from the stressed individual in terms of race, socioeconomic status, and personality. In light of these findings, some researchers contend that under the right circumstances, stress may aid the cause of achieving interracial, intergenerational, or interclass affiliations (Kulik and Mahler 1990; Kulik, Mahler, and Moore 1996; Latané and Glass 1968; Leary 2010; Rofé 2006; Schachter 1959).

The links between stress and affiliation hold true at the collective level as well. Where disasters strike communities at large—be they hurricanes, floods, or terrorist attacks—victims' most common response involves affiliation. While stereotypes of disaster suggest that victims succumb to mass panic and hysteria, studies repeatedly show that disaster victims are much more likely to bond together and work toward mutual survival and recovery (Mawson 2005; Quarantelli 2008; Tierney, Bevc, and Kuligowsky 2006).

It is interesting to note that stress that proves detrimental to the well-being of individuals sometimes may prove productive for societies at large. Consider one such example in the area of chronic stress. Chronic stress can emerge from a particular type of long-term situation, a condition sociologists refer to as role conflict. **Role conflict** occurs when social group members occupy two or more social positions that carry opposing demands. Military chaplains, working parents, and student teachers all provide examples of potentially conflicting role combinations. Roles that carry opposing demands create a tug of war within individuals—a persistent strain characteristic of chronic stress.

Role conflict
Conflict that occurs when individuals occupy two or more social positions that carry opposing demands

Although role conflict can take its toll on an individual's well-being, it sometimes proves the source of positive social change. For example, when role conflict is routinized by changing cultural or economic demands, the resulting stress can actually trigger needed social restructuring. Routinized role conflict can lead societies to institute changes that positively alter the playing field of social interaction. For example, the conflict and stress that emerged from the working-parent role combination served to revolutionize America's work environment. Methods such as flex time, in-house day care, and work-at-home options—originally antidotes to the stress of role conflict—are now a productive dimension of work in the United States.

We can take this analysis of the positive consequences of stress one step further. The phenomenon of stress need not be confined to individual-level inquiries; societies as a whole also can experience stress. Sociologists refer to this type of stress as social strain. **Social strain** develops when a social event or trend disrupts the equilibrium of a society's social structure. For example, an economic depression may generate social strain by forcing increases in unemployment and exacerbating poverty. In essence, the depression event disrupts expected patterns of resource distribution. Similarly, a large increase in a society's birth rate may place strain on various social institutions. Schools, hospitals, or prisons may suddenly be presented with more clients than they were designed to serve.

Social strain
A social event or trend that disrupts the equilibrium of a society's social structure

Sociologists such as Talcott Parsons (1951/1964) or Lewis Coser (1956), although coming from different perspectives, both suggested that social strain creates an opportunity for societal change and growth. (Note that Coser refers to the phenomenon of strain as "social conflict.") With their writings, both Parsons and Coser established a view of society that continues to guide contemporary social thinkers. Social strain is important because it disrupts the status quo. Thus, it can force societies to work at reestablishing smooth operations. For example, the strain placed on the U.S. stratification system by the civil rights movement of the 1950s and 1960s resulted in positive strides toward racial and ethnic equality in America. Similarly, consider the growing demands that U.S. entitlement programs currently place on the nation's economic system. Many credit such strain with prompting much-needed public discourse on major budget reforms.

Is stress harmful to our well-being? Taken as a whole, current research on stress paints a less dismal picture than that promoted by conventional wisdom. To be sure, stress is frequently harmful, and it is rarely a pleasant experience. Yet its consequences do not necessarily jeopardize personal health and happiness. In fact, when one views stress in context or at the level of societies at large, it sometimes proves to be a useful social resource.

LEARNING MORE ABOUT IT

For a good summary of current findings and controversies within the social science literature on stress, see Leonard Pearlin and colleagues' 2005 article "Stress, Health, and the Life Course: Some Conceptual Perspectives" (*Journal of Health and Social Behavior* 46 (2): 205–219), or William C. Cockerham's chapter "Social Stress" in his book *Medical Sociology* (11th edition, Upper Saddle River, NJ: Prentice Hall, 2009).

Several interesting experiments document the positive consequences of stress for individuals. Schachter's *The Psychology of Affiliation* (Stanford, CA: Stanford University Press, 1959) represents a classic among such studies. Mark R. Leary offers a good review of current literature in the field. See his chapter "Affiliation, Acceptance, and Belonging: The Pursuit of Interpersonal Connection." Pages 864–896 in S. T. Fiske, D. T. Gilbert, and G. Lindzey (eds.), *Handbook of Social Psychology,* 5th ed., Vol. 2, Hoboken, NJ: John Wiley and Sons, Inc., 2010.

Stress and conflict within social systems are wonderfully addressed in Lewis Coser's classic yet still relevant theoretical treatise, *The Functions of Social Conflict* (Glencoe, IL: Free Press, 1956).

Want to learn something about the history of stress in the marketplace and its "commerce" value? Visit Stressfocus.com at http://www.stressfocus.com for information, quizzes, and amusing tidbits on the subject.

GOING ONLINE FOR SOME SECOND THOUGHTS

The following organizations maintain websites that can help you learn more about stress:

The American Institute of Stress http://www.stress.org

National Mental Health Association http://www.nmha.org

NOW IT'S YOUR TURN

1. Research suggests that stress can increase an individual's likelihood of affiliating with others. Can stress function in a similar way at the societal level? It's your turn to test the hypothesis that social stress increases social solidarity. See if periods of economic recession or a nation's involvement in a major war are associated with any indicators of increased group cohesion. Using a source such as the Information Please Almanac, track membership rates in five national organizations for 3 years before and after the economic recession of 2008 or World War II.

2. It's your turn to collect some data. Compare 2 weeks' worth of letters to the editor prior to the September 11, 2001, terrorist attacks in America; Hurricane Katrina; the BP oil spill, or a well-publicized murder or accident in your hometown with 2 weeks' worth of letters to the editor after such an event. Analyze the content of letter writers' remarks concerning their personal feelings and reactions to these events. What does your analysis show regarding the social consequences of stress?

ESSAY 6

Conventional Wisdom Tells Us . . . The "Golden Years" Are Tarnished Years

Growing old—no one looks forward to it. Yet this essay illustrates that our worst fears about growing old may be largely unfounded, simply products of a "master status" for which we have been inadequately prepared.

Aging is a curious phenomenon. When we are young, we can't wait to be older—or at least old enough to drive, get a good job, and make our own decisions. When we finally reach adulthood, many of us continue to yearn for a later stage in life, a time when we can begin to capitalize on the lessons of youth, a time when we can enjoy the fruits of our labors. Retirement looks like a pretty good deal from the vantage point of youth. Indeed, one national study found that 88% of Americans 6 to 15 years away from retirement believe that retirement will be a very happy period in their lives, and 76% believe it will be a time when they will achieve their dreams (Ameriprise Financial 2006). This optimism appears to have weathered the great recession of 2007. A recent retirement survey found that 54% of workers are still somewhat (38%) or very (16%) optimistic that they will live comfortably in retirement (Employee Benefit Research Institute 2010).

Eventually, however, there comes a time when the benefits of aging seem less clear cut. We begin to view age as a liability, perhaps even as a thing to be feared. For many, the dread comes with the appearance of their first gray hairs (Graham 2011), and suddenly, we become more attuned to the very negative conventional wisdom on aging. The young are warned never to "trust anyone over 30." Those in their 40s and 50s frequently are characterized as "over the hill." Children and adolescents see their parents as "old fogies." And children aren't alone on this matter. In general, Americans tend to perceive men and women as "old" once they have reached the age of 63 (Abramson and Silverstein 2006). Those 65 and older, however, associate old age with the mid-70s (Pew Research Center 2009a). Advanced age is often viewed as a liability for many practices and occupations. After all, "you can't teach an old dog new tricks." And retiring is often equated with "being put out to pasture." Interestingly, these images are aided by the law. The Age Discrimination in Employment Act protects employees *aged 40 and over* from age discrimination! Organizations such as AARP (Association for the Advancement of Retired Persons) also fuel the fires of conventional wisdom. Note that this organization opens its membership ranks to those *50 and older!*

Thus, despite early desires to "be older," many Americans ultimately develop a rather negative view of growing old. Many picture old age as a time of loneliness, vulnerability, sickness, and anger. Many overestimate senility and the number of accidents that befall the elderly (Abramson and Silverstein 2006). And there is a widespread perception that the aging of America, especially the aging of the Baby Boomers, poses a social and economic problem. Are such images accurate? Is conventional wisdom's negative stance on growing old justified? Research suggests that the negative press on aging is not fully supported by the facts. In reviewing several studies on the elderly, one finds many inconsistencies between the public perceptions and the social realities of old age in America.

Consider, for instance, the image of old age as a lonely, isolated existence. It is true that the elderly make up a relatively large portion of single-occupant households in the United States, but in 2008, only 31% of the elderly lived alone (8.3 million women and 2.9 million men.) Nearly 72% of older men and 42% of older women lived with their spouses in 2008. Twenty percent of those 65 and over live in multigenerational family households (Pew Research Center 2010a). Many elderly are still caregivers. In 2008, 681,000 grandparents 65 and older maintained households for their grandchildren and 471,000 had primary care responsibility for those grandchildren (Administration on Aging 2010).

For those elderly who do live alone, such living arrangements seem to reflect a personal preference rather than a forced choice. Of those 50 and older who live in homes that will meet their changing physical needs, 95% indicate a desire to remain in those homes. Indeed, even when homes don't meet the physical needs of an aging population, most residents (62%) still express a desire to keep living in them (Kochera and Guterbock 2005). A recent survey of a nationally representative sample of the 65+ populations found that 87% indicated that if they needed any help in caring for themselves, they would want that help delivered in their current homes (Barrett 2008). To be sure, the elderly value their independence. Again, survey research documents that fully 95% of the 65+ age group report that it is very important to them to have the ability to do things for themselves; 82% consider it very important to have the freedom to do what they want (Barrett 2008).

In addition to living arrangements, several other behavioral patterns contradict the loneliness stereotype. Today's seniors are connected. A recent AARP survey found that 68% of the 65+ population currently have computers in their homes; 35% report having Internet access (Barrett 2008). Among those 65+ who use the Internet, 60% report using it daily (AARP 2009). In a nationally representative sample of seniors, 83% report highly valuing their friends and family and 67% report being completely satisfied with their level of involvement on this front (Barrett 2008). A recent study of Baby Boomers (who are now starting to join the ranks of 65+ population) found that roughly half of them want to travel and see the world (AARP Services, Inc. and Focalyst 2007). Another recent survey of "social engagement and giving" finds that while these activities decrease with age, the older generations are still quite active in these arenas. Consider for instance that the rate of self-reported volunteering in the previous year was 71% for Baby Boomers (those aged 45–63) and 69% for members of the "Silent Generation" (those aged 64–80). When the focus turns to civic engagement, the older generations can actually outperform younger age groups in certain activities. For instance, 59.3% of the Greatest Generation (those aged 81+) attended community activities sponsored by religious organizations in the last year, while only 43% of those in Generation X (age 29–44) reported doing so. Fifty percent of Baby Boomers and 44.4% of the Greatest Generation worked to fix a problem in their neighborhoods; 42.2% of Generation Xers reported doing so (AARP 2010).

In general, research shows that parents and adult children do have a high sense of mutual normative obligation (Johnson 2000; Rossi and Rossi 1990; Umberson 1996). Caregiving for the elderly is clearly a family affair. (The "typical" caregiver is a 46-year-old daughter who works outside the home but who also spends more than 20 hours a week providing unpaid care to her mother.) In recent years, reliance on paid caregiving has decreased; family caregiving is credited with improving the quality of health for the elderly and with keeping the elderly out of nursing homes. In 2007, the value of unpaid family caregiving was estimated to be $375 billion (Houser and Gibson 2008). And as previously mentioned, many elderly continue in the caregiver role for their children and grandchildren. In 2008, 6% of all children lived with their grandparents. In more than half of these families, adult parents were also present (Child Trends Data Bank 2010). And a majority of grandparents report that spending time with their grandchildren is one of the great benefits of aging (Pew Research Center 2009a).

Loneliness is only one of the many misconceptions we have about old age. Conventional wisdom also paints the elderly as frequent victims of violent crime. Yet violent crime against the elderly has declined drastically over the past 35 years. In fact, age and violent victimization are inversely related—as age goes up, victimization goes down. So while the violent victimization rate for those 65 and older was 3.1 per 1,000 in 2008, it was 37.8 for those aged 20 to 24 (Bureau of Justice Statistics 2010a)! And note that when the elderly are victims of crime, they are more likely to be victims of property crime rather than violent crime, although their victimization rates are still lower than those of other age groups. Consider, for instance, that in 2007, the victimization rate for property crimes involving the elderly was 75; the victimization rate for property crimes involving those aged 20 to 34 was 343 (Bureau of Justice Statistics 2010b). Yet the "crime myth" that surrounds old age is not without its consequences; the myth generates a great deal of anxiety among the elderly (although less today than previously; see Abramson and Silverstein 2006).

Another fear of aging centers on increased poverty. At one point in U.S. history, there was a firm basis for this fear. Until the 1970s, the elderly were more likely than any other age group to live in poverty. While the poverty rate for the population at large was only 13%, a full 25% of those 65 and older were classified as poor. But changes in the Social Security system—in particular, changes linking benefits to cost-of-living increases—have helped reduce the percentage of elderly living in poverty. Today, the poverty rate for those over 65 years of age hovers around 10%. Thus, current poverty rates for the elderly are below the national poverty rate of 13.2% (U.S. Census Bureau 2009b: Table POV1).

BOX 6.1 SECOND THOUGHTS ABOUT THE NEWS: KEEPING SOCIAL SECURITY SOLVENT

It is a fear that may be with us as long as the Baby Boomers are with us: the Social Security program running out of money. For the last several election cycles, candidates have had to weigh in on how they plan on keeping the system afloat. Recently, House leader John Boehner has indicated that the time has come for Americans to have an "adult" conversation about Social Security. And just what would that conversation sound like? For some, it would entail the idea of privatizing Social Security and/or raising the retirement age to 70. And for some, both of these ideas are nonstarters for any productive exchange about Social Security.

(Continued)

(Continued)

The George W. Bush administration tried to push the privatization angle, arguing that Americans should be free to decide what to do with their retirement savings. Traditionally, however, Social Security has been thought of as the "third rail" of the government–touch it at your own peril. And while privatization typically gets some traction when the country is enjoying good economic times (and bullish markets), our recent economic downturn will surely make it harder for many Americans to see the wisdom in casting their retirement fates to the fickle stock market. As it stands now, many retirement nest eggs were ravaged during the recent recession. Older Americans have had to seriously consider the idea of working longer than they initially planned. Consequently, whether mandated or not, we are likely to see the effective retirement age go up by default.

To be sure, we have already seen some movement in this direction. Initially, Social Security set 65 as the retirement age for collection of full benefits. Today, the age cutoff for full benefits depends on one's birth year–currently it is 67 for those born in or after 1960. The United States is not alone in trying to advance the conversation about adjusting the retirement age upward; several European nations are pursuing this route as a way to fight spiraling deficits. During the fall of 2010, millions of French people took to the streets to protest France's decision to raise its retirement age to 62! (Currently, the French retire on average at age 58 and are "in" retirement for about a quarter of a century! In contrast, Americans spend about 17 years in retirement.) A bipartisan Presidential commission on reducing the U.S. debt has re-energized the Social Security debate with the release of its December 2010 report and its recommendation that the retirement age be increased to 69 by 2075. By then, it will be the Baby Boomers' *grandchildren* who may be needing to have the Social Security conversation.

Images of physical and mental deterioration also pervade public perceptions of old age. To be sure, aging does result in some changes on the physical front. Arthritis, hypertension, hearing impairment, and heart disease are frequently occurring chronic conditions of the elderly. About 17% of those 65 and older report trouble seeing; more (32%) report trouble hearing. Yet it is important to note that health, not disease, is the norm for older Americans. People are living longer and healthier lives and are disabled for fewer years than was true for older people of the past (a trend referred to as compression of morbidity; MacArthur Foundation 2009). Only 26.8% of those 65 years and older self-assessed their health as fair or poor (National Center for Health Statistics 2010). With regard to specific acts of daily living (ADLs) such as bathing, dressing, and walking, the need for assistance does increase with age, but overall this need is relatively low for those between 65 and 84 years of age (see Figure 6.1; Administration on Aging 2010). And in recent years, the rate of functional limitations experienced by the elderly has been decreasing. Indeed, our functional capacity appears to be driven more by factors such as race, ethnicity, and education than by age (Federal Interagency Forum on Aging-Related Statistics 2008; MacArthur Foundation 2009).

The elderly's desire for independence and their general good health help explain another important fact: Only 4% of the 65+ population live in an institutional setting such as a nursing home. While the percentage of nursing home residents does increase with age, it is still the case that only 15% of those 85 and older are residing in nursing homes (Administration on Aging 2010).

In terms of mental health, the elderly fare quite well. Most enjoy good mental health—and

Figure 6.1 Percentage of Persons With ADL Limitations, by Age Groups: 2006

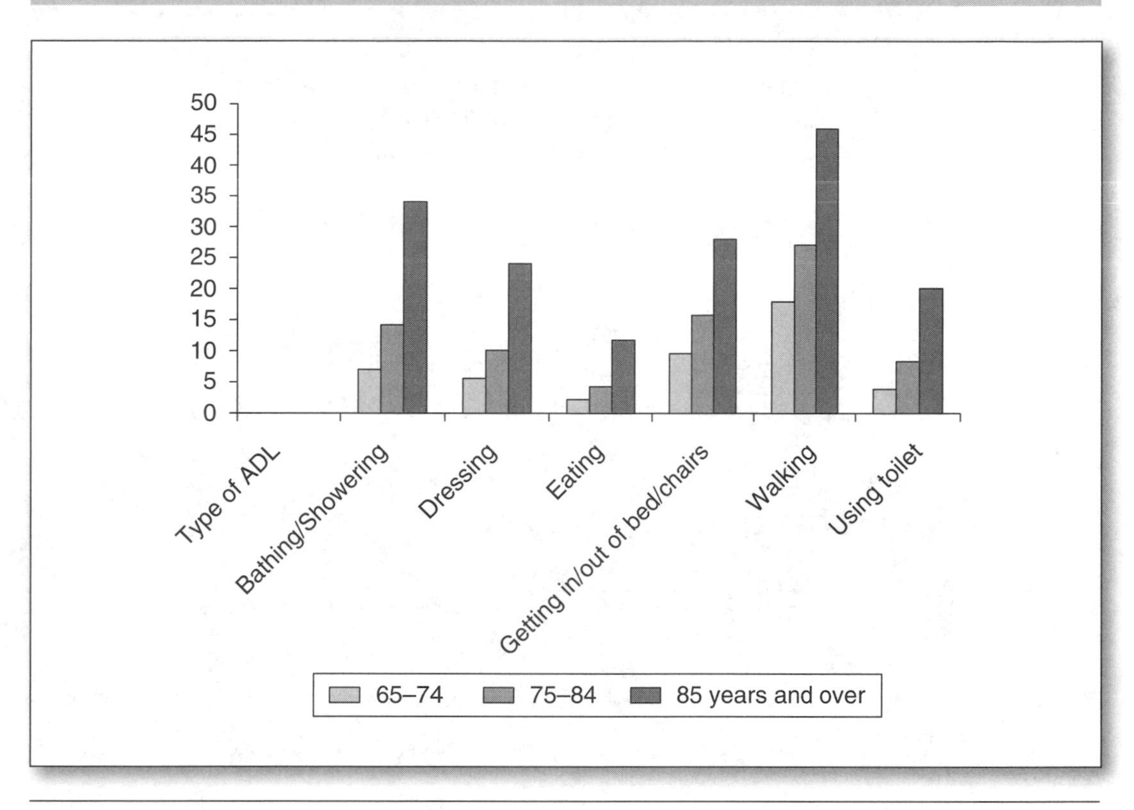

Source: Administration on Aging. (2010). A Profile of Older Americans: 2009. Available at http:www.aoa.gov/AoARoot/Aging_Statistics/Profile/2009/16.aspx.

the mental health problems that do arise are *not* part of the normal aging process per se. For many major mental disorders (schizophrenia; bipolar, obsessive-compulsive, panic, and anxiety disorders), the average age of onset is in adolescence and early adulthood (National Institute of Mental Health 2010). Moderate to severe memory loss affects only 5% of those 65 to 69, 8% of those 70 to 74, and only about a third of those aged 85 and over (Federal Interagency Forum on Aging-Related Statistics 2008). And although Alzheimer's disease is frequently associated with old age, it is not a normal or inevitable part of old age. Indeed, it is estimated that some 500,000 people in their 30s, 40s, and 50s are afflicted with Alzheimer's disease or a related dementia (Alzheimer's Association 2010). Overall, aging appears to bring an elusive mental benefit: The elderly, even with the changes associated with aging, are more likely to report higher levels of life satisfaction than are younger age groups (Administration on Aging 2010). Older Americans also report less mental distress than do younger age groups. In 2008, the mean number of mentally unhealthy days was 2.2 and 2.0 per year for those aged 65 to 74 and 75 and over. For those 18 to 24, the mean number of mentally unhealthy days was 4.0 (Centers for Disease Control and Prevention 2009e; see Figure 6.2). A growing body of research is finding that negative emotions decline with age and that **emotional regulation**, or the ability to maintain a positive affect,

Emotional regulation
The ability to maintain a positive affect

Figure 6.2 Percent Experiencing Mental Distress by Age Groups: 1993–2008

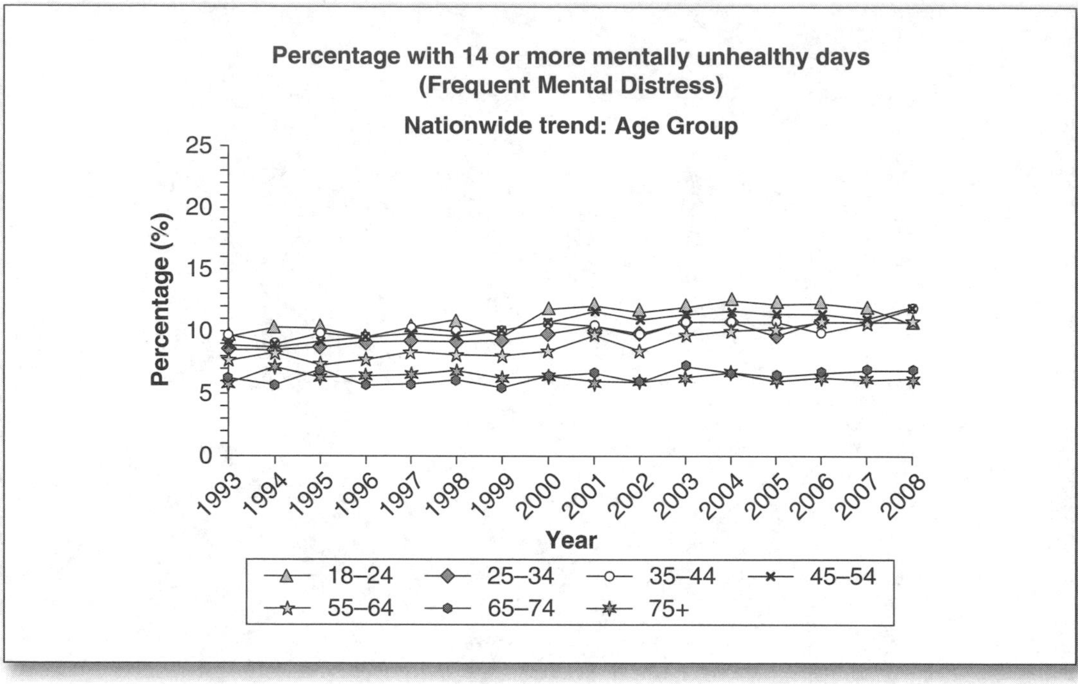

Percentage with 14 or more mentally unhealthy days (Frequent Mental Distress)

Nationwide trend: Age Group

Percentage (%)

Year

18–24 25–34 35–44 45–54 55–64 65–74 75+

Source: CDC Health Related Quality of Life: Nationwide Trend. Available at http://apps.nccd.cdc.gov/HRQOL/TrendV.asp?State=1&Measure=7&Category=3&submit1=Go

improves with age (American Psychological Association 2005; Jayson 2008). Surely, some elderly must already know what research is confirming: A positive attitude toward aging is one of the things that keeps us going (Kennedy, Mather, and Carstensen 2004; Levy et al. 2002).

The facts about old age in America seem to contradict "common knowledge." Why do such misconceptions exist? The concept of master status provides one flash of insight in the matter. A **master status** refers to a single social status that overpowers all other social positions occupied by an individual. A master status directs the way in which others see, define, and relate to an individual.

Master status
A single social status that overpowers all other social positions occupied by an individual; it directs the way in which others see, define, and relate to an individual

Master statuses are powerful identity tools because they carry with them a set of qualities or characteristics that they impose on those who occupy the status. For example, those who occupy the master status of doctor are assumed to be knowledgeable, wealthy, rational, and usually white and male. Similarly, those who occupy the master status of mother are presumed to be caring, nurturing, stable, and female.

In short, a master status and the traits and characteristics that accompany it have a tremendous capacity to influence what others "see" or assume to be true in their social interactions. Thus, although I can look at a doctor and see that she is not a male, the story doesn't end there. Expectations that stem from the master status "doctor" might nonetheless continue to influence my interaction with my doctor. I may question whether she possesses other key traits of the master status, such as knowledge or rationality. I may doubt her ability to diagnose.

During two periods in our lives, childhood and our senior years, age serves as a master status. In childhood, the master status of age is equated with such qualities as dependency, unbounded energy, innocence, inquisitiveness, irresponsibility, and the ability to be uninhibited. As senior citizens, the master status of age is associated with such characteristics as dependency, frailty, loneliness, stubbornness, and the lack of creativity.

In essence, age becomes a master status early and late in life due to a lack of competition. We begin to accumulate statuses more powerful than age only as we move out of childhood and through adolescence, young adulthood, and our middle-aged years. During life's middle stages, we embark on careers, take spouses, raise children, join clubs and associations, become homeowners, and pursue leisure-time or self-fulfilling interests. During life's middle stages, occupational and family statuses typically assume the master status position.

In our later years, we exit many of our occupational and family statuses. Children leave home, people retire, spouses die, and homes are sold. When such status losses occur, age and the characteristics associated with it once again return to the forefront of our identities.

Anticipatory socialization
Socialization that prepares a person to assume a future role

Misconceptions about our golden years may also result from a lack of anticipatory socialization. **Anticipatory socialization** refers to socialization that prepares a person to assume a role in the future. Consider the fact that as children, we play at being mommies and daddies, teachers, and fire fighters. As we move through the early stages of our lives, anticipatory socialization provides a road map to the statuses of young adulthood and the middle years. High school and college put us through the paces via internships, apprenticeships, and occupational training. We receive on-the-job instruction when we are initiated into the workforce. Such preparation is simply not given to the tasks involved in senior citizenship. At no time in our lives are we schooled in the physiological changes and social realities that surround retirement, widowhood, or other events of old age.

This lack of anticipatory socialization should not surprise us. Preparing for old age would be inconsistent with typical American values and practices. We are an action and production oriented society. We generally don't prepare for doing less. As a society, we don't encourage role playing for any statuses that carry negative traits and characteristics, such as being old, criminal, terminally ill, widowed, and so on.

Switching our focus to a macro-level analysis provides additional insight regarding the misconceptions on aging. A **macro-level analysis** focuses on broad, large-scale social patterns as they exist across contexts or through time. Consider aging as a historical phenomenon. Old age in America is a relatively new event. In the first census of 1790, less than 2% of the U.S. population was 65 or older; the median age was 16. (Historians suggest that these statistics probably characterized the population from the early 1600s to the early 1800s; see Fischer 1977.) Thus, old age was an uncommon event in preindustrial America. Those of the period could not reasonably expect to live into old age. Indeed, individuals who did reach old age were regarded as exceptional and often were afforded great respect. Life-earned experience and knowledge were valuable commodities in a preindustrial society (Fischer 1977).

Macro-level analysis
Social analysis that focuses on broad, large-scale social patterns as they exist across contexts or through time

The youthful age structure of early America meant the absence of retirement. Preindustrial societies consisted of home-based, labor-intensive enterprises. Thus, the ability to produce, not age itself, was the relevant factor for working. Before the 20th century, most Americans worked until they died.

With the rise of modern society and industrialization, this pattern changed. The skills, experience, and knowledge of older workers did not resonate with the demands and innovations of factory work. Younger and inexperienced (that is, cheaper) workers were the better economic choice for employers. Such changes in the knowledge and economic base of American society had a profound impact on social and cultural views on aging (Watson and Maxwell 1977). With this shift, old age ceased to be exalted, and a youth-oriented society and culture began to develop.

The youth mentality of the industrial age is clearly articulated in one of the 20th century's most influential pieces of legislation: the Social Security Act of 1935. This act can be credited with setting the "old-age" cutoff at 65. (Note that older workers lobbied for a higher age cutoff and younger workers lobbied for a lower age cutoff at the time of the act's passage.) Furthermore, the act legally mandated that older workers must make way for younger ones. The directive has been successful. In 2008, less than 17% of older Americans were still in the workforce (Administration on Aging 2010).

Your Thoughts . . .

Despite great public protest, France will raise its retirement age from 60 to 62 in 2018 in order to control public deficits. (Germany will raise its retirement age from 65 to 67 as of 2012.) Many Baby Boomers are already facing a retirement age of 67. Should future generations of Americans be made to wait even longer for retirement?

The forced retirement instituted by the Social Security Act contributed to the negative image of old age. Retirement signifies both a social (occupational) and an economic "loss." Despite the reality of Social Security benefits, the elderly as a whole have a lower income than all other age groups except 15- to 24-year-olds (U.S. Census Bureau 2011, Table 701). In 2008, the median income for the 65+ age group was $18,377 (Administration on Aging 2010). When one couples these losses with the natural physiological and social changes that accompany aging (increasing risk of chronic diseases, some vision and hearing impairment, relinquishment of parental and spousal roles, and so on), it becomes easier to understand the development of old age's negative image.

Before leaving this discussion, it is important to note that the analysis of **old age in America** must be qualified with reference to social context. **Social context** refers to the broad social and historical circumstances surrounding an act or an event. Those in certain social circumstances can find the aging experience to be a greater hardship than do others. So while poverty among the elderly *as a whole* has diminished over the past 30 years, some segments of that population still experience high poverty rates (see Figure 6.3). Elderly women, elderly minority members, and, in particular, elderly Hispanic women living alone continue to suffer rates of poverty that exceed the national average (Administration on Aging 2010). Similarly, *overall*, the elderly experience high rates of emotional calm or peace of mind. However, such rates can vary widely among subgroups of the elderly population. For instance, while older

Social context
The broad social and historical circumstances surrounding an act or an event

Figure 6.3 Percent in Poverty for Various Sub-Groups of Elderly, 2008

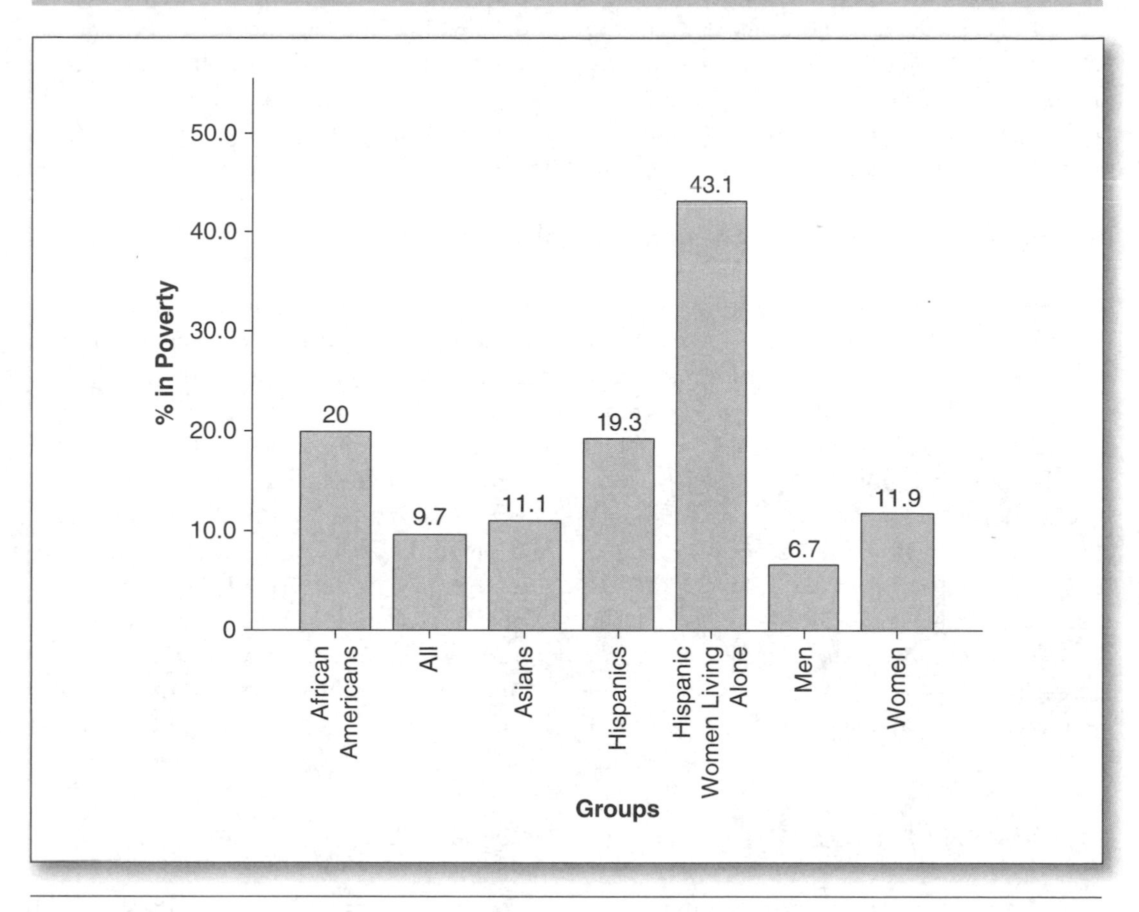

Source: Administration on Aging. 2009. A Profile of Older Americans: 2009. Available at http://www.aoa.gov/AoARoot/Aging_Statistics/Profile/2009/10.aspx.

black females have suicide rates well below the national average, white men over 85 years of age have the highest suicide rate of any age group (National Institute of Mental Health 2009; U.S. Department of Health and Human Services 2009a: Table 42). These high rates have been attributed to the difficulty older males experience in coping with serious illness or the loss of a spouse (National Center for Injury Prevention and Control 2007).

Despite the multiple sources for our misconceptions on aging, all of our views on the matter may soon be significantly revised. The changing age structure of the U.S. population helps explain why such a shift may occur. **Age structure** refers to the distribution that results from dividing a population according to socially defined, age-based categories: childhood, adolescence, young adulthood, middle age, and old age.

Since the early 1900s, the median age in America has risen steadily. So too has the proportion of our population that is 65 or older. Today, nearly 13% of the population is over 65. With

Age structure
The distribution that results from dividing a population according to socially defined, age-based categories: childhood, adolescence, young adulthood, middle age, and old age

the aging of Baby Boomers, these percentages will increase significantly (on the order of 36% between 2010 and 2020). Figure 6.4 shows that by 2050, those 65 and over will account for approximately 20% of the population while those 75 and older will make up 11% of the overall population (Administration on Aging 2010). Now let's move the bar a little bit higher. Those 85 years or older are members of the fastest-growing cohort in both the United States and the world (Shalala 2003). In just a few short years, the 85+ age group will number 6.6 million (Administration on Aging 2010). Such fundamental changes to the age structure of our society will produce major consequences for both the image and the reality of old age.

The 65+ age group has the highest voter registration and turnout rates of any age group—70.3% of older Americans reported voting in 2008 (U.S. Census Bureau 2011, Table 406). With such high civic-mindedness and with their increasing numbers, the elderly represent a powerful voting bloc. Senior voters will increase in number from 40 million in 2010 to 72 million in 2030 (and represent 25% of the voting-age public; MacArthur Foundation 2009).

The political clout of the elderly already has resulted in a significant reduction in the percentage of elderly living in poverty. Such clout also has figured in positive changes with regard to medical care (via Medicare) and lifestyles of the elderly (tax breaks for people over 65). Indeed, the voting power of the old and "near-old" has helped make Social Security and some of its related assistance programs the "third rail" of American politics. To be sure, Social Security is the major source of income for the elderly. It is received by 87% of older persons and provides at least 75% of the total income for 60% of the elderly (Administration on Aging 2007; Employee Benefit Research Institute 2010). As the number and political power of the elderly continue to grow with the aging of the Baby Boomers, we should expect to see significant "corrections" in our social views of old age in America.

Indeed, we may be seeing signs of the change already. There is a new phenomenon being

Figure 6.4 Population Distribution by Age Groups Over Time

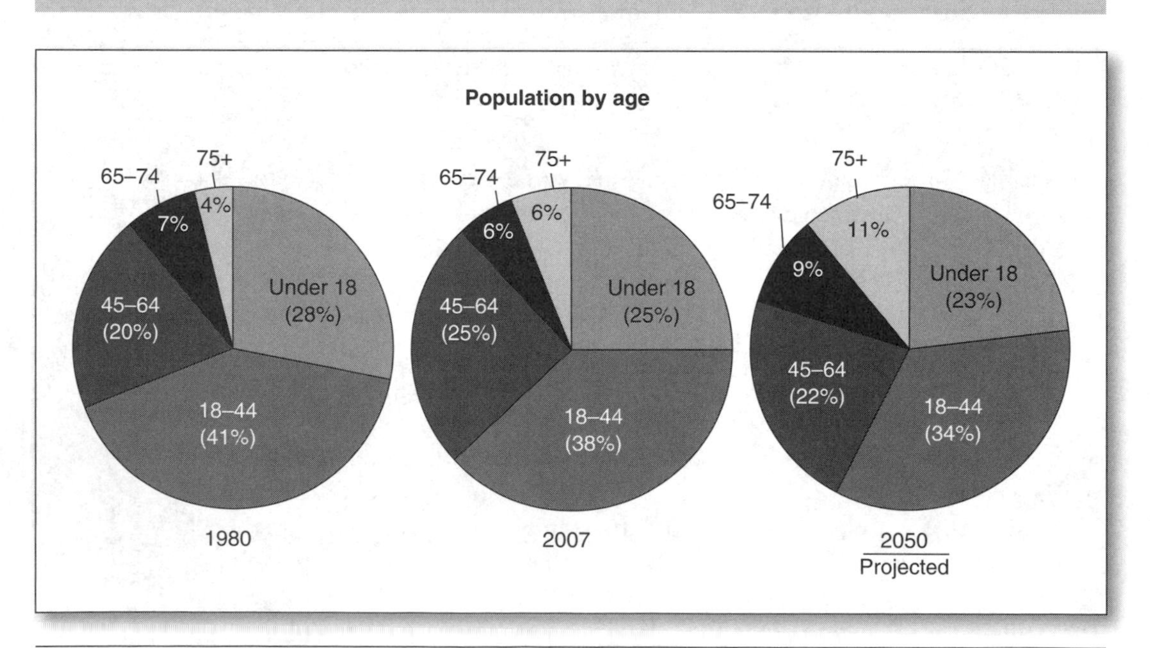

Source: CDC/NCHS, Health, United States, 2009, Figure 1B. Data from the U.S. Census Bureau.

noted by students of pop culture: "elder cool." It seems that the twenty-something crowd are particularly apt to grant celebrity status to older individuals who are "cool" enough to speak their own minds and be their own authentic selves: Muhammad Ali, Clint Eastwood, and Bob Barker (that's right, come on down) are but a few of the original cool oldies (Dudley 2002). Most recently, elder cool was on display in the widely popular command performance by Betty White as she served as the oldest host (at 88) on *Saturday Night Live.* Elder cool may also be influencing the beauty industry. There is some evidence that more women are foregoing hair coloring and instead are intentionally going gray (Harvey 2007). In 2007, DOVE launched "Beauty Comes with Age" as the third phase of its Campaign for Real Beauty. Christy Brinkley, age 56, and Ellen DeGeneres (52) are cover girls for the company's "ageless" line of skin products. And let's not overlook how the boys of summer are aging. The 2010 season saw nine MLB teams with active rosters containing players who were 40 years of age or older—pitcher Jamie Moyer is pushing 50 (ESPN.com 2010)!

Are the golden years tarnished? Research suggests not. We have used several sociological tools—master status, anticipatory socialization, macro historical analysis, and contextual analysis—to understand the discrepancy between the myths and realities of aging.

However, the story on aging in America is hardly complete. Perhaps more than any other social phenomenon, aging is an extremely dynamic process. Given recent and continuing population changes, it is more and more difficult to talk about a single elderly population. On any number of fronts (health, social, and financial), we need to differentiate the "young-old" (65–75) from the "old" (75–85) and the "old-old" (85+). It is estimated today that roughly one-third of Americans will live to 85 and older (President's Council on Bioethics 2005). With these changes, many of us will be facing futures in which one-third of our lives will be spent in "old" and "old-old" age.

Think of the changes this will bring to family, economic, and social relations. As the elderly age into (and beyond) their 80s, their adult children (especially their daughters) will face greater demands for (and conflicts over) elder care (President's Council on Bioethics 2005). Today, the typical caregiver is a 46-year-old daughter who works outside the home but who also spends more than 20 hours a week providing unpaid care to her mother (Houser and Gibson 2008). In the near future, a possible scenario will be a "young-old" daughter (who can't afford to retire) taking care of an "old-old" parent. Consider, too, the fact that Baby Boomers are fueling the projected increases in our elderly population, but this group may pose special problems with regard to caregiving. Currently, spouses and children are the primary caregivers for the elderly. Baby Boomers might be at risk on these fronts with their high divorce rates and smaller families. Neither of these trends bodes well for the caretaking possibilities of the next cohort of the elderly. Indeed, it is estimated that by 2020, there will be 1.2 million people aged 65 and older who will have no living children, siblings, or spouses (Johnson, Toohey, and Wiener 2007; President's Council on Bioethics 2005).

What the future holds for a society experiencing increases in both average age and the absolute numbers of the elderly is unclear. Without breakthroughs in gerontology, living longer will most certainly mean living with chronic, disabling diseases currently associated with the "old-old." For instance, it is estimated that by 2050, the percentage of those 85+ suffering with Alzheimer's disease is expected to increase fourfold if no new cures or treatments are achieved (President's Council on Bioethics 2005). The likelihood of such breakthroughs, however, is undermined by current funding practices for aging-related medicine: In 2006, the National Institutes of Health allotted only 0.1% of its $28 billion budget to such studies (Matousek 2007). To be sure, medical advances are "keeping us young": Who hasn't heard that the 60s are the new 50s? But there is also a paradox of modern aging that deserves mentioning: We may indeed be younger longer, but we are also aged longer. Health care services and practices will surely need to be revamped and expanded to accommodate these changes. A recent Rand study estimates that 4 in 10 of us will die after an extended period of worsening debilitation. Future care of the elderly may entail giving many more than the eight years typically devoted to caregiving at present (President's Council on Bioethics 2005).

Finally, consider that when the Social Security Act was first passed, approximately 50 workers "supported" each Social Security recipient. Today, the burden of support falls on fewer than three workers for every retiree (see Figure 6.5). What will happen to this ratio in the not-too-distant future when the aging Baby Boomers hit retirement age and continue to live long, healthy lives? It is currently estimated that the ratio will fall to just two workers for every one retiree by the year 2034 (U.S. Social Security Administration 2009). But with as dynamic a process as aging, any and all projections might very well be shaky. The recent recession has also thrown a curve to workers thinking about retirement. In a 2010 Retirement Confidence Survey, 25% of workers indicated that they now plan on staying in the workforce longer than initially planned. The poor economy and change in employment are the two most frequently cited reasons for the delay. Indeed, about a quarter of workers now indicate that they will *not* retire until they are 70 or older (in

Figure 6.5 Ratio of Covered Workers to Social Security Beneficiaries

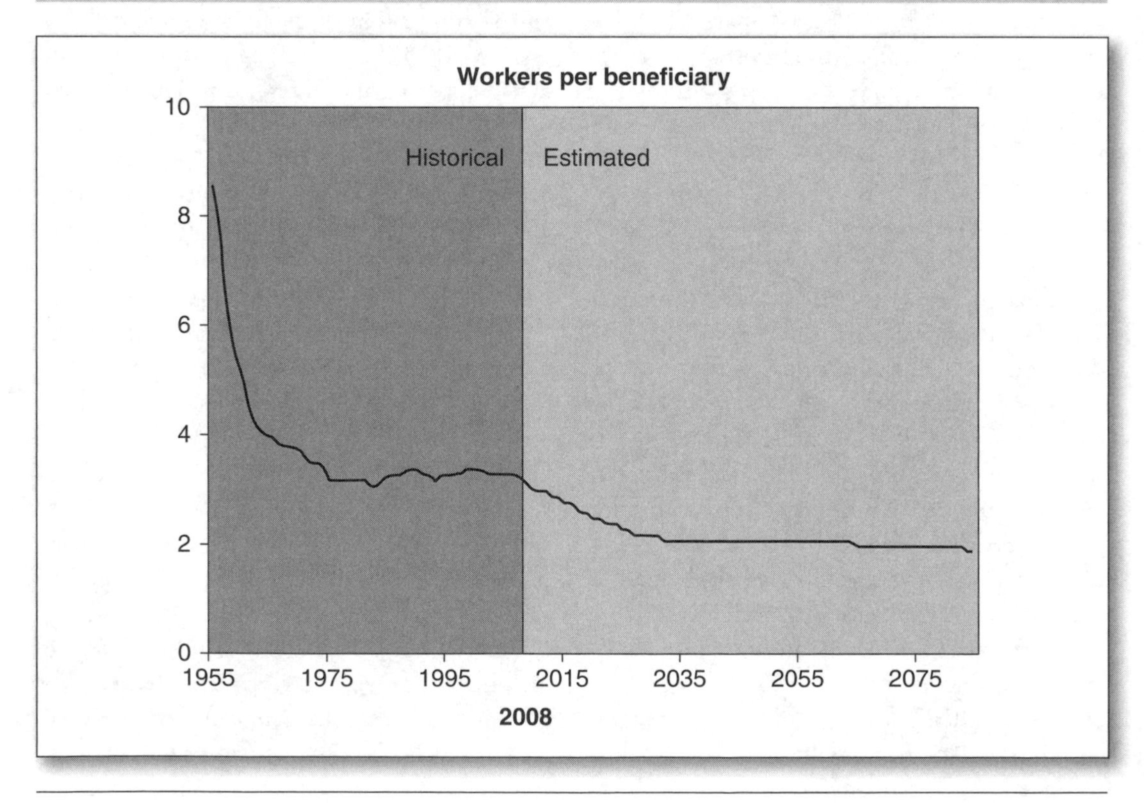

Source: Fast Facts and Figures about Social Security: 2009. Available at http://www.ssa.gov/policy/docs/chartbooks/fast_facts/2009/fast_facts09.html#highlights

the early 1990s, only 9% of workers planned on delaying retirement this long). Furthermore, 70% of workers indicate that they plan on working for pay *after* retirement (Employee Benefits Research Institute 2010)!

To be sure, the "aging" story is one that will continue to evolve; it is a story that will continue to demand our attention. Thus, old age in America is an area to which all of us must apply careful second thoughts.

LEARNING MORE ABOUT IT

An informative collection of articles on aging is available in the fall 2009 and winter 2010 special issues of *Contexts* Vol. 8, Issue 4 and Vol. 9, Issue 1, American Sociological Association: http://contexts.org/aging/.

To learn more about the concept of master status, consult Everett C. Hughes's "Dilemmas and Contradictions of Status" (*American Journal of Sociology* 50 (5): 353–359, 1945); Howard Becker's *The Outsiders* (Glencoe, IL: Free Press, 1963); or J. L. Simmons's "Public Stereotypes of Deviants" (*Social Problems* 13: 223–232, 1966).

To probe more deeply into the process of anticipatory socialization, check Robert K. Merton's *Social Theory and Social Structure* (Glencoe, IL: Free Press, 1957) or "Adult Socialization," by Jeylan T. Mortimer and Roberta G. Simmons (*Annual Review of Sociology* 4: 421–454, 1978).

Todd Nelson's edited volume, *Ageism: Stereotyping and Prejudice Against Older Persons* (Cambridge, MA: MIT Press, 2002) examines the origins and consequences of negative stereotyping of the elderly.

For an informative and "in their own words" account of making one's home in assisted living facilities, see Jacquelyn Frank's *The Paradox of Aging in Place in Assisted Living* (Westport, CT: Bergin and Garvey, 2002).

Sociologist Robert Weiss presents "before and after" interviews with 89 retirees and gives us a chance to hear subjects' experiences of the retirement process in his book *The Experience of Retirement* (Ithaca, NY: Cornell University Press, 2005).

GOING ONLINE FOR SOME SECOND THOUGHTS

To learn more about cultural adaptations to our aging population, visit Stanford's Longevity center: http://longevity.stanford.edu/.

A list of data sources on the elderly that are available through the Federal Interagency Forum on Aging-Related Statistics (Forum) as well as other Federal agencies can be found at Data Sources on Older Americans 2009: http://www.aoa.gov/agingstatsdotnet/Main_Site/Data/2009_Documents/Final_DSOA2009_508.pdf.

The Administration on Aging offers one-stop shopping for an array of statistics on aging at http://www.aoa.gov/AoARoot/Index.aspx.

For more on the perceptions and misperceptions of aging, see Pew Research Center's "Growing Old in America: Expectations vs. Reality." http://pewsocialtrends.org/pubs/736/getting-old-in-america.

NOW IT'S YOUR TURN

1. Try a little fact-finding yourself. Ask some friends and family members what they believe to be true about the financial and social situations of those 65 and older. For instance, what social characteristics do your subjects associate with the elderly? What percentage of the elderly do your subjects believe are still employed? Are the elderly financially secure? Are the elderly happy after retirement? After you've solicited several opinions, compare your results with the latest census figures or with the results of a recent survey executed by the AARP. How do the social locations of your respondents influence their knowledge of the elderly?
2. It is not always possible to know in advance which status will emerge as one's master status. Consider the following individuals and try to identify the master status for each: Hillary Clinton, Tiger Woods, Barack Obama, Oprah Winfrey, Mother Theresa, Martha Stewart, Lindsay Lohan, and yourself. Be prepared to discuss your selections.
3. We've offered the concepts of master status and anticipatory socialization as useful devices for understanding conventional wisdoms regarding old age. Consider some of the other sociological ideas introduced in previous essays. Select and discuss two of them that you feel also offer insight into the misconceptions about old age. (Hint: Do the dynamics of a self-fulfilling prophecy feed a negative view of old age?)
4. Visit the AARP website (http://www.aarp.org/). Take note of the lead articles and the featured page sections/links. Do the articles/topics confirm or challenge stereotypical views of the elderly?

SOCIALIZATION AND IDENTITY

ESSAY 7

Conventional Wisdom Tells Us . . . What's in a Name? That Which We Call a Rose by Any Other Name Would Smell as Sweet

This essay explores the power of names, highlighting the central role of symbols and labels in the construction of identity.

Shakespeare's verse argues for substance over labels. It is a sentiment common to much of the conventional wisdom on names. You've heard the adages: "Sticks and stones may break my bones, but names will never hurt me" or "The names may change, but the story remains the same." These truisms consistently downplay the importance of names. Are the labels we give to people and things as inconsequential as conventional wisdom suggests? Do names really lack the power to influence or the force to injure? Would a rose really command such deep respect and awe if it were known as a petunia or a pansy?

A large body of social science literature suggests that conventional wisdom has vastly underestimated the significance of names. Indeed, names in our society function as powerful symbols. **Symbols** are arbitrary signs that come to be endowed with special meaning and, ultimately, gain the ability to influence behaviors, attitudes, and emotions.

Symbols
Arbitrary signs that come to be endowed with special meaning and gain the ability to influence behaviors, attitudes, and emotions

The symbolic nature of names makes them much more than a string of alphabetics. Rather, names function as calling cards or personal logos. They signify important aspects of one's history and heritage; they pinpoint an individual's social location and group affiliations. Based on our names, others make important decisions regarding our nature and temperament. In this way, names serve as important markers of identity. **Identity** refers to those essential characteristics that both link us to and distinguish us from other social players and thus establish who we are.

Identity
Essential characteristics that both link us to and distinguish us from other social players and thus establish who we are

The link between names and identity helps explain why forgetting someone's name often is viewed as a social faux pas. Similarly, we view situations that preclude the linking of name to identity with great pathos. Consider the sadness that surrounds every tomb of an unknown or unnamed soldier. In settings where one's name remains unknown, it is not unusual for an individual to feel alienated or disconnected. Think of those large lecture courses in which neither the professor nor other students know your name.

Names contribute to the construction of identity in a variety of ways. A family surname, for example, provides instant knowledge of an individual's history. Surnames, worldwide, serve as road maps to the past; they guide us through an individual's lineage and archive one's traditional group affiliations and cultural ties. Thus, historically, children who were denied their father's surnames were denied legitimate social locations. Without this signifier to chronicle their paternal pasts, such children were considered faceless and anonymous, with no rightful place in their social environments. This is especially true for daughters (Brunet and Bideau 2000; Colantonio, Fuster, Ferreyras, and Lascano 2006; Isaacs 1975; Lloyd 2000; Lundberg, McLanahan, and Rose 2007; Nagata 1999; Sanabria 2001; Sapkidis 1998; Stevens 1999; Tait 2006).

In the modern world, many states manipulate surnames in ways that forward their current national agendas. Some governments, China's being one recent example, try to squelch unusual surnames in the cause of bureaucratic efficiency (LaFraniere 2009). In other cases, governments may create official policies that use surnames to include or exclude particular communities in the national identity (Alcoff 2005; Scassa 1996; Stevens 1999; Tait 2006; Wang 2002). Consider the case of Japan. After annexing Korea in the 1900s, the Japanese government forced the first wave of Korean immigrants to adopt Japanese surnames. In this way, the government placed Koreans under both physical and symbolic control (Fukuoka 1998). In the white- (British-) controlled Jamaica of the 18th century, immigrants were an issue as well. But here, the immigrants were African slaves. Rather than assimilating the group, the government wished to mark that population's separation from whites. Hence, upon their arrival in Jamaica, slaves were stripped of ancestral forenames and surnames. Furthermore, they were renamed using only common English forenames. These truncated labels marked slaves as a distinctive group. Their "missing" surnames also made it clear that slaves lacked the lineage of whites (Burnard 2001; Prabhakaran 1999). Note too that certain ethnic surnames place their "owners" in negative social space. In some European countries, experimental work showed that Arab job applicants who changed their surnames received more lucrative job offers than those who retained their Arab surnames (Arai and Thoursie 2009). And in post–9/11 America, research documents that those whose surnames suggested Arab lineage experienced a period of increased harassment, violence, and workplace discrimination. This milieu proved especially dangerous for pregnant Arab women. Indeed, the risk of a poor birth outcome (i.e., a stillbirth, a premature birth, a problematic delivery, etc.) proved significantly higher for Arab-named women than it did for non-Arab-named women (Lauderdale 2006).

A name's ability to pinpoint personal histories leads some individuals to abandon their family surnames and adopt new ones. A well-chosen replacement name can bring one closer to groups or social histories that seem more in vogue, more powerful, or more in tune with one's future aspirations and endeavors. The entertainment industry is rife with such examples. Many performers see name changes as the building blocks of successful careers. Consequently, Stefani Germanotta understood that "Lady Gaga" would better capture a fan base. Dwayne Johnson decided "The Rock" was better suited to his "tough guy" image. And credit singer/actress "Queen Latifah" for understanding that names such as Dana Owens are not the stuff of which pop idols are made. In particular, personalities with ethnic surnames often feel the need for more mainstream, English-sounding names. Thus, fans know Winona Horowitz as Winona Ryder and Carlos Irwin Estevez as Charlie Sheen (www.brighternaming.com/; http://www.people.com/people/quiz/0,,20210827,00.html#20275124-1; www.famousnamechanges.net). When careers encounter dry spots or take a new direction, a name change can become the perfect jump start. Perhaps that explains why singer Prince evolved from "Prince" to "the artist formerly known as Prince" and back to "Prince" or why Sean Combs moved from "Puff Daddy" to "Puffy" to "P Diddy" to "Diddy." Of course, name changing is not restricted to the famous. In many African nations, for example, one's name is *expected* to change with every significant life event—for example, puberty, marriage, or entering the life of a soldier (Brinkman 2004). Often, newly naturalized citizens choose to change their names, selecting something familiar within their adopted nation. Such name changing can be political in nature. For example, after the September 11, 2001, terrorist attacks on the United States, some Muslims began legally changing their names to avoid bias and discrimination (Donohue 2002; see *Indian Express* 2008 for similar findings on Indians and Thompson 2006 for work on Koreans).

Impression management
Process by which individuals manipulate or maneuver their public images to elicit desired reactions

In each of these cases, a name change became a tool for impression management. **Impression management** is a process by which individuals manipulate or maneuver their public images so as to elicit certain desired reactions. In the process of impression management, new surnames become the foundation upon which broadly targeted identities are constructed.

First or common names function as powerful symbols as well. No doubt, you have witnessed the care and consideration most parents-to-be take as they set about naming a newborn. Many parents start the process months before the child is born. Such care may represent a worthy investment, for research suggests that the selection of a personal name can have long-term consequences for a child. Several studies show that individuals assess others' potential for success, morality, good health, and warmth on the basis of names. Thus, when asked to rate other people on the basis of first names alone, subjects perceived "James" as highly moral, healthy, warm, and likely to succeed. In contrast, "Melvin" was viewed as a potential failure, lacking good character, good health, or human caring. Similarly, individuals with names that correspond to contemporary norms of popularity—currently these include Jacob, Ethan, and Michael for boys and Isabella, Emma, and Olivia for girls—are judged to be more intelligent and better liked than individuals with old-fashioned names such as Arnold, Earl, Fred, Betty, Judy, and Phyllis (Christopher 1998; Feinson 2004; Gueguen, Dufourcq-Brana, and Pascual 2005; Karylowski et al. 2001; Mehrabian 2001; Mirsky 2000; Ofovwe and Awaritefe 2009; Rosenkrantz and Satran 2004; Twenge and Manis 1998; Wattenberg 2005; Young et al. 2006).

Personal names can influence more than just the perception of performance and ability. Some studies show a significant association between uncommon, peculiar, undesirable, or unique names and actual outcomes such as low academic performance, low professional achievement, and psychological maladjustment (Banse 1999; Bruning, Polinko, Zerbst, and Buckingham 2000; Coffey and McLaughlin 2009; De Schipper, Hirschberg, and Sinha 2002; Gueguen, Dufourcq-Brana, and Pascual 2005; Insaf 2002; Liddell and Lycett 1998; Marlar and Joubert 2002; Rosenkrantz and Satran 2004; Twenge and Manis 1998; Willis, Willis, and Grier 1982).

BOX 7.1 SECOND THOUGHTS ABOUT THE NEWS: YOU CAN CALL ME ANYTHING . . . *NOT*

The influence of personal names is thought to be so strong that some nations actually regulate the process of naming. French law, for example, allows officials to reject any name deemed at odds with a child's well-being (Besnard and Desplanques 1993). The Canadian government has set similar standards, refusing to register unusual names for babies. So after trying to register the name "Ivory" for their daughter, two Quebec parents were told, "No! Ivory is only a brand of soap" (McLean 1998).

To be sure, naming concerns are a worldwide phenomenon. The New Zealand government rejected the request of Pat and Sheena Wheaton when they petitioned the government registry to name their son "4Real" (Reuters 2007a); the court felt the name was unfair to the child. Similarly, an Italian court ruled that a couple could not name their baby "Friday" and ordered that he be named "Gregory"—the patron saint for the day on which he was born (BBC News 2008a). In China, government officials discouraged one couple from naming their baby "@" (Reuters 2007b). And in the United States, the Marine Corps pulled the plug on Sgt. Cody C. Baker when they discovered that the soldier planned to change his name to one suggested by the highest bidder on his website, ChooseMyName.com. According to Marine officials, U.S. federal law prohibits military personnel from making such commercial endorsements (*New York Times* 2006).

Your Thoughts . . .

Given what we know about the impact of certain names, do governments have an obligation to protect their citizens from making bad naming choices?

While laws on naming are somewhat rare, there are strong social norms that govern the practice of naming. **Norms** are social rules or guidelines that direct behavior; they are the "shoulds" and "should nots" of social action, feelings, and thought. In the United States, norms tell us that naming is a very personal affair. That may explain why Jason Black and Frances Schroeder drew the wrath of the public when they tried to cash in on the naming of their son. In 2001, the couple announced they were willing to sell a corporate buyer the right to name their child. People were aghast when they posted the naming rights on eBay and started the bidding at $500,000 (Goldiner 2001)! And it is worth noting that the couple got no takers. Not a single corporation wanted to get involved in the controversial scheme. Of course, there are always some who are willing to break naming norms. In 2008, for example, David Partin sold the name of his unborn child for a $100 gas card. The buyers? Radio station WHTQ. The station bought the right to name Partin's child "Dixon and Willoughby" after their two "morning drive" DJs. It is worth noting that Partin tried to "up the stakes" and get a corporation to beat WHTQ's bid. But no bites. Apparently, corporations are more conservative than the media when it comes to

Norms
Social rules or guidelines that direct behavior, they are the "shoulds" and "should nots" of social action, feelings, and thought

breaking well-entrenched naming rules (*Central Florida News 13*, 2008).

In charting the role of personal names, it is interesting to note that child name selection follows some predictable patterns. For example, naming patterns initiate and reinforce certain gender scripts. Parents quite frequently select trendy or decorative names for their daughters, for example, Chloe, Destiny, Lily, or Jade. In contrast, parents prefer traditional or biblical names for their sons—such as David, Ethan, Jacob, or Michael. (Think about it. How often have you met a man named after a flower, a season, or a concept?) Furthermore, although little boys frequently are given the names of their fathers or grandfathers, little girls rarely share a name with any family member (Lieberson 2000; Satran and Rosenkrantz 2008). Indeed, among some immigrant groups, straying from ethnic or family names (especially where daughters are concerned) is seen as a powerful means of achieving family assimilation within the host country (Gerhards and Hans 2009; Sue and Telles 2007).

In addition to establishing identity, names often demarcate shifts in identity or changes in social status. In this way, names facilitate a process sociologists refer to as boundary construction. **Boundary construction** is the social partitioning of life experience or centers of interaction. When we cross the boundary from childhood to adulthood, for example, we often drop childlike nicknames—Mikey, Junior, or Princess—in favor of our full birth names. Similarly, when acquaintances become close friends, the shift is often signaled by a name change. Mr. or Ms. becomes "Bob" or "Susan"; William becomes "Bill," or Alison becomes "Ali" (Tait 2006).

Boundary construction
Social partitioning of life experience or centers of interaction

A shift from singlehood to marriage, an occupation change, and a religious conversion are marked by name changes. Thus, in the occupational arena, "Ike," "the Gipper," and "W" all became "Mr. President" when they moved into their new status. In the religious realm, Siddhartha's conversion was signaled by his new name, "Buddha." With a similar experience, Saul became "Paul." In the modern day, a change in religion transformed Cassius Clay into "Muhammad Ali." Of course, name changes can also be used to separate a difficult past from one's future trajectory. That was the hope of the scandal-ridden Blackwater USA when owners changed the company name to Xe. The new name, owners hoped, would forever partition public memories of the company's illegal activities in Iraq (Sauer and Chuchmach 2009). Similar hopes fueled Susan Madoff's name-changing bid. This daughter-in-law of the infamous Ponzi schemer Bernie Madoff petitioned New York courts to change the surname of her and her children to "Morgan." The Madoff name, she argued, brought shame and risk to her family, including several death threats wielded by some of Bernie Madoff's victims (Ansari 2010).

Name changes accompany the shifting identities of places as well. With the reemergence of Russia's nationhood, for example, the city of Leningrad reverted to "St. Petersburg." Similarly, with political reorganization, the plot of land once known as Czechoslovakia was renamed as the "Czech Republic" and "Slovenia." Political reorganization also resulted in renaming the place once known as Yugoslavia as "Bosnia," "Croatia," and "Serbia." Currently, Felix Camacho, the Governor of Guam, seeks to change the country's official name to Guahan. The name Guahan was used for three centuries prior to the 1898 Treaty of Paris, and Camacho argues that readopting Guahan will help to preserve the Chamorro language and culture (Sennitt 2010). Of course, the physical terrain of these areas remains the same. But new names serve to reconfigure each location's political identity. So important are names to place identities that naming struggles can be long lasting. For example, after more than 60 years of peaceful protest by its citizens, the Japanese government agreed to reinstate the prewar name for the island of Iwo Jima (now the famous site of one of World War II's bloodiest battles). The island's original inhabitants argued that Allied forces hijacked their identity when they renamed the land. With the island's name returned to Iwo Tu, the Japanese say that they have "reclaimed their identity" (Greimel 2007).

Changing place names often does more than mark an identity shift. Such changes can also function as tools of social control. Gonzalez Faraco and Murphy (1997) illustrate this condition as they trace the rise and fall of three socially transformative regimes in 20th-century Spain. The authors note that extensive changes to the street names of a town called Almonte served as a strategy by which each ruling body announced its relationship to the ruled. Indeed, the street names chosen by each regime proclaimed each government's intentions, methods, philosophy, and ethos. Thus, the Second Republic (1931–1936) chose street names that promoted the regime's educational agenda. In contrast, Franco's oppressive dictatorship (1936–1975) employed intimidating street names. And the Socialist Democracy that followed the Franco regime adopted a clever set of symbolic compromises designed to heal the rifts between the nation's opposing camps. Azaryahu and Kook (2002) tell a similar story in their study of street names in pre-1948 Haifa and post-1948 Umm el Fahm. The researchers show that the naming of streets in these locations represents variations on the theme of Arab-Palestinian identity. For example, street names in pre-1948 Haifa define Arab identity in the broadest meaning of the term, celebrating culture and politics, Catholics and Muslims, heroes, and occasions that traverse locality and time. In contrast, street names in Umm el Fahm project a much narrower version of Arab-Palestinian identity. The overwhelming majority of these names highlight persons and events critical to early Islamic history. The differences one finds in Haifa and Umm el Fahm street names are important, for they reveal the interests and attitudes of local political elites. In studying them, one can concretize a political shift and track a period of changing ideologies. Naming can serve a similar function with regard to institutions. In 1993, the New Orleans school board launched an effort to lead the population toward greater racial tolerance. The school board mandated that slave owners' names be removed from all city schools. Schools that bore the names of slave owners were renamed after racially tolerant individuals. The seriousness of the school board's intentions was dramatically illustrated in 1997 when President George Washington's (a slave owner) name was removed from one of New Orleans' schools. School board officials renamed the institution after Charles Drew, a pioneering black doctor. Spain, Palestine, the United States: These examples illustrate one common point. By reading the symbolic tapestry created by place names, we learn something about official identity formation in a region. We learn something as well about the ideological premises upon which such identities are built. (For a similar study on street naming in New York City, see Rose-Redwood 2008; for one on naming neighborhoods in the White Mountain Apache Reservation, see Nevins 2008.)

Beyond person and place, names can illustrate changing collective identities. In this regard, consider the experience of African Americans in the United States. Note that when the name "Negro" appeared in the United States, it was synonymous with the status of slave. To distance themselves from slavery, free African Americans of that period elected to call themselves "African" rather than "Negro." However, when a movement developed in the 1830s encouraging slaves and their descendants to return to Africa, free African Americans renamed themselves "Colored" or "People of Color." The color term was adopted to underscore disapproval of the "return to Africa" movement. Interestingly, "Black" was repeatedly rejected as a name for this collective and appeared on the scene only with the social and legal changes of the 1960s and 1970s (Isaacs 1975). Currently, the name "African American" is favored, with many arguing that this change will emphasize African Americans' cultural heritage and help address problems such as racial disparity and poverty (Philogene 1999; Sangmpam 1999).

Your Thoughts . . .

What were the founders of the Tea Party trying to accomplish with their name? In what ways did they succeed . . . fail?

Postmodern theory
An approach that destabilizes or deconstructs fixed social assumptions and meanings

Postmodern theorists suggest that collective identities generated by shared group names can sometimes prove more harmful than helpful. **Postmodern theory** represents an approach that destabilizes or deconstructs fixed social assumptions and meanings. Collective names imply a unity of identity—a sameness—among all members of a group. In this way, collective names can mask the diversity that exists within groups. Collective names can lead us to conclude that all "Hispanics," "women," or "senior citizens" think or act in identical ways by virtue of their shared classifications. Postmodernists also warn that collective names can give a false sense of distinctiveness to groups. Labeling collectives suggests that "whites" are profoundly different from "blacks," "men" irreconcilably different from "women," and "nations" unique unto themselves (Collins 1990; Foucault 1971; Hacking 1995, 1999; Riley 1988; Smith 1991; Wong 2002).

Just as name changes symbolize shifts and movement, they can also function to immortalize certain identities. The name often becomes the tool of choice in poignant and permanent commemorations of extraordinary human efforts. Consider the American Immigrant Wall of Honor located at Ellis Island or the AIDS Memorial Quilt. Special individuals often are honored by attaching their names to buildings or streets. War memorials elicit heightened emotions by listing the names of those they honor. Witness the deeply moving response elicited by the Vietnam War Memorial in Washington, D.C., or the ceremonies for the September 11, 2001, terrorist attacks that have involved reading the victims' names.

Extraordinary athletes have their "numerical names," or numbers, retired, indicating that there will never be another Yankee "Number 7" (Mickey Mantle), another Jets "Number 12" (Joe Namath), or another Red Sox "Number 22" (Roger Clemens). The retirement and return of the Chicago Bulls' Michael Jordan provides an interesting example of this phenomenon. Recall that upon Jordan's 1993 exodus from basketball, "Number 23" was ceremoniously retired by his team. When he returned to the Bulls in 1994, his new beginning was signaled by the assignment of a new "name": "Number 45." These numerical names served to distinguish the old, proven Jordan from the new, mysterious Jordan. Indeed, the National Basketball Association viewed the boundary protected by these symbols to be so sacred that it fined Jordan heavily the first few times he tried to wear his old number during one of his "second-life" games. The power of the symbol 23 continues. In 2010, LeBron James—though on a completely different team—petitioned to have his number changed from 23 to 6 in an effort to retain the sacred milieu surrounding the number linked to Jordan's hey-day (Huffington Post 2010).

Names can be used to indicate possession or ownership. It is not unusual for valuable belongings such as homesteads, boats, aircraft, cars, or pets to be named by their owners. In the same way, conquerors reserved the right to name the continents they discovered or acquired, as well as the indigenous people living there. Columbus, for example, named the indigenous people he met "Indians," a term that came to be used generically for all native peoples. Similarly, colonial populations were frequently renamed by those controlling them so as to reflect the cultural standards of the ruling power. In one case, a mid-19th-century Spanish governor replaced the Philippine surnames of his charges with Spanish surnames taken from a Madrid directory as a method of simplifying the job of Spanish tax collectors (Isaacs 1975). Scientists, too, use names to mark their discoveries. Most of us are familiar with "Lucy," the name given to some of the oldest human remains known to contemporary scientists. Indeed, the naming of scientific discoveries is so important that scientists fight hard for the right to name. Witness the multiyear controversy that surrounded heavy elements 104 (discovered in the 1960s), 105 (discovered in the 1970s), and 107 through 109 (discovered in the early 1980s). These elements went nameless, in some cases for decades, because researchers in the field disagreed as to the parties responsible for their discovery. Similar controversies now surround the human genome project as companies and universities fight for the right to patent and trademark thousands of genes and gene fragments (Browne 1997; *Information Please Almanac* 1997; Pollack 2000).

Family names function as a sign of ownership as well. In bestowing their surnames on children, parents identify the children as "theirs." And historically, wives were expected to take the names of their husbands to indicate to whom the women "belonged" (Arichi 1999; Finch 2008; Johnson and Scheuble 2002; Suarez 1997). These examples highlight a normative expectation: That which we name belongs to us. This expectation may help explain why adopted children are more likely to be named after a parent or relative than are biological children. In the absence of shared genes, names become a mode of establishing familial connection. And, indeed, research shows that namesaking generally strengthens the bond between father and child (Sue and Telles 2007). This link among males, surnames, and possession has been a difficult one to challenge. The American Community Survey shows that only 6% of native-born married women opted to keep their own names or adopt a hyphenated surname (Gooding and Kreider 2010). And interestingly, in places and times in which women have won the right to keep their surnames upon marriage, the large majority choose to name their children in accord with the surnames of their husbands (Auerbach 2003; Finch 2008; Furstenburg and Talvitie 1980; Johnson, McAndrew, and Harris 1991; Johnson and Scheuble 2002; Lloyd 2000).

Labeling theory Maintains that the names or labels we apply to people, places, or circumstances influence and direct our interactions and the emerging reality of the situation

Perhaps the importance of the power of naming is best revealed in research on labeling. **Labeling theory** is built around a basic premise known to sociologists as Thomas's theorem: If we define situations as real, they are real in their consequences. In other words, the names or labels we apply to people, places, or circumstances influence and direct our interactions and thus the emerging reality of the situation.

Thomas's theorem was well documented in a now-famous study of labeling practices in the classroom. After administering intelligence tests to students at the beginning of the academic year, researchers identified to teachers a group of academic "spurters"—that is, children who would show great progress over the course of the approaching school year. In fact, no such group really existed. Rather, researchers randomly assigned students to the spurter category. Yet curiously enough, when intelligence tests were readministered at the end of the academic year, the spurters showed increases in their IQ scores over and above the "nonspurters." Furthermore, the subjective assessments of the teachers indicated that the spurters surpassed nonspurters on a number of socioeducational fronts. The researchers credited these changes to the power of labels: When teachers came to define students as "spurters," they began to interact with them in ways that guaranteed their success (Anderson-Clark et al. 2008; Rosenthal and Jacobson 1968; Schulman 2004).

A famous study in the area of mental health also demonstrates the enormous power of labels. David Rosenhan engaged colleagues to admit themselves to several psychiatric hospitals and to report symptoms of schizophrenia to the admitting psychologists. (Specifically, Rosenhan's colleagues were told to report "hearing voices.") Once admitted to the hospital, however, these pseudopatients displayed no signs of mental disorder. Rather, they engaged in completely normal behavioral routines. Despite the fact that the pseudopatients' psychosis was contrived, Rosenhan notes, the label "schizophrenia" proved more influential in the construction of reality than their actual behaviors. Hospital personnel "saw" symptomatic behaviors in their falsely labeled charges. The power of the label of schizophrenia caused some normal individuals to remain hospitalized for as long as 52 days (Rosenhan 1973; Schulman 2004).

The labeling phenomenon is not confined to what others "see" in us. Labels also hold the power to influence what we see in ourselves. Recent emphasis on politically correct (PC) speech is founded on this premise. The PC movement suggests that by selecting our labels wisely, we may lead people to more positive self-perceptions. There is, after all, a difference between calling someone "handicapped" and calling that person "physically challenged." The former term implies a fundamental flaw, whereas the latter suggests a surmountable condition. Many believe that applying such simple considerations to the use of positive versus negative labels can indeed make a critical difference in self-esteem levels. Others feel the shift to PC language is far more critical. Psychologists Brian Mullen and Joshua Smyth contend that negative ethnic labels are among several direct causes of increased suicide among the targets of such hate speech (Mullen and Smyth 2004).

Similar logic can be found within the literature on social deviants. Some contend that repeated application of a deviant label—class clown, druggie, slut, troublemaker, and so on—may lead to a self-transformation of the label's "target." Sociologists refer to this phenomenon as secondary deviance. **Secondary deviance** occurs when labeled individuals come to view themselves according to what they are called. In other words, the labeled individual incorporates the impressions of others into his or her self-identity. Thus, just as positive labels such as "spurter" can benefit an individual, negative or deviant labels can help ensure that an individual "lives down" to others' expectations (Lemert 1951; Schulman 2004).

Secondary deviance
Occurs when labeled individuals come to view themselves according to what they are called. The labeled individual incorporates the impression of others into her or his self-identity.

The power of names and labels may be best demonstrated by considering the distrust and terror typically associated with "the unnamed." Namelessness is often synonymous with invisibility and exclusion. One can note several examples of this phenomenon. Historically, Christian children were considered nonpersons until they received a name. Indeed, Christian children were not granted full rights to heaven until they were baptized and "marked" by a Christian name. The souls of children who died before such membership cues could be bestowed were believed to be barred from heaven and condemned to perpetuity in limbo. Indeed, the bodies of such children could not be buried in sacred ground (Aries 1962). Things that are unnamed can strike terror in social group members because, in a very real sense, such things remain beyond our control. The most feared diseases, for example, are those that are so new and different they have not yet been named. The lack of a name implies unknown origins and thus little hope for a cure. In contrast, the mere presence of a diagnosis, even one that connotes a serious condition, often is viewed as a blessing by patients. Think of the number of times you've heard a relieved patient or family member say, "At least now I know what the problem is."

Alzheimer's disease also illustrates the terror that accompanies namelessness. For many people, the most frightening aspect of Alzheimer's disease is its ability to steal from us the names of formerly familiar people and objects. Generally, our life experiences are rendered understandable via insightful naming and labeling.

In another realm, note that anonymous callers and figures can strike dread in their targets. The namelessness of these intruders renders them beyond our control. Wanted "John Does" are frequently perceived as greater threats than known criminals because of their no-name status. Recall the intensive search efforts for the suspect "John Doe II" following the Oklahoma City bombing. Similarly, note the frantic aura that surrounded the hunt for the unknown "Unabomber." And no time was wasted in putting names to faces after the September 11, 2001, attacks. Fears of namelessness plague us in cyberspace as well. Anonymous communication forums or situations in which people use fake names and identities make Internet users uncomfortable and unsure of who they're dealing with. In these settings, namelessness can threaten the development of trust (Marx 1999).

When we experience disruptive behaviors that appear new or unusual, our first step toward control involves naming. We coined the label "rumble," for example, to characterize the violent and frightening gang fights that began to erupt on urban streets in the 1950s. We applied the label "wilding" to the new and shocking acts of violence by packs of juveniles that emerged as a phenomenon of the 1980s. And we use the term "hacking" to describe those who harm websites and machinery by releasing unwanted computer viruses and worms. In essence, naming such phenomena provides us with a sense of control. In addition to control, naming people, places, objects, and events seems to make them more appealing. Thus, marketers work long and hard to get the right brand name, one that will hook potential consumers. Nothing can hurt the image of a product more than locating it in the generic or no-name arena (Achenreiner and John 2003; Grassl 1999).

What's in a name? Obviously more than conventional wisdom implies. Names and labels can effectively reshape an individual's past, present circumstance, or future path. Indeed, research seems to leave little doubt: A rose by any other name . . . would somehow be different.

LEARNING MORE ABOUT IT

For a fascinating examination of the social patterns of naming, see Stanley Lieberson's award-winning book *A Matter of Taste* (New Haven, CT: Yale University Press, 2000). For a more lighthearted discussion, consult Linda Rosenkrantz and Pamela Redmond Satran's *Beyond Ava and Aiden: The Enlightened Guide to Naming Your Baby* (New York: St. Martin's Press, 2009).

Harold Isaacs offers some interesting reflections on collective naming in his now-classic treatise, *Idols of the Tribe* (Cambridge, MA: Harvard University Press, 1975). For an interesting discussion of postmodern perspectives on collective naming and identity, see James Wong's article, "What's in a Name: An Examination of Social Identities" (*Journal for the Theory of Social Behavior* 32 (4): 451–464, 2002).

For the classic work on impression management, see Erving Goffman's *The Presentation of Self in Everyday Life* (New York: Anchor, 1959). Howard Becker provides a highly readable discussion of labeling in *The Outsiders* (Glencoe, IL: Free Press, 1963). An informative discussion of labeling as it pertains to women comes from Edwin Schur in *Labeling Women Deviant* (Philadelphia: Temple University Press, 1984).

GOING ONLINE FOR SOME SECOND THOUGHTS

Want to trace the popularity of a name or discover the most popular names of the day? Visit the Social Security website on names at http://www.ssa.gov/OACT/babynames/.

For some interesting reflections on the business of naming and name changes, visit brighternaming.com at http://www.brighternaming.com/.

NOW IT'S YOUR TURN

1. Choose five friends or acquaintances and intentionally call them by the wrong name several times over the course of a day. Record your friends' reactions. It's your turn to review the data. What do these data tell you about the power of personal symbols?
2. It's your turn to review some recent naming events—for example, the naming of new buildings on your campus, sport complexes, or airports. See if you detect any trends. Are the norms of naming time bound? Class bound? Gender bound?
3. Visit the website "Name your Pet" (http://www.nameyourpet.co.uk/index.html) and click on the link for "Your Ideas." Record the suggested names for various types of pets. It's your turn to analyze the data. Do you notice any gender, regional, or historical patterns to the names people suggest for pets? Do names vary by the type of pet? How do pet naming patterns compare to the naming norms discussed for people?

ESSAY 8

Conventional Wisdom Tells Us . . . Beauty Is Only Skin Deep

This essay documents the social advantages enjoyed by physically attractive individuals–tall, slim, and beautiful or handsome women and men. We also discuss the powerful role physical attractiveness can play in the construction of self-identity.

"Beauty is only skin deep," goes the old adage. It's a lesson we learn very early in our lives. From youth to old age, we are promised that, ultimately, we will be judged on the basis of our inner qualities and not simply by our appearance.

The conventional wisdom on beauty is echoed on many fronts. Religious doctrines teach us to avoid the vanity of physical beauty and search for the beauty within. Popular Broadway shows such as *Phantom of the Opera* or *Beauty and the Beast*, fairy tales such as "The Ugly Duckling," or songs such as Billy Joel's "I Love You Just the Way You Are" or Prima J's "Inside Out" promote the notion that appearances are too superficial to seriously influence our fate. All in all, our culture warns us not to "judge a book by its cover," for "all that glitters is not gold."

The conventional wisdom on beauty is reassuring, but is it accurate? Do social actors really look beyond appearance when interacting with and evaluating one another?

One finds considerable **cultural inconsistency** surrounding the topic of beauty. Cultural inconsistency refers to a situation in which actual behaviors contradict cultural goals. Cultural inconsistency depicts an imbalance between ideal and real cultures. Although we say that appearances don't matter, our actions indicate something quite to the contrary. Indeed, a large body of research suggests that an individual's level of attractiveness dramatically influences others' assessments, evaluations, and reactions.

Cultural inconsistency
A situation in which actual behaviors contradict cultural goals; an imbalance between ideal and real culture

Several studies show that attractive individuals—tall, slim, and beautiful or handsome women and men—are better liked and more valued by others than individuals considered unattractive (Gueguen et al. 2009; Langlois et al. 2000; Lemay, Clark, and

Greenberg 2010; Ludwig 2005; Patzer 2006, 2008; Zuckerman, Miyake, and Elkin 1995). Interestingly, these preferences begin shortly after birth and are amazingly widespread (Ramsey et al. 2004). In seeking friends, individuals prefer the companionship of attractive versus unattractive people (Marks, Miller, and Maruyama 1981; Patzer 2006, 2008; Reis, Nezlek, and Wheeler 1980; Sprecher and Regan 2002). In the workplace, attractive people are more likely to be hired and promoted than their unattractive competitors, even when an experienced personnel officer is responsible for the hiring (Cash and Janda 1984; Haas and Gregory 2005; Marlowe, Schneider, and Nelson 1996; Patzer 2006, 2008; Tews, Stafford, and Zhu 2009; Watkins and Johnston 2000). In courts of law, physically attractive defendants receive more lenient sentences than unattractive defendants (Abwender and Hough 2001; DeSantis and Kayson 1997; Erian et al. 1998; Vrij and Firmin 2002), and attractive victims are more credible with juries than unattractive victims (Madera, Podratz, King, and Hebl 2007). Within the political arena, attractive candidates regularly garner more votes than unattractive candidates (Lutz 2010; Ottati and Deiger 2002; Sigelman, Sigelman, and Fowler 1987). And some studies suggest that students rate the performance of their physically attractive professors higher than that of their less attractive professors (Freng and Webber 2009; Riniolo, Johnson, Sherman, and Misso 2006)!

Only in the search for a lifetime mate does the influence of physical attractiveness wane. People tend to choose long-term partners whom they judge to be of comparable attractiveness (Kalick and Hamilton 1986; Kalmijn 1998; Keller and Young 1996; Wong, McCreary, Bowden, and Jenner 1991). However, "comparable" can be a pretty big category, and within it people shoot for the most attractive person they can get (Carmalt, Cawley, Joyner, and Soba 2008; Takeuchi 2006)!

The link between physical attractiveness and being liked and rewarded exists at all stages of the life cycle, including infancy and childhood. Studies show, for example, that attractive babies are held, cuddled, kissed, and talked to more frequently than unattractive babies. This pattern holds true even when one restricts the focus to mother–child interactions (Badr and Abdallah 2001; Berscheid 1982; Leinbach and Fagot 1991; Weiss 1998; Yamamoto et al. 2009). It is worth noting that babies apparently feel the same way about attractiveness. In research settings, newborns and young infants spend more time looking at pictures of attractive faces than they do pictures of unattractive faces (Ramsey et al. 2004; Rhodes et al. 2002; Riniolo et al. 2006; Rubenstein, Kalakanis, and Langlois 1999; Slater et al. 2000). When attractive children make their way to school, they tend to be more frequently praised and rewarded by teachers than their less attractive counterparts (Clifford and Walster 1973; Kenealy, Frude, and Shaw 1988; Parks and Kennedy 2007; Patzer 2006, 2008). Furthermore, studies show that children themselves come to equate attractiveness with high moral character (Dion 1979; Dion and Berscheid 1974; Langlois and Stephan 1981; Ramsey and Langlois 2002). The typical children's fairy tale is one source of this lesson. Remember Cinderella and her evil stepsisters? Or Snow White and her wicked stepmother, who is disguised as an ugly witch? And how about Oz's beautiful, "good" witch of the North versus the ugly and "wicked" witch of the West? The stories of our youth regularly couple beauty with goodness, whereas ugliness is usually indicative of wickedness (Ramsey and Langlois 2002).

In addition to issues of liking, reward, and moral character, physically attractive individuals are perceived as having a host of other positive and highly desirable characteristics. "Beautiful people" are assumed to possess pleasing personalities, personal happiness, great intelligence, mental and physical competence, high status, trustworthiness, and high success in marriage. Furthermore, these perceptions persist even when the facts contradict our assumptions (Andreoletti, Zebrowitz, Leslie, and Lachman 2001; Brewer and Archer 2007; Chia and Alfred 1998; Dion 2001; Dion, Berscheid, and Walster 1972; Feeley 2002; Grant, Button, Hannah, and Ross 2002; Jones, Hansson, and Phillips 1978; Ludwig 2005; Meier et al. 2010; Patzer 2006, 2008; Perlini, Marcello, Hansen, and Pudney 2001; Yerkes and Pettijohn 2008; Zaidel, Bava, and Reis 2003; Zebrowitz, Collins, and Dutta 1998).

Your Thoughts . . .

Some neuroscientists believe that our positive reactions to physical attractiveness are based in human evolution. They argue that humans' attraction to beauty ensures the survival of the species. Can you think of a more sociologically based explanation for the impact of physical attractiveness as described in this chapter?

Some researchers feel that our perceptions of attractive people and their lifestyles may create a **self-fulfilling prophecy** (Leonard 1996; Snyder 2001; Zebrowitz et al. 1998). A self-fulfilling prophecy is a phenomenon whereby that which we believe to be true, in some sense, becomes true for us. Thus, when we expect that handsome men or beautiful women are happy, intelligent, or well placed, we pave the way for expectation to become reality. This may explain why attractive individuals tend to have higher self-esteem and are less prone to psychological disturbances and display more confidence than do unattractive individuals (Cash and Pruzinsky 1990, 2004; Figueroa 2003; Garcia 1998; Gergen et al. 2003; Judge, Hurst, and Simon 2009; Patzer 2006, 2008; Phillips and Hill 1998; Rudd and Lennon 1999; Sanderson, Darley, and Messinger 2002; Taleporos and McCabe 2002). Some studies suggest, however, that this connection may be race specific. African Americans, for example, are far less likely than whites to link their appearance to feelings of self-worth (Hebl and Heatherton 1998; Jefferson and Stake 2009; Kelly et al. 2007; Lovejoy 2001; Milkie 1999; White et al. 2003).

Self-fulfilling prophecy
A phenomenon whereby that which we believe to be true, in some sense, becomes true for us

By contributing to a self-fulfilling prophecy, social reactions to physical appearance may endow handsome men and beautiful women with valuable cultural capital. **Cultural capital** refers to attributes, knowledge, or ways of thinking that can be converted or used for economic advantage. Cultural capital is a concept that was originally introduced by contemporary theorist Pierre Bourdieu. According to Bourdieu (1984), one accumulates cultural capital in conjunction with one's social status. **Social status** refers to the position or location of an individual with reference to characteristics such as age, education, gender, income, race, religion, and so on. The more privileged one's status, the better one's endowment of cultural capital.

Cultural capital
Attributes, knowledge, or ways of thinking that can be converted or used for economic advantage

Social status
The position or location of an individual with reference to characteristics such as age, education, gender, income, race, and religion

Bourdieu argues that an individual's cultural capital works like a good investment. The capital itself—typically defined as family background, education, communication skills, and so on—has inherent value and gains for the individual's entry into "the market." "Working" one's cultural capital enables its "owner" to "buy" or accumulate additional social advantages.

The many studies reviewed in this essay suggest that physical attractiveness also forms another type of cultural capital that operates according to the same dynamic as the one described by Bourdieu. Physical attractiveness provides individuals with an extra resource in meeting life's demands. Beauty places individuals in a preferred or more powerful position (Espino and Franz 2002; Haas and Gregory 2005; Hunter 2002; Kwan and Trautner 2009; Lynn and Simons 2000; Mulford et al. 1998; Patzer 2006, 2008). As such, appearances are frequently converted to economic gain.

Thinking of beauty as cultural capital helps to explain Americans' propensity for physical alterations. In the United States, we are "lifted," "augmented," "tucked," "Botoxed," and liposuctioned more than any other country in the world. In 2009, for example, Americans had more than 12.5 million cosmetic procedures (up 69% from the year 2000)—all of them performed for aesthetic reasons rather than reasons of necessity. Further, in 10% of those procedures, the patient was between the ages of 13 and 29, indicating that our concerns with beauty start early (American Society of Plastic Surgeons 2010). Some of us, of course, choose less drastic measures. National figures indicate that more than $170.2 billion is spent in the United States each year on personal care products and services (i.e:., cosmetics, perfumes, hair care, health clubs, etc.). This figure has more than doubled in the past 15 years. And note that such "beauty spending" surpasses the dollar amounts Americans devote each year to other socially central concerns such as nursery, elementary, and secondary schools combined ($38.5 billion); gasoline and other fuels ($26.6 billion), public transportation ($86.2 billion), electricity ($157 billion); and heath insurance ($161.8 billion; *World Almanac* 2010). It also is worth noting that some impose their obsession with beauty on their pets! In 2003, controversy swirled through the prestigious Crufts Dog Show in England. It seems that Danny, a Pekingese who won the show, underwent facial surgery to improve his prospects of victory (Trebay 2003)! The American Kennel Club has since prohibited any plastic surgery beyond tail cropping and ear docking for show dogs. But for everyday pooches, the surgeon's knife is still an option. Owners invest in pet face lifts, tummy tucks, nose jobs, breast reductions, and the newest craze, "neuticles"—testicular implants designed to give neutered dogs a more "masculine" look (Brady and Palmeri 2007).

BOX 8.1 SECOND THOUGHTS ABOUT THE NEWS: ZEROING IN ON FASHION

In the fashion world, it pays to be zero–*size* zero. Yes, a zero dress size (that's a 23-inch waist) is currently the industry standard for runway models and fashion photo shoots. You may recall the flap in 2009 over a Ralph Lauren ad featuring a touched-up photo of one of his long-standing models–Filippa Hamilton. The photo (which went viral) depicted the model's head as wider than her hips! (Check it out on the Web.) The model later claimed that she was fired by Ralph Lauren for being too fat. During fashion week in 2010, model Coco Rocha, currently a size 4, grabbed headlines with claims that her career took a hit once she abandoned industry standards. The $100 million fashion industry has yet to deliver on its promise to abandon its unrealistic and dangerous weight rules. In 2006, Ana Carolina Reston, an 88-pound model from Brazil, died of complications from anorexia. Coco Rocha (who is 5 feet 10 inches tall) says it was her own ill-fated experience with dieting when she was 108 pounds that convinced her to ignore the industry's "pin-thin" demand.

In 2006, Madrid issued the first ban on rail-thin models at its fashion week, arguing that the industry had a responsibility to portray healthy body images. The mayor of Milan indicated that she would follow suit in her city. In 2009, the editor of the German magazine *Brigette* announced that the magazine would no longer use professional models, opting instead for using real, everyday women.

Here in the United States, *Glamour* magazine caused quite a media stir (and received favorable feedback from women) when it featured a photo spread of plus-size model Lizzi Miller. (Actually, Miller is a size 12 to 14, but the fashion industry regards size 6 and over as "plus.") And while Kate Moss believes that "nothing tastes as good as skinny feels," Heidi Klum says the secret to her looking young is keeping on extra weight. "Weigh" to go, Heidi.

"Buying" beauty is not strictly an American phenomenon. In many Asian and African cultures, for example, the increasing value placed on American facial and skin features has resulted in a dramatic upsurge in eyelid, nose, and facial reconstruction surgery as well as skin-lightening procedures. Reports also show massive increases in cosmetic sales throughout rural China and other Asian nations and long lines in Moscow as Muscovites fight to purchase Estée Lauder and Christian Dior cosmetics (Aginsky Consulting Group 2007; Branigan 2001; Kovaleski 1999; Mok 1998; *New Zealand Herald* 2004; Pitman 2010; Poblete 2000, 2001; Savacool 2009; Shaffer, Crepaz, and Sun 2000). All in all, human behavior may confirm Aristotle's ancient claim: Beauty may be better than all the letters of recommendation in the world.

The effects of physical attractiveness go beyond our interactions with others. An individual's "attractiveness quotient" also proves to be one of the most powerful elements in the construction of one's self-identity. **Identity** refers to those essential characteristics that both link us to and distinguish us from other social players and thus establish who we are.

Identity
Essential characteristics that both link us to and distinguish us from other social players and thus establish who we are

When considering attractiveness and its impact on identity, body weight proves a particularly crucial factor. Each year, Americans spend more than $35 billion on weight-loss programs, diet aids, and low-calorie foods in an effort to shed those extra pounds (Landers 2009). We trim down, pump up, tan, tattoo, and even surgically reshape our bodies, all in the hopes that a "new" and more beautiful body will boost our sense of self.

In theory, connections between body weight and identity should be quite straightforward. Throughout the socialization experience, we are exposed to what sociologists call appearance norms. **Socialization** refers to the process by which we learn the norms, values, and beliefs of a social group, as well as our place within that social group. **Appearance norms** refer to a society's generally accepted standards of appropriate body height, body weight, distribution or shape, bone structure, skin color, and so on.

Socialization
The process by which we learn the norms, values, beliefs, and symbols of a social group and our own place in that social group

Appearance norms
Society's generally accepted standards of appropriate body height, body weight, distribution or shape, bone structure, skin color, etc.

When individuals conform to appearance norms, they enjoy positive feedback from intimates, peers, and social group members at large. These reactions enable one to develop a "normal" body image and a heightened sense of self. In contrast, individuals who deviate from appearance norms are likely to be negatively sanctioned. As such, those who stray from average body weight may develop deviant or negative self-identities (Goffman 1963; Millman 1980; Schur 1984; for recent empirical work, see Carr and Friedman 2005; Gergen et al. 2003; Gimlin 2002; Hurd 2000; Neumark-Sztainer et al. 2002; Wang et al. 2009). This phenomenon begins in childhood, and some worry that with increases in childhood obesity, Americans' collective self-esteem will drop precipitously in coming decades (Wang et al. 2009).

In the everyday world of experience, body weight and its connection to identity can be quite complex. For example, several studies, as well as testimony found on countless weight-loss websites, show that when certain individuals move from thin to fat (in American society, a shift from a normal to a deviant body), such individuals nevertheless maintain a slim, and hence "normal," body image. This sense of normalcy often persists even in the face of objective evidence to the contrary, such as scale readings or clothing size (Berscheid 1981; Degher and Hughes 1992; Gettleman and Thompson 1993; Kuchler and Variyam 2003; Millman 1980; Stanley, Sullivan, and Wardle 2009). Similarly, some individuals who achieve "normal" bodies via diet, eating habits, illness, or surgery continue to identify themselves as overweight, disproportioned, or disfigured (Kuchler and Variyam 2003; Rubin, Shmilovitz, and Weiss 1993; Stenson 2009; Waskul and van der Riet 2002; Williamson et al. 2001).

What explains the failure to incorporate a "new" body into one's identity? Some believe the phenomenon is social in nature. This is because misperception of one's body varies with one's social location. Thus, overweight men are more likely than overweight women to underestimate their weight. Average or underweight women are more likely than average or underweight men to overestimate their weight. Racial minorities—especially African Americans—are likely to underestimate their weight, as are those who are highly religious. And the elderly as well as those with low levels of education are more likely than other age and education groups to underestimate their weight (Kim 2007; Kuchler and Variyam 2003; Martin, Frisco, and May 2009; Standley, Sullivan, and Wardle 2009).

But the misperception of one's body may also be a function of one's childhood years—in particular, the "first impressions" such individuals formed of their bodies during their primary socialization. Sociologists define **primary socialization** as the earliest phase of social "training," a period in which we learn basic social skills and form the core of our identities.

> **Primary socialization**
> The earliest phase of social training wherein we learn basic social skills and form the core of our identities

Children who develop "slim and trim" images of their bodies often succeed at maintaining that image as they build their adult identities. In essence, that skinny kid of an individual's past can cover her or his adult eyes and obscure the portly grown-up in the mirror. In contrast, children who are labeled as "fat" or ridiculed during their early years seem never to fully embrace the notion of a normal or thin body, even when they achieve body weight within or below national weight guidelines (Laslett and Warren 1975; Millman 1980; Pierce and Wardle 1997; Rubin et al. 1993; Sands and Wardle 2003; Thompson and Stice 2001; Wang et al. 2009).

Can those affected by first impressions of their bodies ever synchronize their identities with their current physical condition? Research shows that certain rituals prove helpful in this regard. Sociologists define **rituals** as a set of actions that take on symbolic significance.

> **Rituals**
> Set of actions that take on symbolic significance

When body transitions are marked by some sort of "rite of passage," individuals are more likely to adjust their identities to reflect their new weight. So, for instance, patients opting for surgical weight loss may request a "last meal," write a will, or burn old clothing and photographs. Such rituals prove quite powerful in signaling the death of one's "old" body. Similarly, dieters often engage in rituals such as clothing shopping sprees or body-boasting beach vacations to mark the achievement of a target weight. Dieters report the power of these rituals in signifying a physical "rebirth" (McCabe and Ricciardelli 2003; Rubin et al. 1993, 1991).

Intense social feedback also appears critical to synchronizing identity with body weight.

Repeated reaction to one's actual weight can eventually alter faulty self-perceptions. Thus, although the overweight individual may be able to neutralize the numbers that appear during his or her morning weigh-in, that same individual proves unable to ignore repeated stares or blatant comments on weight gain by family, friends, or strangers. Similarly, the newly thin often report the wide-eyed gasps, exclamations, and smiles of those viewing their new bodies for the first time as the factors most significant to their adoption of a true sense of body size (McCabe and Ricciardelli 2003; Rubin et al. 1993, 1994).

Note, however, that some sources of social feedback can hinder the synchronization process. For example, when individuals use TV images to measure their own appearance, they tend to overestimate their body weight. Such overestimations, in turn, have a negative impact on self-identity. Women appear particularly susceptible to such media influence. Although the media present the "acceptable" male in a variety of shapes and sizes, "acceptable" females rarely deviate from the thin standard (Bartlett, Vowels, and Saucier 2008; Dittmar 2005; Dohnt and Tiggemann 2006; Grave, Ward, and Hyde 2008; Harper and Tiggemann 2008; Harrison 2000; Harrison and Cantor 1997; Martin 2010; Myers and Biocca 1992; Myers et al. 1999; Van den Buick 2000; Vartanian, Giant, and Passino 2001).

Work by communication researchers Philip Myers and Frank Biocca (1992; Myers et al. 1999) demonstrates that daily exposure to as little as 30 minutes of TV programming may contribute to the self-overestimation of body size typical among women. Furthermore, these same short periods of TV viewing may indirectly increase the incidence of anorexia nervosa and bulimia among women and steroid use among men (see also Grave, Ward, and Hyde 2008; Groesz, Levine, and Murnen 2002; Harper and Tiggemann 2008; Martin 2010; Park 2006; Stice, Spangler, and Agras 2001; Thomsen et al. 2002).

Before leaving this discussion of the ideal body, it is important to note that definitions of that ideal can change dramatically as one moves through history or across different racial and ethnic groups. As recently as the 1950s, for example, a Marilyn Monroe-ish figure—5 feet 5 inches and 135 pounds—was forwarded as the American ideal. But just four decades later, the ideal body has slimmed down considerably. Images of the 1990s depict that same 5-foot 5-inch female at under 100 pounds (Greenfield 2006; Killbourne 2000; Wolf 1991)! The movement toward thinness is also illustrated in longitudinal studies of beauty icons such as Miss America winners and Playboy centerfolds. Current figures suggest that these icons are approximately 20% thinner than a woman of average size. Indeed, an increasing number of these icons fall within the medically defined range of "undernourished" (Rubinstein and Caballero 2000; Wiseman et al. 1992). But note that such research projects illustrate race-specific standards of attractiveness. For example, African Americans—both men and women—associate fewer negative characteristics with overweight bodies than do Anglo-Americans. Furthermore, in cross-sex relationships, African American males are nearly twice as likely as Anglo-American males to express a preference for heavier females (Hawkins, Tuff, and Dudley 2006; Hebl and Heatherton 1998; Jackson and McGill 1996; Lovejoy 2001; Martin 2010; Milkie 1999; Thompson, Sargent, and Kemper 1996).

Your Thoughts . . .

Think of two public figures that capture today's "ideal" male and female body. How do these images compare to the ideal bodies of earlier eras? Has the historical cultural trajectory changed in any way?

Social feedback on weight and the use of such feedback in identity construction illustrate the utility of Charles Horton Cooley's concept, the looking-glass self. The **looking-glass self** refers to a process by which individuals use the reactions of other social group members as mirrors by which to view themselves and develop an image of who they are. From Cooley's perspective, individuals who seem unable to "see" their current bodies may be using reactions of the past as their mirrors on the present. Similarly, the use of TV "mirrors" in the definition of self may lead to fun house–type distortions. The key to accepting one's current body type is collecting appropriate contemporary mirrors and elevating them over those of the past.

Looking-glass self
Process by which individuals use the reactions of other social members as mirrors by which to view themselves and develop an image of who they are

If we could wipe the slate clean—start over in a new context—would we change our thinking on physical appearance? Some researchers have asked just that in exploring virtual worlds such as those created in Second Life, an online community. Despite all the hype that promises such sites will create new and better contexts of interaction, the sociocultural patterns surrounding physical attractiveness remain the same. People prefer to date, befriend, and interact with other users whose profiles present physically attractive faces and bodies (Chow et al. 2009; O'Brien and Murnane 2009). Interestingly, however, people are unwilling to "steal" this valuable form of cultural capital. Studies repeatedly show that those who create online images of themselves such as "avatars" tend to create images that match their actual physical appearance or only minimally improve it (McCue 2008; Messinger et al. 2008; O'Brien and Murnane 2009; Taylor 2002).

Thus far, we have discussed the various effects exerted by an individual's physical appearance. But it is interesting to note that the influence of physical appearance goes beyond the realm of the person. Appearances influence our evaluation of objects as well. Often, we judge the value or goodness of things in accordance with the way they look. Some researchers have discovered, for example, that the architectural style of a home can affect the way in which others describe the atmosphere within the structure. Farmhouses, for instance, are generally identified with trustworthy atmospheres. Colonial-style homes are perceived to be the domains of "go-getters." And Tudor-style homes are associated with leadership (Freudenheim 1988). And, of course, the great chefs have taught us a similar lesson: Looks equal taste. Thus, great chefs underscore the beauty of food. For the connoisseur, the presentation of the food—the way it looks on the plate—is as important as the flavor.

Such links between an object's appearance and notions of quality or identity are at the heart of the marketing industry. Indeed, in the world of advertising and public relations, "packaging" a product so as to convey the right image is truly the name of the game. A product must be more than good. Its appearance and "story" must lure the consumer. The importance of packaging holds true even when the object is a living thing! Indeed, research shows that "good marketing" can change our perception of an animal's attractiveness and, thus, its desirability. In one set of studies, labeling an animal as an endangered species increased subjects' perceptions of the animal's attractiveness. Once subjects learned that an animal was endangered, animals routinely thought to be ugly were reassessed as cute, majestic, or lovable. This shift in perception is important, because the same study showed that people are more sympathetic to campaigns designed to save physically attractive animals. Like people, being attractive affords an animal with more cultural capital (Gunnthorsdottir 2001; Tisdell, Wilson, and Nantha 2005).

Is beauty only skin deep? After reviewing research findings on physical attractiveness, we cannot help but view this conventional wisdom with some skepticism. When it comes to evaluating and reacting to others, ourselves, and even inanimate objects, beauty matters. The more attractive the proverbial cover of the book, the more likely we are to value its story.

LEARNING MORE ABOUT IT

For a very readable discussion of physical attractiveness and its impact on people's lives, see Gordon Patzer's book *Looks: Why They Matter More Than You Ever Imagined* (New York: AMACOM, 2008). Jeffrey Sobal and Donna Maurer provide some compelling articles that use symbolic interactionism to examine the ways in which people deal with body image and construct, define, and sanction fatness and thinness. See *Interpreting Weight: The Social Management of Fatness and Thinness* (New York: Aldine de Gruyter, 1999).

In *The World Has Curves: The Global Quest for the Perfect Body,* Julia Savacool provides us with cultural snapshots from around the world of what the "ideal body" looks like in different countries (New York: Rodale Books, 2009).

In *Thin,* Lauren Greenfield explores the intersection of femaleness and weight in America. Her photographs are paired with extensive interviews and journal entries from 20 girls and women who are suffering from various weight-related afflictions (San Francisco: Chronicle Books, 2006). On the other side of the spectrum, Sander L. Gilman's book *Fat: A Cultural History of Obesity* takes us to the crossroads of fact and fiction about obesity, tracing public concern from the mid-19th century to the modern day (Cambridge: Polity Press, 2008).

Abigail Saguy and Kjerstin Gruys (2010) forward a fascinating study of the news media's portrayal of weight and its connection to morality. See "Morality and Health: Mews Media Constructions of Overweight and Eating Disorders" (*Social Problems* 57 (2): 231–250).

Check out the five-part documentary *Skin Deep: The Business of Beauty,* directed by Michael Katz, Maryellen Cox, Laura Fleury, and Tamar Hacker and published in 2008 by A&E Home Video. These five cutting-edge episodes reveal the dark side of our obsession with beauty.

GOING ONLINE FOR SOME SECOND THOUGHTS

The Media Awareness Network provides some interesting discussions of beauty and body image in America. Visit them at http://www.media-awareness.ca/english/issues/stereotyping/women_and_girls/women_beauty.cfm.

The Renfrew Center provides information and educational resources on body image, self-esteem, and weight control. Visit its website at http://www.renfrew.org.

NOW IT'S YOUR TURN

1. Choose approximately 10 bridal pictures from your local paper. Using conventional cultural standards, choose brides of varying attractiveness. Remove any identifying names and show the pictures you've selected to five "judges." Supply the judges with a 5-point scale, where 5 equals "just right" and 1 equals "inadequate," and have the judges rate the brides on the following standards:

attractive	sensual	good-humored	sophisticated	happy
successful	intelligent	trustworthy	pretty	wealthy

Check the judges' ratings. Is there any relationship between the answers addressing physical attractiveness and those pertaining to personality characteristics? Now, repeat Exercise 1 using pictures of men from your local newspaper. In choosing your pictures, be sure to select men who are similarly dressed and of similar ages.

2. For this exercise, you will need to gather 20 to 30 ads that feature both products and people. In making your selections, choose ads for "glamorous" products (perfume, clothing, vacations, and the like), as well as ads for nonglamorous products (antacids, cleansers, insecticides). Analyze the patterns you find (if any) between the type of product being marketed and the attractiveness of the people used in the product's ad.

3. Review the personal ads in three newspapers: the *Village Voice,* your local town newspaper, and your college newspaper. Content-analyze 3 days' worth of ads that feature people. Record all the information about their physical appearance—weight, height, facial characteristics, and so on. What do your data tell you about current appearance norms? Using your data, discuss the similarities and differences in the appearance norms that govern each of these three contexts. Now look at the personal descriptions offered on 10 Facebook sites. Are the appearance norms for cyberspace different from those stressed in the print media?

STRATIFICATION

ESSAY 9

Conventional Wisdom Tells Us . . . The More We Pay, the More It's Worth

If so, our garbage collectors are worth more than our teachers, and baseball players are worth more than those searching for a cure for AIDS. This essay addresses the inconsistencies often found between what we pay for work and the value we place on it.

Price tags mean a lot to consumers. With time and experience, most consumers come to embrace the notion that "you get what you pay for." To be sure, many shoppers are frequently driven to find a good bargain. But on the whole, Americans seem to equate high price with quality.

Antonio Rangel and colleagues at the California Institute of Technology demonstrated this tendency quite dramatically. The researchers asked 20 people to sample wine while undergoing MRIs that would map their brain activity. The subjects were told they were tasting five different Cabernet Sauvignons, each selling for a different price. In reality, however, the "tasters" received only three wines. Two of those wines were offered twice—each time marked with a different price. For example, a $90 bottle of wine was presented with its real price, and then later as a wine costing only $10 a bottle. Similarly, a $5 bottle of wine was presented with its real price, and then later as a wine costing $45 a bottle. In each instance, tasters were drinking the same wine. But their brains showed more pleasure at the higher price than at the lower one (Schmid 2007).

In the real world, Americans' willingness to equate high prices with quality has led to some ingenious marketing strategies. The founders of Häagen-Dazs ice cream, for example, readily admit to conscious price inflation in introducing their product on the market. Given consumer tendencies to gauge product value and attractiveness by price, the owners of Häagen-Dazs correctly perceived a high price tag as the best path to high sales (Cowe 1990). Thus business economists Michael J. Silverstein and Neil Fiske (2003) ask this question: Does that high-priced Starbucks coffee really taste that much better than its Dunkin' Donuts counterpart, or does Starbucks' "double-the-price" cost influence our taste buds?

The more we pay, the more it's worth. The art world has certainly embraced the conventional wisdom. Indeed, the willingness of Mexican financier David Martinez to pay more than $140 million for Jackson Pollock's *No. 5, 1948* (a record price at this writing) drastically changed the value, not only of that single painting but also of master art works in general (Vogel 2006).

Conventional wisdom seems on the money with regard to patterns of product consumption. However, it is important to note that the adage falls short when we apply it to other economic arenas. For example, in the area of human effort or work, what we pay is not always a signal of the worth of people's work. Determining the social worth of work requires that we look far beyond an individual's paycheck.

We might begin our inquiry by asking these questions: What do we pay for work? What occupations draw the biggest paychecks in the United States?

Chief executive officers (CEOs) of large corporations earn the highest yearly wages in the United States. In 2008, the average CEO of a Standard & Poor's 500 company received $9.25 million in total compensation. (Due to the recession, this figure was 9% lower than that of the previous year. However, salary decreases were more than offset by a 23% increase in executive retirement benefits. See AFL-CIO 2010.) What do these figures mean in relative terms? Currently, the CEOs of America's major corporations earn nearly 42 times the salary of the Chief Justice of the Supreme Court, more than 55 times the average salary of a family doctor, about 263 times the average salary of a U.S. elementary schoolteacher, and just about 279 times the average construction worker's wage (U.S. Bureau of Labor Statistics 2009).

Close on the heels of the CEOs are major league baseball players. Along with long winter vacations and great adulation, the "boys of summer" enjoy an average yearly wage of $2.99 million per year (Major League Baseball Players Association 2009).

Not surprisingly, physicians also fall near the top of the nation's pay scale. But note that within the profession, the distribution of salaries is somewhat varied. A family practitioner (average yearly salary: $168,550) or a psychiatrist (average yearly salary: $163,660), for example, earns significantly less than an anesthesiologist (average yearly salary: $211,750) or a radiologist (average yearly salary: $234,065; see SalaryList.com 2010; U.S. Bureau of Labor Statistics 2009).

American lawyers also earn a hefty paycheck. But like physicians, attorneys' financial rewards vary across employment settings. The average salary for partners in a large law firm can range from $500,000 to $850,000, whereas associates in the same firm average only about $136,000. And lawyers employed by energy manufacturing firms (median yearly salary: $196,190) earn nearly twice as much as those employed by state and local governments (median yearly salary: $82,750 and $91,040, respectively; see the Connecticut Law Tribune 2010; U.S. Bureau of Labor Statistics 2009).

We have reviewed some of the highest-paid occupations in America. What occupations generate the smallest paychecks in the United States? The average yearly salaries of sewing machine operators ($22,250), home health aides and nurses' aides (both at $21,620), farm workers ($19,780), cashiers ($19,030), child care workers ($18,900), and fast food cooks ($18,230) make these some of the lowest-paid occupations in the United States. And it is worth noting that those who care for our children earn more than 15% less than those who care for our pets! (Animal caretakers average $21,830 per year.) Those who care for our children fare even worse when compared to those who care for our appearance. Note that the average yearly salaries for manicurists ($22,150), hairdressers ($27,070), and skin care specialists ($31,990) are 17 to 69% higher than the salaries of child care workers (U.S. Bureau of Labor Statistics 2009)!

Are members of highly paid occupations worth the paychecks they collect? Is income the true measure of worth in the United States? **Income** refers to the amount of money earned via an occupation or investments during a specific period of time. One theoretical position in sociology, the Davis-Moore thesis, supports the connection between income and worth. The **Davis-Moore thesis** asserts that social inequality is beneficial to the overall functioning of society. According to Davis and Moore, the high salaries and social rewards attached to certain occupations reflect the importance of these occupations to society. Furthermore, high salaries and social rewards ensure that talented and qualified individuals are well motivated to pursue a society's vital jobs. Inequality, then, is an important source of occupational motivation; income variation ultimately works to the benefit of society as a whole (Davis and Moore 1945; Jeffries and Ransford 1980).

The Davis-Moore thesis represents a functional analysis of society. A **functional analysis** focuses on the interrelationships among the various parts of a society. The approach is ultimately concerned with the ways in which such interrelationships contribute to social order. But not all sociologists share this functionalist view.

Income
The amount of money earned via an occupation or investments during a specific period of time

Davis-Moore thesis
The view that social inequality is beneficial to the overall functioning of society

Functional analysis
Analysis that focuses on the interrelationships among the various parts of a society; it is ultimately concerned with the ways in which such interrelationships contribute to social order

Proponents of conflict theory question the social benefits of salary discrepancies. **Conflict theorists** analyze social organization and social interactions by attending to the differential resources controlled by different sectors of a society. Conflict theorists note that certain occupational salaries far outweigh the occupation's contribution to society. Furthermore, negative attitudes and bias can prevent some people from occupying jobs for which they nonetheless are qualified. Thus, conflict theorists suggest that salary variations reflect discrepancies in wealth and power, discrepancies that allow a select group of individuals to determine the financial rewards of various occupations (Tumin 1967). **Wealth** refers to the totality of money and resources controlled by an individual (or a family). **Power** is the ability of groups and/or individuals to get what they want even in the face of resistance. Consider, for example, that in 2009, the average CEO salary was 319 times higher than the salary of the average U.S. worker (AFL-CIO 2010). The conflict perspective suggests that these salary increases are not reflective of the CEOs' social contributions. Rather, the increases occurred because the CEOs had the capacity, or the power, to command them.

Similar reasoning is used to explain the multimillion dollar salaries of some baseball stars. From the conflict perspective, these megasalaries do not reflect the contributions of these athletes. Rather, wealthy team owners pay the salaries because they are convinced that they will reap the

Conflict theorists
Sociologists who analyze social organization and social interactions by attending to the differential resources controlled by different sectors of a society

Wealth
Totality of money and resources controlled by an individual or a family

Power
The ability of groups and/or individuals to get what they want even in the face of resistance

financial benefits of such investments. Baseball stars can generate huge baseball revenues for club owners by attracting paying customers to the stadium gates and to home TV screens.

Your Thoughts . . .

What is a life worth? After 9/11, and now in the wake of the BP oil spill, independent agencies are calculating the appropriate value of lost life or lost careers. Generally, these formulae involve considerations of presumed salary loss, expected benefits, likely duration of a victim's career, trauma experienced by one's survivors, and so forth. Thinking sociologically, what factors would you add or subtract from this equation?

Note that income, wealth, and power do not tell the whole story when it comes to defining the social worth of one's work. Worth is also a function of occupational prestige. **Occupational prestige** refers to the respect or recognition one's occupational position commands. Occupational prestige is determined by a variety of job-related factors: the nature of the job, the educational requirements for the job, honors or titles associated with the job, the job's use of brainpower versus brute strength, and the stature of the organizations and groups affiliated with the job.

Occupational prestige
The respect or recognition that one's occupational position commands

Periodically, Americans give insight into the prestige factor by rating hundreds of U.S. occupations. Researchers then use such ratings to form an occupational prestige scale. The **occupational prestige scale** provides relative ratings of select occupations as collected from a representative national sample of Americans. In reviewing these ratings, we can quickly see that prestige complicates the road to worth. High income, power, and prestige do not always travel together (Gilbert 2003).

Occupational prestige scale
Relative ratings of select occupations as collected from a representative national sample of Americans

CEOs, for example, enjoy great wealth and can wield immense power. Yet the prestige associated with this occupation is comparatively weak: CEOs score only 71 when rated on the 100-point prestige scale. Doctors and dentists earn similar average salaries, yet doctors enjoy significantly more prestige for their work, earning a rating of 86 versus a 72 rating for dentists.

Bank tellers and secretaries find themselves near the bottom of the income scale. Yet their prestige ratings are more moderate in magnitude; these occupations receive rankings of 43 and 46, respectively. And a rating of 64 suggests that if prestige were currency or power, U.S. elementary school teachers would take home much larger paychecks.

Occupational prestige ratings also can take us beyond the workplace and into the realm of general American values. Often, such insight presents a disturbing commentary. Consider that the lifesaving acts of a firefighter (rated 53) are given no more social recognition than the cosmetic acts of a dental hygienist (rated 52). Similarly, police officers, our legally sanctioned agents of power (rated 59), appear only slightly more valued than the actors who entertain us (rated 58). The "information highway" seemingly has bulldozed the heartland, for the farmers who grow our food appear equal in prestige to the telephone operator who answers our information questions (both rated 10). Trades that played a central role in building the nation—carpenters, masons, and miners—no longer command our favor when it comes to prestige (their ratings are 39, 36, and 26, respectively). And, interestingly, the midwife who delivers a baby (rated 23) fares slightly worse than the waitress who delivers food (rated 28), the bellhop who delivers bags (rated 27), and the bartender who delivers drinks (rated 25; Nakao and Treas 1992).

When we note that income, power, and prestige are not always a package deal, we come to realize the complexity of the U.S. stratification system. The **stratification system** ranks individuals hierarchically with regard to their control of a society's resources, privileges, and rewards.

Stratification system
The hierarchical ranking of individuals with regard to control of a society's resources, privileges, and rewards

Those on the highest rungs of the stratification ladder enjoy a critical combination of wealth, power, and prestige. Knowing this helps explain why electricians or plumbers are rarely considered members of the upper crust. Their incomes may be high, but their prestige levels are relatively low because of a lack of higher education and title and the use of manual labor in their jobs. Similarly, major league baseball players rarely are classified among the elite. Although their incomes are high, their level of prestige is moderate (rated 65). Furthermore, the historic baseball strike of 1994 to 1995 suggests that the high income associated with this occupation does not always translate into power.

Individuals who enjoy high income and prestige but are barred from the inner circles of power will never gain full entry to the American upper class. Indeed, many argue that it is this very condition that impedes the progress of political minorities—African Americans, Hispanics, women, youth, and so on—in our nation. Our public rhetoric suggests open access to advanced education, good jobs, and thus high incomes, but our behaviors often block members of minorities from entering the professional and social networks through which power is brokered.

The more we pay, the more it's worth. Conventional wisdom needs some qualification here. When it comes to the value of certain objects, conventional wisdom may be accurate, but with regard to other aspects of the economy—such as human effort or work—research suggests that the more we pay simply means the more we pay.

LEARNING MORE ABOUT IT

For a good review of occupations, compensation, and value, see Steven Vallas, William Finley, and Amy S. Wharton's *Sociology of Work: Structures and Inequalities* (New York: Oxford University Press, 2009).

The definitive work on the three dimensions of stratification—wealth, prestige, and power—can be found in Max Weber's classic work *Economy and Society* (New York: Bedminster, 1968; original work published 1922).

For a classic review of the functionalist versus conflict perspectives on stratification, see Arthur Stinchcombe's article, "Some Empirical Consequences of the Davis-Moore Theory of Stratification" (*American Sociological Review* 28(5):805–808, 1963).

GOING ONLINE FOR SOME SECOND THOUGHTS

Visit the U.S. Bureau of Labor Statistics website and access the *Occupational Outlook Handbook*. Learn about hundreds of occupations and find information on training, earnings, working conditions, and job outlooks: http://stats.bls.gov/oco/.

How many years would it take you to earn the same amount of money that either Coca-Cola's, Exxon's, or Walt Disney's CEO earns in a year? How much would you be earning if your salary increased at the rate of CEOs? *Executive Paywatch* allows you to compare your salary with that of nearly 300 U.S. CEOs. Visit http://www.aflcio.org/corporatewatch/paywatch/ceou/index.cfm.

NOW IT'S YOUR TURN

1. Make a list of all the occupations mentioned in this essay. Classify each occupation with regard to the gender, race, and ethnicity of those typically associated with the occupation. What patterns can you determine with reference to income and prestige as the occupations vary by gender, race, and ethnicity?

2. Ask 10 of your relatives and/or friends to list their occupations. Then ask them to rate the prestige of the occupation on a 100-point scale. Compare the occupation ratings given by your subjects with the national ratings; you can get a copy of the national ratings at this link: http://publicdata.norc.org:41000/gss/DOCUMENTS/REPORTS/Methodological_Reports/MR070.pdf. Did your subjects underestimate, overestimate, or pinpoint their prestige levels? If errors were made, were there any patterns to these errors that might be related to the age, ethnicity, gender, or race of your subjects?

ESSAY 10

Conventional Wisdom Tells Us . . . Money Is the Root of All Evil

This essay documents the impact of income on issues of mortality and life chances. Money, with all its alleged downfalls, can still mean the difference between life and death.

When it comes to issues of wealth and poverty, conventional wisdom spins a compelling tale. On the one hand, we are warned of money's ills. Money is touted as the "root of all evil," an intoxicating drug with the power to enslave us. (Charles Dickens's Scrooge could tell us something about that!) Biblical scripture contains similar cautions, noting that one "cannot serve God and money." And adages of popular culture warn that "money can't buy happiness or love."

In conjunction with admonitions regarding the perils of wealth, conventional wisdom often paints a rather comforting picture of poverty. From Shakespeare, one hears that "poor and content is rich, and rich enough." In the modern era, Gershwin promoted a similar sentiment, writing that "plenty o' nuttin" is plenty enough. These messages reflect a more general belief that poverty brings serenity and simplicity to one's life. The poor are lauded as free of the possessions that can cloud the mind and tempt the spirit. Indeed, the conventional wisdom on poverty suggests that it can breed great character. Such beliefs may explain why politicians—Abraham Lincoln, Richard Nixon, and Bill Clinton among them—love to remind us of their humble beginnings.

Is money the root of all evil and poverty a blessing in disguise? The everyday world of wealth and poverty contradicts such conventional wisdom. Indeed, when we review the connections between one's wallet and one's well-being, it becomes quite clear that the difference between wealth and poverty can literally have life-and-death consequences.

Consider, for example, the issue of mortality. **Mortality rates** document the number of deaths per each 1,000 (or 10,000 or 100,000) members of the population. Such rates suggest that the length of

Mortality rates
Document the number of deaths per each 1,000 or (10,000 or 100,000) members of the population

Socioeconomic status
A particular social location defined with reference to education, occupation, and financial resources; *see also* **Poverty and wealth**

Infant mortality rates
The number of deaths per 1,000 live births for children under 1 year of age

one's life is greatly influenced by one's socioeconomic status. **Socioeconomic status** refers to a particular social location defined with reference to education, occupation, and financial resources.

Those in America who have the highest socioeconomic status live significantly longer than those who have the lowest status. In fact, some sources suggest that a privileged person's life-span can exceed that of a disadvantaged person by as much as 3 to 7 years. Patterns of infant mortality paint a similar picture. **Infant mortality rates** gauge the number of deaths per 1,000 live births for children under 1 year of age. Rates of infant mortality are twice as high among the economically disadvantaged as they are among the privileged (Centers for Disease Control and Prevention 2010c; Cockerham 2009; Ezzati et al. 2008; Houweling et al. 2005; Lindstrom and Lindstrom 2006; Manchester and Topoleski 2008; Ram 2006; Robert and House 2000).

The link between poverty and mortality stems, in part, from issues of health care. The economically disadvantaged have less access to health care than do members of any other socioeconomic status. Furthermore, the quality of care received by the disadvantaged is significantly worse than that enjoyed by those with higher incomes. Thus, people at the bottom of the U.S. economic hierarchy face the greatest risk of contracting illness and disease. When the disadvantaged get sick, they are more likely to die from their ailments than are those who are more economically privileged (Centers for Disease Control and Prevention 2010c; Cockerham 2009; Manchester and Topoleski 2008).

Poor individuals, for example, are much more likely to suffer fatal heart attacks or fatal strokes or to die from cancer than members of any other socioeconomic status. Interestingly, these economic patterns of health hold true even for diseases nearly eradicated by modern medicine. Disadvantaged patients are several times more likely to die of tuberculosis, for example, than are their more privileged counterparts. Similarly, the poor are more likely than are members of any other socioeconomic strata to die from generally nonfatal illnesses, such as influenza, stomach ulcers, and syphilis. These trends led former U.S. Surgeon General C. Everett Koop to remark, "When I look back on my years in office, the things I banged my head against were all poverty" (Centers for Disease Control and Prevention 2010c; Cockerham 2009; Manchester and Topoleski 2008; Robert and House 2000).

Your Thoughts . . .

Pretend for a moment that you are a member of the U.S. Congress. Given the links between poverty and health, what points might you argue should become the central components of any health care bill?

Poverty's relationship to life and death, to health and well-being, is a worldwide phenomenon. According to the World Health Organization, 1.2 billion people around the world—just over 20% of the world population—suffer from serious illnesses attributable to poverty (World Health Organization 2010a, 2010b). Poor sanitation, unvaried diet, and malnutrition all set the stage for poor health. The lack of medical care also greatly contributes to the high rates of death and disease among the poor. Note that in the world's most disadvantaged nations—for example, Angola, Burundi, Guinea-Bissau,

Indonesia, Malawi, Mozambique, Niger, Rwanda, Togo, or Uganda—there are fewer than 10 doctors for every 100,000 of the country's inhabitants (World Health Organization 2010b, 2010c)!

The effects of world poverty seem especially harsh when one considers the plight of children. Despite the technological advancements of the 20th century, nearly 9 million children worldwide will never see their fifth birthdays. In the poorest nations of the world, such as Angola, Chad, or Sierra Leone, approximately 1 in 5 children face this sad plight. And it is important to note that about one third of the children who die before the age of five will not succumb to incurable diseases or tragic accidents. Rather, their deaths will be linked to malnutrition, a problem that is clearly solvable (World Health Organization 2010b, 2010c).

Life expectancy The average number of years that a specified population can expect to live

Poverty also helps explain the short life expectancies of those living in the poor nations of the world. **Life expectancy** refers to the average number of years that a specified population can expect to live. For example, although a U.S. citizen can expect to live approximately 78 years, individuals in many African nations can only expect to live for roughly 50 years. For some nations, life expectancy is significantly lower. In Afghanistan, Swaziland, or Zimbabwe, for example, the average citizen lives for 42 years (World Health Organization 2010b, 2010c)! Clearly, for many parts of our world community, poverty can be viewed as the leading cause of death.

Poverty's link to mortality goes beyond issues of health and hygiene. Simple membership in a society's lower economic ranks, regardless of one's health, increases the risk of premature death. The sinking of the *Titanic* in 1912 offers a stark illustration of this phenomenon. Among passengers on that ill-fated ship, socioeconomic status was a major determinant of survival or death. When disaster strikes on the high seas, norms dictate that women and children should be the first evacuated. On the *Titanic*, however, that norm apparently applied only to wealthier passengers. Forty-five percent of the women in third class met their deaths in contrast to the 16% death rate of women in second class and the 3% death rate of women in first class. What explains the discrepancy? Historians tell us that first-class passengers (both male and female) were given the first opportunity to abandon ship, while those in third class were ordered—sometimes forced at gunpoint—to stay in their rooms. It was only when the wealthy had been safely evacuated from the ship that third-class passengers were permitted to leave. Thus, for many aboard the *Titanic*, mere membership in the ranks of the poor proved to be a fatal affiliation (Hall 1986; Lord 1981; Zeitlin, Lutterman, and Russell 1977).

Of course, we don't have to travel very far in time to see a repeat of the lessons learned from the *Titanic*. It was the poor in New Orleans and throughout Mississippi who bore the brunt of Hurricane Katrina's death and destruction. Homes in wealthier neighborhoods fared better than those in poor areas; injuries to those in wealthy neighborhoods were lower than those suffered by people in poor neighborhoods. Perhaps one statistic more than any other drives home the hopelessness of poverty. Nearly 134,000 of the people in Katrina's path simply could not evacuate before the storm because they could not afford transportation (Faw 2005).

BOX 10.1 SECOND THOUGHTS ABOUT THE NEWS: TAX BREAKS FOR THE WEALTHY; BAD BREAKS FOR THE POOR

As we approached the end of the 111th Congress, political pundits started gearing up for the great tax-cut debate. With their victories in the 2010 midterm elections, Republicans maintained that Americans want to see a change in President Obama's fiscal recovery strategy. And the change

(Continued)

(Continued)

Republicans were pushing was the extension of the Bush-era tax cuts beyond their December 31, 2010, expiration date. (Their push was successful.)

While President Obama agrees that extending middle-class tax cuts is a good idea, he (along with some but not all fellow Democrats) questions the wisdom of maintaining the tax cuts for the wealthiest 2 to 3% of Americans. In terms of actual income levels, the "wealthiest" Americans would include individual filers with incomes of $200,000 or more and families with incomes of $250,000 or more. (Proposed changes in tax-rate brackets, however, would mean that those close to these cutoffs will not see any increase in their taxes at all and may even see some reductions [NPR 2010]).

To be sure, extending the tax cuts will result in a significant loss of tax revenues. It is estimated that permanently extending *all* the Bush-era tax cuts will cost the government $3.6 trillion over the next 10 years. On the other hand, extending the tax cuts to just the top 2% of wealthiest Americans would add $700 billion to our national debt.

Can the United States afford such revenue loss given the current historically high deficit? Republicans argued that letting the tax cuts expire on the wealthiest 3% would hurt the economy by depriving these Americans of tax dollars they could otherwise pump back into the economy. (Consider, for instance, that millionaires stand to gain on average an extra $100,000 a year via extended tax cuts.) Those critical of this logic argue that the last 10 years of tax cuts did nothing to help stimulate the economy. Independent senator Bernie Sanders of Vermont maintains that the demand for extending tax cuts for the wealthiest Americans is just one more instance of unbridled greed–greed that comes at the expense of the overwhelming majority of Americans.

To be sure, the tax revenues foregone with the tax-cut extension merely force the government to look elsewhere for needed revenues. Politicians will be forced to turn their attention to cutting federal spending–an idea that fuels the current Tea Party movement and current congressional debate. And if the bipartisan committee on deficit reduction has its way, major spending cuts will be seen in areas traditionally considered sacrosanct: Social Security and Medicare. With these kinds of spending cuts, we can expect to see the gap between the wealthiest and the rest of us continue to grow.

The link between poverty and mortality, so dramatically witnessed on the decks of the *Titanic* and in the winds of Katrina, haunts every corner of American life. In the United States, poverty doubles one's chances of being murdered or raped and quadruples one's chances of being assaulted. Similarly, members of the lower economic strata are more likely than others to die as a result of occupational hazards—that is, from diseases such as black lung, from machinery injuries, and the like. Among children, those of the lower class are more likely to drown, to die in fires, to be murdered, or to be killed in auto accidents than their more affluent counterparts. And during wars, the sons of the poor are most likely to serve in the military and therefore most likely to be casualties (Cockerham 2009; Kriner and Shen 2010; Reiman and Leighton 2009; U.S. Department of Justice 2010).

Physical health, life, and death—poverty influences all of these. But the negative effects of low socioeconomic status extend beyond physical well-being. Many studies document that poverty can also negatively influence mental and emotional states. For example, the poor are more likely than those in middle and upper economic groups to report worrying all or most of the time that their household incomes will be insufficient to meet their basic family expenses. Similarly, the poor are less likely to report feelings of happiness, hope, or satisfaction than their more wealthy counterparts. As a result, the poor are more likely to greet the day with trepidation, despair, and depression than with enthusiasm, drive, and stamina (American Psychological Association 2010; Centers for Disease Control and Prevention 2009h; Cockerham 2009, 2010).

Negative life events
Major and undesirable changes in one's day-to-day existence, such as the loss of a spouse, divorce, or unemployment

Negative life events also befall the disadvantaged more frequently than those of any other socioeconomic status. **Negative life events** refer to major and undesirable changes in one's day-to-day existence, such as the loss of a spouse, divorce, or unemployment. For example, divorce occurs most frequently among the poor, with rates steadily decreasing as one moves up the socioeconomic hierarchy. Similarly, job loss and unemployment are most common among those of the lower socioeconomic strata. The frequency with which the poor experience such events affects their mental and emotional well-being as well. Negative life events have been linked to increases in depression, low self-esteem, and use of drugs and alcohol (Cockerham 2009, 2010).

Socioeconomic status can also influence the ability to cope effectively with life's struggles; the poor again appear at a disadvantage in this regard. Consider that family members typically constitute the support networks of the poor. This stands in contrast to the networks of the privileged, which typically are made up of friends, neighbors, colleagues, and Internet-based support networks. The restricted outlets of the poor are not without cost. Research indicates that the poor experience less security in social exchanges with nonfamily members and greater distrust and fear of the "outside world" than those in more privileged segments of the population (Cockerham 2009, 2010; Knowlton and Latkin 2007).

Poverty's links to premature death, physical disease, and poor mental and emotional health suggest that membership in the lowest socioeconomic strata can severely limit an individual's life chances. **Life chances** are the odds of one's obtaining desirable resources, positive experiences, and opportunities for a long and successful life.

Life chances
Odds of obtaining desirable resources, positive experiences, and opportunities for a long and successful life

Poverty damages the general quality of life. The condition also limits the ability to improve or change one's circumstances. In the face of disease, depression, unrest, or danger, it becomes hard to summon the motivation necessary for upward mobility.

Given the debilitating consequences of poverty, why have societies been so ineffective in combating it? Sociologist Herbert Gans (1971) suggests that poverty may serve some positive social functions for society. In this regard, he offers a functional analysis of poverty. A **functional analysis** focuses on the interrelationships among the various parts of a society. The approach is ultimately concerned with the ways in which such interrelationships contribute to social order.

Functional analysis
Analysis that focuses on the interrelationships among the various parts of a society; it is ultimately concerned with the ways in which such interrelationships contribute to social order

Consider the economic benefits afforded by the existence of poverty. The poor constitute an accessible pool of cheap labor. They fill jobs that are highly undesirable yet completely necessary to a functioning society: garbage collector, janitor, poultry processor, and so on. The existence of poverty also generates jobs for those in other socioeconomic strata. Social workers, welfare agents, and public defenders, for example, occupy positions created either to service the poor or to isolate them from the rest of society. A society's poor also provide a ready market for imperfect or damaged goods. By consuming products that others would not consider, the poor help many manufacturers avoid financial loss.

At a social level, the poor provide a measuring rod against which those of other socioeconomic statuses gauge their performance. In this way, the continued existence of a poor class reassures the more privileged of their status and worth. Finally, the poor often function as social scapegoats, symbols by which the larger society reaffirms its laws and values. The poor are more likely to be arrested and convicted of crimes than are members of any other socioeconomic strata (Reiman and Leighton 2009). By focusing the social audience on the "sins" of the poor, societies can effectively convey the message that crime doesn't pay.

Reviewing the realities of money and poverty and their place in a society casts serious doubt on conventional wisdom. Money may not guarantee happiness; it may not buy love. Money may trigger greed and, ultimately, personal pain. Yet the disadvantages of money pale in comparison to the absence of money and its effects. Poverty has clear, negative consequences for social actors. In fact, it can be argued that poverty has been a more destructive force in this nation than any medical disease or any international threat. Yet poverty also has clear, positive social functions for society as a whole. Perhaps this point best explains a harsh fact of our times: Despite society's "war on poverty," poverty has proven a tenacious opponent. The battle wages on, with casualties growing in number. Yet victory over poverty may come at a cost too high for the nonpoor to embrace.

LEARNING MORE ABOUT IT

For a comprehensive look at how America's working families are faring in today's economy, see Lawrence Mischel, Jared Bernstein, and Heidi Shierholz's *The State of Working America 2008–2009* (Ithaca: Cornell University Press, 2009).

Douglas S. Massey offers a compelling examination of the roots of inequality in America and the potential to create a fairer system in *Categorically Unequal: The American Stratification System* (New York: Russell Sage, 2008).

Jeffrey Reiman and Paul Leighton offer a highly readable look at the ways in which poverty influences justice in *The Rich Get Richer and the Poor Get Prison* (Saddle River, NJ: Prentice Hall, 2009).

GOING ONLINE FOR SOME SECOND THOUGHTS

The World Health Organization provides a wealth of information on global health patterns. Its website provides a variety of reports and surveys that fully document the effects of poverty on health and life chances: http://www.who.int/whosis/whostat/EN_WHS10_Full.pdf.

Want to calculate your life expectancy based on your personal and social characteristics? Visit the Life Expectancy Calculator: http://moneycentral.msn.com/investor/calcs/n_expect/main.asp.

The following organizations can help you learn more about poverty:

Center for Community Change http://www.communitychange.org

National Center for Children in Poverty http://www.nccp.org

National Coalition for the Homeless http://www.nationalhomeless.org/

NOW IT'S YOUR TURN

1. Essay 6, on aging, introduced the concept of master status. Consider the ways in which an individual's financial position can function as a master status in our society. What auxiliary traits or characteristics are presumed to accompany the status of rich? Of poor? Under what conditions does one's financial status fail to operate as a master status?

2. Choose one of the lowest-paid occupations listed on page 98. Using federal and state income tax tables, calculate your salary after taxes. (If you cannot find the tax tables, settle for deducting 15% of your salary.) Now take that figure and plan a monthly budget. Remember to budget for rent or a mortgage, food, gas and auto maintenance, medical expenditures, clothing, insurance, and so forth. (Use the classified section of your local newspaper and Internet sites to gather the estimates you need.) How far will your salary take you each month? What things that you now enjoy would you have to give up?

ESSAY 11

Conventional Wisdom Tells Us . . . You've Come a Long Way, Baby

In the past 40 years, women have made great strides toward equality with men. But have they journeyed far enough? Here, we focus on gender relations in the home, the schools, and the workplace, illustrating the gains and losses faced by women and men in the current era.

Drop in on any historical period—past or present—and chances are great that you will find a story filled with gender inequality:

- *Dateline, Preindustrial Europe:* Artisan guilds limit their apprenticeships to men, thereby ensuring the exclusion of women from the master crafts (Howell 1986).
- *The Shores of Colonial America:* Colonists adopt "the Doctrine of Coverture" from British common law, thus subsuming women's legal identities and rights to those of their husbands (Blackstone 1765–1769/1979).
- *United States, circa 1870:* The "conservation of energy" theme is used to support the argument that education is dangerous for women. The development of the mind is thought to occur at the expense of the reproductive organs (Clarke 1873).
- *The State of Virginia, 1894:* In reviewing a case on a Virginia state regulation, the U.S. Supreme Court rules that the word *person* is properly equated with *male*, not *female.* The decision upholds the state's decision to deny a law license to a "nonperson" female (Renzetti and Curran 1989).
- *Turn-of-the-century America:* Twenty-six U.S. states embrace the doctrine of "separate spheres" and pass laws prohibiting the employment of married women. The doctrine asserts that a woman's place is in the home, while a man's is in the public work sphere (Padavic and Reskin 2002; Skolnick 1991).
- *Sharpsburg, Maryland, 1989:* A female participant in a historical re-creation of the Civil War battle of Antietam is forced to leave the event. She is evicted by a park ranger, who tells her that women were not

allowed to portray Civil War soldiers at reenactments.[1]

- *March 2010:* The Women Airforce Service Pilots (WASPs) were granted the Congressional Gold Medal for service rendered *nearly seven decades* earlier during World War II. These 1,074 women flew more than 60 million miles, and 38 WASPs were killed in service. They were deactivated in 1944, their records classified and sealed. They were never granted military rank and were denied veteran benefits until 1977.

"You've come a long way, baby." No doubt, you have heard this phrase used to acknowledge the dramatic change in women's social roles and achievements. Today, much has improved for women. Thousands of women have moved into traditionally male jobs. Marital status is no longer a legal barrier to the employment of women. Court rulings have struck down gender-based job restrictions. Women participate in higher education at rates equal to or greater than those of men, and the law has made concerted efforts to advance and protect the legal rights of women. Even the historical record is slowly but surely being corrected. Yet despite the long way that "baby" has traveled, a careful assessment of gender relations in the United States indicates that "baby" has a long haul ahead.

Obstacles to gender equality begin with gender socialization. **Gender socialization** refers to the process by which individuals learn the culturally approved expectations and behaviors for males and females. Even in a child's earliest moments of life, gender typing, with all its implications, proves to be a routine practice. **Gender typing** refers to gender-based expectations and behaviors. Several early studies documented parents' differential treatment of male and female infants. An observational study by Goldberg and Lewis (1969), for example, revealed mothers unconsciously rewarding and reinforcing passivity and dependency in girls while rewarding action and independence in boys. In another early study by Lake (1975), researchers asked 30 first-time parents to describe their newborn infants. The exercise revealed that parents' responses were heavily influenced by prominent gender stereotypes. **Stereotypes** are generalizations applied to all members of a group. Thus, daughters were most often described using such adjectives as *tiny*, *soft*, and *delicate*. In contrast, boys were most frequently described with adjectives such as *strong*, *alert*, and *coordinated* (also see Karraker, Vogel, and Lake 1995; Rubin, Provenzano, and Luria 1974; Sweeney and Bradbard 1988). Other studies on gender typing in infancy uncovered similar patterns. For example, when the infants were dressed in blue clothing and identified as boys, women participating in the study described the infant in masculine terms and engaged in more aggressive (bouncing and lifting) play. When the *very same* infants were dressed in pink and identified as girls, women participating in the study described the infants in feminine terms, handled them more tenderly, and offered the "girls" a doll (Bonner 1984; Will, Self, and Dalton 1976; see also Bevans 2009). Similarly, when asked to assess the crawling ability of their babies, mothers overestimated the ability of their sons and underestimated the ability of their daughters. In actual performance, infant boys and girls displayed identical levels of crawling ability (Mondschein, Adolph, and Tamis-LeMonda 2000). Gender typing in infancy is a widespread phenomenon. Even in Sweden, a society that actively promotes gender equality, there is nonetheless evidence of differential treatment of male and female infants by mothers (Heimann 2002). And to be sure, gender typing begins early. One

Gender socialization
Process by which individuals learn the culturally approved expectations and behaviors for males and females

Gender typing
Gender-based expectations and behaviors

Stereotypes
Generalizations applied to *all* members of a group

recent study documents that prenatal talk by mother to the child in utero varies by the gender of the baby (Smith 2005).

Often, the gender typing of infants occurs in subtle ways. Several studies, for example, have focused on gender differences in vocalizations of both infants and parents. In one study, both mothers and fathers perceived their crying infant girls more negatively as the crying increased. Increased crying by sons, on the other hand, led their mothers to rate them as "more powerful" (Teichner, Ames, and Kerig 1997). Another study found that babies who "sounded" like boys (i.e., babies with less nasal vocalizations) received higher favorability ratings by adults (Bloom, Moore-Schoenmakers, and Masataka 1999)! Gender typing can even occur in the realm of naming. In many traditional cultures (e.g., Iranian, Japanese, or Jewish), the naming of boys involves more elaborate public rituals than the naming of girls. These differences suggest that boys' identities are viewed as more central to the society's well-being. Similarly, immigrant Hispanic couples are more likely to give sons Spanish names while giving daughters names with no Spanish referents, a practice that sends a very early message about the perceived keepers of family heritage (Sue and Telles 2007).

Gender typing is not restricted to the infancy period; it continues into and beyond the toddler years. Observation studies of parents and toddlers reveal that parents are rougher and more active with their sons than with their daughters. Studies also show that parents teach their toddlers different lessons about independence. For example, fathers teach boys to "fend for themselves," while encouraging daughters to "ask for help." These distinctions occur even among parents who claim to use identical child-rearing techniques with their sons and daughters (Basow 1992; Lindsey and Mize 2001; Lips, 1993; Lytton and Romny 1991; Richardson 1988; Ross and Taylor 1989; Witkin-Lanoil 1984). When toddlers play with other children, the gender typing continues. At the playground, for example, fathers' supervision of sons is more lax than their supervision of daughters, suggesting different expectations with regard to risk-taking and injury (Kindleberger Hagan and Kuebli 2007). And daughters seem to model their parents' fear and avoidance reactions to a greater degree than sons (Gerull and Rapee 2002). When it comes time to discipline a toddler, gender typing remains. Misbehaviors from sons elicit anger from mothers while misbehaviors from daughters elicit disappointment. This is because mothers expect more risk-taking behaviors from sons but think there is less they can do to prevent it. (You know the old saying: Boys will be boys!) On the other hand, mothers think they can modify risk-taking behaviors among daughters (Morrongiello and Hogg 2004).

Of course, not all gender typing is quite so blatant. Studies show that parents of young children engage in more implicit gender scripting as well. In storytelling about their own pasts, for example, studies show that fathers tell stories with stronger autonomy themes than do mothers, and sons hear these stories more than daughters (Fiese and Skillman 2000). Research also suggests that the very presence of sons versus daughters can influence general family dynamics. Both mothers and fathers use more affiliative speech with sons and more assertive speech with daughters (Shinn and O'Brien 2008). Fathers invest more and are more likely to stay married in families that have sons. Mothers report greater marital happiness if they are in families with sons (Raley et al. 2006).

To be sure, parents are not the only family members to contribute to the process. Siblings also play a role. Boys with older brothers and girls with older sisters engage in more gender-typed behaviors than children whose older siblings are of the opposite sex (Rust et al. 2000). *Family system* dynamics can also influence gender role attitudes. For instance, parents with two sons are more likely to exhibit traditional gender role attitudes across the entire unit. But for parents with two daughters, a "divergent" family type is likely where parents are more traditional but the daughters are more egalitarian in their gender role attitudes (Marks et al. 2009).

Children appear to learn their gender lessons well and early. In one study, toddlers were shown photos of male and female adults engaged in gender-stereotyped activities and gender-neutral activities. Toddlers as young as 2 years of age were able to identify "men's work" and "women's work" (Serbin, Poulin-Dubois, and Eichstedt 2002). Indeed, well before their third birthdays, children display knowledge of the ways in which familiar family activities are gender stereotyped (Poulin-Dubois et al. 2002). That knowledge appears to get stronger with age (Owen-Blakemore 2003). Preschool children prove quite aware of gender-typed competencies and occupations (Levy, Sadovsky, and Troseth 2000). In one study, children 3 to 5 years old predicted that their parents would be upset if they were to play with cross-gender toys. This finding was true even for children whose parents claimed to reject common gender stereotypes (Freeman 2007).

Of course, the gender typing of children is not confined to the family but extends to play settings as well. Children who engage in more same-sex pretend play were better liked by their peers and seen as more socially competent by their teachers. Boys who engaged in rough-and-tumble play with girls were less liked by their peers (Colwell and Lindsey 2005). And even when children play in mixed groups, they engage in "borderwork" that reinforces gender norms (Boyle et al. 2003).

And, of course, no discussion of gender typing would be complete without a serious look at gender stereotypes in the media. The mass media contribute to gender inequality by prioritizing the male experience in explicit ways. Consider, for instance, the message delivered via television news. A 2008 study of prime-time cable news programs found that of the 1,700 guests appearing during the month of May 2008, 67% were men. And all but two of the prime-time cable news programs in the study had male program hosts (Media Matters for America 2010). While their numbers have grown over the years, women are still underrepresented in leading movie roles and in prime-time television programming. A recent content analysis of prime-time series found that 60% of characters were male and 40% were females. Female characters were more likely to enact roles involving romance, family, and friends while male characters were more likely to enact work-related roles (Lauzen et al. 2008). Popular prime-time programs (e.g., *Desperate Housewives,* the *Law and Order* series, etc.) and reality shows (e.g., *Survivor* or *The Apprentice*) frequently reinforce negative stereotypes of women (Cuklanz and Moorti 2006; Lauzen, Dozier, and Cleveland 2006; Merskin 2007).

Television commercials present more of the same—despite the fact that women do most of the purchasing for the home, male characters outnumber females and gender stereotypes still fill prime-time commercial spots (Ganahl, Prinsen, and Netzley 2003). When men are shown doing domestic chores in commercials, they are inept or unsuccessful, thus reinforcing traditional gender scripts about housework (Scharrer, Kim, Lin, and Liu 2006). The gender bias that fills the airwaves permeates other media venues as well. Magazine ads tend to portray women in stereotypical ways and as sex objects and as victims—the worst offenders on these fronts were men's magazines but also women's fashion and female adolescent magazines (Lindner 2004; Smith 2006; Stankiewicz and Rosselli 2008). And when magazines and marketers sell gender-neutral products such as computers or cell phones, they favor a hyper-feminine pitch: Don't use that phone for business—get it to keep in touch with the kids (Gannon 2007)!

The reach of gender stereotyping is quite long. Look up at your morning cereal box and chances are you will see more males than females (by a 2 to 1 margin). Males on those boxes are more likely to be shown as adults while females are more likely to be shown as children (Black et al. 2009). If you watch some morning cartoons with your cereal, you are likely to see some subtle and not-so-subtle devaluation of women. In a recent study of animated cartoons produced by major studios between the 1930s and the mid-1990s, only 16.4% of major characters were female.

(Indeed, from the mid-50s to mid-70s, females were symbolically annihilated in short cartoons [Klein and Shiffman 2009]) A study of more current Saturday morning and after school "superhero" cartoons once again found that male characters outnumbered females (by 2 to 1). Furthermore, the female superheroes were depicted as more emotional, attractive, and concerned about their appearances than their male counterparts. The study also noted that superheroics is typically cast in male characteristics (Baker and Rainey 2007). Similar findings emerge in the world of video games. Males are more likely to be cast in the role of heroes and be depicted as able or aggressive and powerful while females are more likely to be cast in supplemental roles and portrayed as sexual objects or as attractive innocents (Dill and Thill 2007; Miller and Summers 2007). Music videos also deliver clear gender scripts that reinforce traditional gender views (Ward, Hansbrough, and Walker 2005).

Reviewing the places and ways in which gender typing occurs is important because such stereotypes have tangible and important outcomes. Gender stereotypes, for example, have resulted in strikingly different educational experiences for boys and for girls. Boys' noise and activity levels are effective at getting a teacher's attention (Gordon et al. 2005). Numerous studies document that teachers give more attention, questions, direction, evaluation, and praise to male students (Beaman et al. 2006; Chira 1992; Smith et al. 2007; Spencer et al. 2003; Thorne 1995). This gender imbalance is further exacerbated with the use of classroom technologies such as interactive whiteboards or interactive classroom communication systems (Kay 2009; Smith et al. 2007). Social scientists contend that such differential treatment can have long-term consequences. Teacher response patterns send an implicit message that male efforts are more valuable than female efforts. More importantly, teachers' gender-driven responses also appear to perpetuate stereotypes of learning. Consider that gender stereotypes suggest that boys are more skilled at math and science than girls. Yet more than 100 studies document that during the elementary and middle school years, girls actually perform equal to or better than boys in math and science. Some suggest that the decline in girls' math skills and interest during the high school years occurs because teachers begin tracking boys and girls in drastically different directions.

Peer and family support also influences boys' and girls' intellectual preferences and future aspirations. Girls with interests in the sciences, for example, enjoy less support from their friends than do their male counterparts. Yet such support appears to be essential for gifted female students contemplating a future in science (Riegle-Crumb et al. 2006; Stake and Nickens 2005). Parents (and adults in general) contribute to the mix by endorsing gender stereotypes about math and science (Bhanot and Jovanovic 2005; Dai 2002; Kurtz-Costes et al. 2008). The divergence in female/male interest in math/science can also be linked to children's out-of-school activities, that is, boys' involvement with computers and math/science clubs versus girls' involvement with reading and dance/music/art classes (Downey and Yuan 2005). To be sure, female students' own gender biases about their perceived competencies and their interests also influence their choice of education and career plans (Kurtz-Costes et al. 2008). In order to overcome negative stereotypes about women in science, female students need extra supportive measures—that is, seeing counterstereotypical images (e.g., female scientists), receiving parental encouragement, having a strong network of academically successful female friends, or participating in high school sports can all aid the development of the skills and attitudes needed for entering and persevering in traditionally masculine domains (Good et al. 2010; Pearson et al. 2009; Riegle-Crumb et al. 2006; Sonnert 2009; Ziegler and Stoeger 2008).

Teachers,' parents,' and students' perceptions and actions have tangible costs. **Longitudinal data**—data collected at multiple points in time—show that 7th- and

Longitudinal data
Data collected at multiple points in time

10th-grade boys and girls have a similar liking for both math and science. But by the 12th grade, boys are more likely than girls to report enjoying math and science (U.S. Department of Education 1997). Gender differences in actual performance increase over time as well. A study of high-scoring male and female math students found that, despite a similar starting point in elementary school, the male students' math performance accelerated faster as years in school progressed (Freeman 2004; Leahey and Guo 2001). And to come full circle, such performance differences have been attributed to pedagogical approaches that are male friendly rather than female friendly (Strand and Mayfield 2000).

Given these dynamics, it should not surprise us to learn that junior high school students today express career interests that fall along traditional gender paths (Lupart and Cannon 2002). Furthermore, the lack of training in math and science also serves to keep females out of lucrative career paths in engineering and the sciences (Mitra 2002). While females earned 51% of *all* doctoral degrees in 2007 to 2008, they earned only 31% of the doctoral degrees in math and statistics, 30% of the degrees in the physical sciences, and 21.5% of the doctoral degrees in engineering. And consider some noteworthy developments in the area of computers. Male and female students appear equal in their access to and use of computers (Freeman 2004). Yet in 2008, men earned 73% of the master's degrees and 88% of the doctoral degrees in computer/information sciences (Aud et al. 2010, Indicator 42). Gender scripts and stereotypes surely play some role in this outcome.

Perhaps the most telling lesson regarding the relationship between gender and education, however, is that schooling leads to greater financial benefits for males than it does for females. For every level of educational attainment, the average earnings for women are lower than those for men (Aud et al. 2010, Indicator 17). In 2008, a female high school graduate's earnings were only slightly above the average earnings for a male with less than a ninth-grade education. Indeed, it takes a *college degree for a female worker* to exceed the average earnings of a *male with a high school diploma.* The gender gap in earnings grows still larger for those with graduate training. In 2008, American males (aged 25–64) with master's degrees had an average annual income of nearly $92,000 per year, whereas females with the same amount of graduate training averaged just less than $56,000 per year. Males (aged 25–64) with professional degrees had average earnings of nearly $154,000, while their female counterparts averaged just under $88,000 (U.S. Census Bureau 2009c, PINC-03). The lower financial returns of education for women are made more exasperating when one realizes that women are increasingly participating in advanced education. Indeed, since the 1970s, females have been the driving force behind the increases in undergraduate and graduate enrollments. By 2018, female enrollments at four-year colleges are expected to reach 5.8 million while male enrollments are expected to be 4.2 million. Also through 2018, enrollment rates in graduate and professional programs are projected to increase at a faster rate for females than males (Planty et al. 2009).

When boys and girls become men and women, they carry learned gender differences into the domestic sphere. Thus, despite current rhetoric to the contrary, the division of labor on the domestic front is anything but equal. A recent Bureau of Labor study found that on an average day, 83% of women and 64% of men report spending some time on household activities (cooking, cleaning, lawn care, etc.). Women, however, regardless of marital status, spend more time on these activities: 2.6 hours a day for women versus 2.0 hours a day for men. If we restrict the focus exclusively to *housework*, then on an average day 50% of women versus 20% of men report doing some cleaning or laundry (U.S. Department of Labor 2009a). In recent years, there has been a narrowing of the gap between women's and men's contributions to housework. But this "advance" is attributed to the fact that women have been systematically *cutting back* on the number of hours they spend on housework (Bianchi, Milkie, Sayer, and Robinson 2000). And this "advance" comes with a cost: In married and cohabiting couples, women's earnings are more important than men's for covering the costs of outsourcing domestic work (Treas and de Ruijter 2008).

Most sociologists agree that the greatest strides toward gender equality have been made within the workplace. Despite such strides, however, the old industrial practice of separating work along gender lines continues. Sex segregation is common practice in many workplaces and within many occupations. **Sex segregation** in the work sphere refers to the separation of male and female workers by job tasks or occupational categories.

Sex segregation (in the work sphere) The separation of male and female workers by job tasks or occupational categories

When it comes to women and work, it is very clear that sex segregation still thrives. Indeed, you might be surprised to see how many common occupations are still **"nontraditional"** for women (i.e., occupations where women comprise 25% or less of those employed). Take a look at Table 11.1.

Ninety-two percent of registered nurses, 92% of receptionists, 95% of child-care workers, 97% of secretaries, 98% of preschool and kindergarten teachers, and 97% of dental hygienists are female. One of every three female workers can be found in "sales and office occupations" (U.S. Department of Labor 2010a). Table 11.2 lists the top 10 leading occupations for women in 2009.

The histories of many female-dominated occupations suggest an economic motive for such segregation: Employers used female workers to reduce their wage costs. Employers were able to pay female workers lower wages than males. Employers also thought that women were less likely to be susceptible to the organizational efforts of unions. Furthermore, by confining their hiring to young, single women, employers ensured a high worker turnover in their businesses (young, single women left their jobs to marry) and thus a continuous supply of inexperienced, low-wage workers (Padavic and Reskin 2002).

Table 11.1 A Sampling of Nontraditional Occupations for Women, 2008

Occupation	Percent Female
Architects	25%
Chief executives	23%
Computer programmers	22%
Detectives and criminal investigators	19%
Chefs and head cooks	17%
Couriers/messengers	17%
Chiropractors	15%
Clergy	15%
Police patrol officers	15%
Taxi drivers and chauffeurs	13%
Machinists	7%
Home appliance repairers	5%
Truck drivers	5%
Firefighters	5%
Construction workers	3%
Aircraft pilots	2%
Electricians	1%

Source: U.S. Department of Labor, Women's Bureau. "Quick Facts on Nontraditional Occupations for Women," http://www.dol.gov/wb/factsheets/nontra2008.htm.

*Note: A **nontraditional female occupation** is one in which women comprise 25% or less of total employment.*

We may be tempted to think that sex segregation can lead to certain positive outcomes. For example, an abundance of women within certain occupations suggests arenas of power born from numbers. However, it is important to note that there is a negative relationship between the percentage of female workers within an occupation and that occupation's

Nontraditional female occupation One in which women comprise 25% or less of total employment

Table 11.2 The Top 10 Female Occupations, 2009

Occupation	Number	Median weekly earnings
1. Secretaries and administrative assistants	3,074,000	$619
2. Registered nurses	2,612,000	$1,035
3. Elementary and middle school teachers	2,343,000	$891
4. Cashiers	2,273,000	$361
5. Nursing, psychiatric, and home health aides	1,770,000	$430
6. Retail salespersons	1,650,000	$443
7. First-line supervisors of retail sales workers	1,459,000	$597
8. Waitresses	1,434,000	$363
9. Maids and housekeeping cleaners	1,282,000	$371
10. Customer service representatives	1,263,000	$587

Source: U.S. Department of Labor, Women's Bureau. "20 Leading Occupations of Employed Women." http://www.dol.gov/wb/factsheets/20lead2009.htm.

Note: These figures are for full-time wage and salary workers.

earnings. Occupations dominated by women enjoy less pay, less prestige, and less power than occupations dominated by males. Female-dominated industries also fare less well on health insurance coverage than do male-dominated industries (Dewar 2000). Furthermore, once an occupation becomes female dominated, it is effectively abandoned by men.

The opposite trend—male displacement of female workers—is unusual (Padavic and Reskin 2002). Indeed, it is a trend typically limited to instances where immigrant men replaced native-born women, as they did in American textile mills or in the cigar-making industry (Hartman 1976; Kessler-Harris 2003). Men moving into female work has also occurred when there has been a compelling financial incentive. Title IX of the 1972 Higher Education Act, for instance, required salaries of college coaches of female teams to be brought in line with those for coaches of male teams. With this change, there was a marked increase in the number of men taking positions as coaches for women's collegiate programs (Padavic and Reskin 2002). Men in female-dominated professions (e.g., male librarians and nurses) can benefit from presumed leadership skills and careerist attitudes (Simpson 2004). In general, however, men have little motivation to enter lower-paying, lower-status, female-dominated occupations. Those who do are apt to encounter challenges to their masculinity and witness eventual wage erosion in the occupation (Catanzarite 2003; Cross and Bagilhole 2002; Simpson 2004).

In general, male workers dominate in relatively high-paying precision production, construction, repair, and protective service occupations. Only 5% of employed women are found in production, transportation, and material-moving occupations. Less than 1% are found in natural resources, construction, and maintenance occupations. In addition, the most prestigious professions are primarily the domains of men. Only

25% of chief executives, 30% of dentists, 31% of financial analysts, 32% of physicians and surgeons, and 32% of lawyers are female (U.S. Department of Labor 2010b). Women who enter nontraditional occupations are likely to face gender segregation within the occupation. For example, females in medicine are most likely to specialize in pediatrics or obstetrics and gynecology, while anesthesiology and radiology remain the preserve of male physicians (American Medical Association 2006). And in the last decade, medical specialties dominated by women are finding it more and more difficult to recruit new residents (Bienstock and Laube 2005).

Women who enter nontraditional occupations are also underrepresented in leadership positions. Among physicians, for example, women make up 49% of graduating medical students and 42% of residents and fellows. Yet they constitute only 16% of full professors and 11% of medical school deans (Association of American Medical Colleges 2006). Similar patterns are found in the legal profession. A recent study of Massachusetts lawyers found that while men and women enter law firms in equal numbers, women leave law firm practice at much higher rates than men. The primary reason for the departure: the conflict between maximizing billable hours for firms and attending to family needs (Harrington and Hsi 2007). The female exodus from law firms means that fewer women make partner and fewer women lawyers become judges, law school professors, and business executives (Pfeiffer 2007).

In professional occupations, men are much more likely than women to be in the highest-paying professions (e.g., engineers and mathematical and computer scientists). Women are more likely to work in lower-paying occupations, such as teaching. They also tend to take jobs that allow them to move into and out of the labor force in order to accommodate family needs. Such jobs tend to offer lower compensation (Day and Downs 2007; U.S. Department of Labor 2006, 2009c). The picture fails to brighten in service-oriented work. In the world of waiting tables, gender segregation persists as well. Expensive restaurants tend to hire waiters; inexpensive eateries and diners hire waitresses (Padavic and Reskin 2002). Even in "God's work," sex segregation rules the day. Women clergy are overrepresented in low-status, subordinate congregational positions (Sullins 2000).

The gender segregation of jobs and occupations takes a financial toll on women. For example, in 2009, the median weekly earnings for full-time male workers averaged $819; for female workers, weekly earnings averaged $657 (Institute for Women's Policy Research 2010a). This disparity means that women must work about 15 months to earn the 12-month wage of men. Such pay discrepancies are reflected in a statistic known as the **pay gap**. The pay gap refers to a ratio calculated when women's earnings are divided by men's earnings. A pay gap favoring men over women is a well-established tradition. Currently, the pay gap for the annual average of median weekly earnings is approximately 80—that is, for every $10,000 paid the average male worker, the average female worker is paid around $8,000. While the gap did narrow through the 1980s, it has been rather stubbornly consistent over the last decade (U.S. Department of Labor 2009c). Furthermore, review of the Bureau of Labor statistics on weekly median earnings clearly shows the pay gap holds across virtually all occupations (Institute for Women's Policy Research 2010a).

Pay gap
The discrepancy between women's and men's earnings; a ratio calculated when women's earnings are divided by men's earnings

However, the pay gap can vary according to factors such as race, ethnicity, and educational level of workers. For example, the pay gap is actually lower for black and Hispanic female workers. These groups earn 89% and 90% (respectively) as much as their male counterparts. Ironically, it is Asian and white female workers who, though earning higher median wages, nonetheless face larger wage gaps; they earn only 78% and 79% (respectively) as much as their male counterparts.

(Can you understand this pattern?) Females with professional degrees earn 72% of their male counterparts' salaries, while female high school graduates earn 98% of their male counterparts' income (U.S. Department of Labor 2009c). Women hoping to improve their financial status should consider the jobs listed in Table 11.3. These jobs offered the highest median weekly earnings for full-time female workers in 2009 (although as you can see, a pay gap is still present in these occupations.) Women should also consider working in nontraditional occupations, as these tend to offer higher entry-level pay (U.S. Department of Labor 2009b).

Your Thoughts . . .

In spring 2010, the Department of the Navy cleared the way for women to serve on submarines, one of the last military locations to exclude women. Some are celebrating the ruling while others are calling it a huge mistake. What do you think? Are there some places that should be restricted to same-sex workers?

Ironically, one area in which women do appear to be achieving equity is in the realm of disease and mortality. Traditionally, women have enjoyed a health advantage over men. Females display lower rates of infant mortality than males. Females enjoy longer life spans than males. Male death rates generally are higher than female death rates within all age

Table 11.3 Top 10 Occupations With Highest Median Weekly Earnings for Full-Time Female Workers, 2009

Occupation	Median Weekly Earnings for Women	Women's Earnings as % of Men's
1. Chief executives	$1,553	74.5%
2. Pharmacists	$1,475	75.5%
3. Lawyers	$1,449	74.9%
4. Computer and information systems managers	$1,411	78.9%
5. Computer software engineers	$1,311	84.6%
6. Physicians and surgeons	$1,228	64.2%
7. Computer programmers	$1,182	93.3%
8. Management analysts	$1,177	85.8%
9. Computer scientists and systems analysts	$1,167	92%
10. Occupational therapists	$1,155	NA

Source: U.S. Department of Labor, Women's Bureau, Quick Stats on Women Workers, 2009, http:www.dol.gov/wb/stats/main.htm; Institute for Women's Policy Research, Fact Sheet: The Gender Wage Gap by Occupation, http:www.iwpr.org/publications/pubs/the-gender-wage-gap-2009.

categories. But as women embrace more of the behaviors traditionally associated with the male role (such as alcohol consumption and smoking), and as they make inroads into male occupations, their health advantage appears to be waning.

Consider smoking, the leading cause of preventable death in the United States. Currently, 23% of men and 18% of women are smokers (Centers for Disease Control and Prevention 2009i). Since 1984, the incidence rate for lung cancer has *decreased substantially* for men but *increased* for women until 2003, when it finally stabilized. Today, lung cancer accounts for the largest number of cancer-related deaths in both men and women. Since 1987, more women have died from lung cancer than from breast cancer (American Cancer Society 2009).

Similarly, women's increased representation in the workforce has been linked to increases in female heart disease. Although typically thought of as a man's health issue, heart disease is now the leading cause of death for *both men and women*—roughly the same number of men and women die each year from heart disease. Despite this fact, more than a third of women do not perceive themselves as being at risk for heart disease. This lack of awareness can be fatal: About two-thirds of women who die suddenly from heart disease have no previous symptoms (Centers for Disease Control 2010d).

Despite women's increasing representation in cancer and heart disease rates, several studies show that the female experience receives only secondary consideration by medical researchers. For instance, physicians are less likely to prescribe drugs for treating angina, blood clots, or heart disease for their female than male patients. They are less likely to counsel women about key risk factors and lifestyle changes relevant to heart disease. Physicians are less likely to hospitalize their female patients than their male patients for heart attacks or heart disease. After the first heart attack, women are less likely to receive diagnostic, therapeutic, and rehabilitative procedures. Consequently, women are more likely to die or suffer a second heart attack (Agency for Healthcare Research and Quality 2005, 2008, 2009).

Clearly, the social and economic contexts of women's lives are related to their health and health care. During the 1990s, activists aggressively lobbied Congress to obtain a more equitable share of funding for women's health issues. In the last decade, the Institute of Medicine (IOM) and the Office of Research on Women's Health (ORWH) called for studies to examine the ways in which health and disease processes may differ between men and women (Institute of Medicine 2001; Office of Research on Women's Health 2007). In 2009, the ORWH held public hearings in order to formulate recommendations for research priorities on women's health issues for the coming decade (Office of Research on Women's Health 2009).

In general, women's health care reflects many of the gender stereotypes and discrepancies documented throughout this essay. To make this point clear, consider the ways in which the experience of pain differs by gender. **Gender scripts**—the articulation of gender norms and biases—are useful in this exercise. The nurturing and empathic roles supported by female gender scripts make women more likely to see pain in others. As a result, women are more likely to acknowledge and experience pain themselves. In contrast, male gender scripts emphasize courage and strength. Hence, men are slow to acknowledge pain to themselves and even slower to report pain to their doctors. Gender scripts even influence medical protocols on pain treatment. Because women are viewed as overly sensitive, women's pain has been taken less seriously by the medical community. As such, women who complain of pain are too often discounted (Wartik 2002).

> **Gender scripts**
> The articulation of gender norms and biases

It is said that the longest journey begins with the first step. Women have taken that step, but their journey is far from complete. Perhaps the greatest evidence of the distance yet to be covered is found in the area of politics. Governorships,

Senate seats, and House seats are noteworthy for their near absence of women. Only six women currently serve as governors (this is down from a recent all-time high of nine in 2007). Women hold 88 (16.4%) of the 535 seats in the 112th (2011–2013) Congress—17 in the Senate and 71 in the House of Representatives (Center for American Women and Politics 2010). Social psychologist Sandra Lipsitz Bem (1993) contends that the male dominance of political power has created a male-centered culture and social structure. Such an environment works to the clear advantage of men. A male-centered perspective on the world dictates a set of social arrangements that systematically meets the needs of men while leaving women's needs unmet or handled as "special cases."

Witness, for instance, the influence of the male perspective within the legal arena: A case in point is the area of no-fault divorce laws. Such laws treat parties to a divorce as equal players despite their unequal work and occupational histories. Present social arrangements are such that a husband's earning power is enhanced over the course of a marriage. Consequently, in the wake of no-fault divorce laws, ex-wives typically experience a decrease in their standard of living, while ex-husbands typically enjoy an increase (Peterson 1996; Popenoe 2002). A male-centered perspective can also influence many government policies and assistance programs. Consider first women's standing with regard to health insurance. Job-based health benefits are typically not available for part-time or low paying jobs, yet women account for 67% of all part-time workers and 59% of low-wage earners. Consequently, women come up short on health insurance. Less than half of women have the option of obtaining health insurance through their work (Seshamani 2009). (The recent health care reform legislation of 2010 will not significantly impact individual coverage until 2014.) With regard to the unemployment insurance (UI) system, note that many states exclude part-time workers and low-wage earners from eligibility. Again, such policies are particularly harsh on females. It was only with the recent American Recovery and Reinvestment Act (passed as part of President Obama's stimulus program) that these gender disparities in unemployment benefits are being addressed (Institute for Women's Policy Research 2010b). The Temporary Assistance to Needy Families (TANF) program has been criticized as well for forcing mothers to prioritize wage work (in low-paying female jobs) over child-care responsibilities (Oliker 2000; Peterson 2002). Indeed, family support and occupational segregation issues have been systematically neglected as critical elements to any welfare or workforce reform efforts (Jones-DeWeever, Peterson, and Song 2008). Lastly, consider sick day policies. Currently, no state or federal law guarantees paid sick days. This policy oversight hits working women quite hard. Female-dominated jobs are the *least* likely to offer paid sick days. Also, since women are still the primary caregivers in families, they suffer financially when other family members get sick. Two-thirds of low-income working mothers will lose pay when they miss work in order to care for a child (National Partnership for Women & Families 2010).

Male-centered social arrangements also permeate current disability policies. Such policies recognize nearly all "male" illnesses and medical procedures (circumcision, prostate surgery, and so on) as potentially eligible for compensation. In contrast, many insurance companies have long engaged in "gender rating" of applicants—a practice that proved more costly to women. For instance, many insurance plans consider pregnancy to be a "special" or "pre-existing" condition that is not eligible for coverage. While the new health care reform bill bans such discriminatory practices, women are well advised to review insurance plans regarding maternity coverage until the bill is fully implemented. In essence, models or standards of normalcy and behavior are male oriented, a situation that automatically puts women at a disadvantage (Bem 1993; Crocker 1985).

Your Thoughts . . .

Insurance companies have long practiced charging women higher premiums than men by claiming that women use health services more than men and therefore are more costly clients. Following this logic, should women get a tuition cut for postsecondary education because their degrees yield lower financial returns than men's degrees? What do you think?

By increasing their numbers and voice in the political arena, women may achieve an effective check on social inequality. More women than men are registered voters. In recent years, women have made important strides in the area of voter turnout: In every presidential election since 1980, the percentage of female voters exceeded the percentage of male voters. In the 2008 presidential election, women made up a larger portion of the strategically important and much-wooed undecided voter (Center for American Women and Politics 2009a, 2009b). Without these kinds of developments, it will remain far too easy to maintain policies and practices that disadvantage women and risk continuing gender inequality.

LEARNING MORE ABOUT IT

In his work *Out of Play: Critical Essays on Gender and Sport* (Albany: State University of New York Press, 2007), Michael Messner offers an analysis of the dynamics of gender in organized sports.

Claudia Buchmann, Thomas Prete, and Anne McDaniel provide a thorough review of the literature regarding gender gaps in educational performance and attainment in elementary schools and beyond in "Gender Inequalities in Education" (*Annual Review of Sociology*, 2008, 34: 319–337).

Douglas Schrock and Michael Schwalbe explore the ways that men practice acts of manhood and examine how these practices reproduce gender inequality in "Men, Masculinity, and Manhood Acts" (*Annual Review of Sociology*, 2009, 35: 277–295).

In *Mismatch: The Growing Gulf Between Women and Men* (New York: Scribner, 2003a), Andrew Hacker examines the widening divide between men and women as evidenced in marriage patterns, divorce trends, career paths, politics, and so on.

A very readable and interesting discussion of the working woman's disproportional domestic duties is offered by Arlie Russell Hochschild (with Anne Machung) in *The Second Shift: Working Parents and the Revolution at Home* (New York: Penguin, 2003b).

Irene Padavic and Barbara Reskin have constructed a very informative review of gender and its relationship to work. Readers can consult *Women and Men at Work*, 2nd edition (Thousand Oaks, CA: Pine Forge Press, 2002).

GOING ONLINE FOR SOME SECOND THOUGHTS

Center for American Women and Politics http://www.cawp.rutgers.edu/

Institute for Women's Policy Research http://www.iwpr.org

(Click the link for "Status of Women and Girls" and then click on the link for "Status of Women in the States" and then the link for "State Reports" to see how each of the 50 states ranks on indicators such as political participation, earnings, health and well-being, social autonomy, etc.)

For an overview of many of the most pressing legal issues facing women in society, visit the National Women's Law Center site: http://action.nwlc.org/site/PageServer. To find an array of articles addressing the pay gap , click on "Our Issues" and the Equal Pay topic.

NOW IT'S YOUR TURN

1. Using your own experiences and the experiences of friends and classmates, construct a list of paying jobs typically performed by adolescent boys and girls. Be sure to note the activities, duration, and rate of pay that normally characterize these jobs. Discuss the anticipatory socialization (see Essay 6) implications of your findings.
2. Go to the National Women's Law Center website and find which states are doing the best and which are doing the worst at reducing the pay gap (http://www.nwlc.org/resource/wage-gap-persists-all-50-states). Can you offer any insight regarding the composition of the two lists?
3. Observe parents with small children in some public setting. Identify 5 to 10 gender lessons being provided by the nonverbal exchanges you observe.
4. Visit the Institute for Women's Policy Research website and review the information found via the "Status of Women and Girls " link (see above). Do you think that the indicators for assessing the status of women are reasonable ones? Are there areas or issues of life that are overlooked or slighted? Would the same indicators work for assessing the status of men?
5. In this exercise, you will be listening to some TV! For 30 minutes in each of three distinct time periods (morning, afternoon, and evening) turn on the TV and listen to the voiceovers on the commercials that are aired during your listening session. On a coding sheet, note the product being sold, the type of program, the network, the day of the week, and the gender of the voice. Run your listening test for a week (and switch up the networks and type of programs from one day to the next). Analyze your findings with regard to the voice of commercial authority.

NOTE

1. In fact, more than 250 women fought on both sides of the Civil War; five women died at the battle of Antietam (Marcus 2002).

ESSAY 12

Conventional Wisdom Tells Us . . . America Is the Land of Equal Opportunity

Is the United States a level playing field for all Americans? In this essay, we review the many arenas of continued segregation and racism in the United States. Furthermore, we explore the basis for determining one's race, noting that with all of the implications the classification holds, categorizing race is, at best, a tenuous process.

In 2007, the Pulitzer Prize in History went to *The Race Beat*—a book documenting journalists' role in the civil rights movement. The book was 16 years in the making, and the authors, Gene Roberts and Hank Klibanoff, attribute that fact to the complexity of the story (Online Newshour 2007; Roberts and Klibanoff 2006). To be sure, issues of race in America are extraordinarily complex. Some recent news events drive this point home.

Consider the 2008 Presidential election: Barack Obama, a person of mixed racial heritage (a white mother and a black father) was elected the 44th president of the United States. The inauguration generated the largest crowds ever for a D.C. event—an estimated 1.8 million. An additional 38 million watched the inauguration on TV. Political commentators hailed the event as a proud moment in American history. Yet 2 years later, the "birther movement" continues to challenge the legitimacy of the Obama presidency. "Birthers" maintain that Obama is not a natural citizen of the United States and therefore cannot hold the office. Incredibly, a 2010 *Washington Post* poll found that 1 in 5 Americans hold the same suspicions that the president was not born in the United States. What keeps this movement alive? Some believe that racist motives are behind the birthers—they can't accept a member of a racial minority holding the office of president (Rich 2009).

To be sure, with the election of Barack Obama, many expected change, especially in the area of race relations. But a 2009 Gallup Poll found that the percentage of Americans thinking that solutions to our race problems would be eventually worked out stood at 56%, almost the exact same level of optimism as found in the mid-1960s (55%). This 56% seeing good things coming is up from the low of 29% in the mid-1990s, but it is down from the 67% who held this positive view the day after Obama's election. Perhaps even more telling, the percentage of blacks who are optimistic about race relations in the United States was only 42% in the fall of 2009 (Newport 2009). Perhaps the low level of optimism among blacks is due in part to the famous "Beer Summit" of July 2009. At that time, President Obama hosted Harvard historian Henry Louis Gates and Cambridge police sergeant James Crowley at the White House for some "thoughtful conversation," conversation prompted by Crowley's cuffing and arresting Gates as Gates was entering his Cambridge home. (Crowley mistakenly thought Gates was an intruder.)

BOX 12.1 SECOND THOUGHTS ABOUT THE NEWS: DR. LAURA'S RANT

As we were nearing the end of the hottest summer ever, radio talk personality Dr. Laura Schlessinger heated things up a bit more. A long-time listener called Dr. Laura for some advice regarding an issue in her interracial marriage stemming from her husband's friends' use of her as a "black" authority as well as their use of the N-word. The caller asked Dr. Laura if the word was ever appropriate. Dr. Laura suggested that the N-word, depending on circumstances, might be OK. She then went on to use the word herself 11 times during her "advice" session. When the caller expressed her dismay over Dr. Laura's response, Dr. Laura accused the caller of being too sensitive and told her that she should not have married outside her race if she couldn't deal with these issues. When the caller continued to express her distress with the direction of Dr. Laura's advice, Dr. Laura told the caller, "Don't NAACP me."

While the good doctor was quick to apologize for her "N-word rant," not all thought the apology was sincere (including the caller). Indeed, within days of the incident, Dr. Laura announced that she was leaving radio at the end of her contract so that she could regain her First Amendment rights. She told Larry King that debate is no longer possible in our country given the presence of special-interest groups wanting to silence dissent. While some applauded Dr. Laura's on-radio rant, others insisted blacks but not whites "owned" the word, and still others voiced their strong objections to *anyone* using the word. The incident has been credited with setting off new debates over the use of the N-word and about the state of race relations in general. Ultimately, though, skeptics fear that people who might benefit from such discussions won't truly engage the issues because they tend to live in segregated realms. Readers interested in learning more about this story should see the CNN video "Is the N-Word Ever OK?" (http://www.youtube.com/watch?v=RTJlfWPc09I) or the video (that went viral) of an 11-year-old's call for the elimination of the N-word from our vocabulary (http://www.youtube.com/watch?v=bMyp8y8SkUM).

Social status
The position or location of an individual with reference to characteristics such as age, education, gender, income, race, and religion

The racial divide witnessed in our country should not surprise us. This is because our race, as well as our other social statuses, greatly influences our perceptions of reality. **Social status** refers to the position or location of individuals vis-à-vis each other with reference to characteristics such as age, education, gender, income, race, religion, and so on. Indeed, racial status is an extremely pertinent location in anyone's **status set**. A status set refers to the total collection of statuses that a social actor occupies.

Status set
The total collection of statuses occupied by a social actor

In survey data, the links between race and perception come through loud and clear. In the 2009 Gallup survey referenced above, 82% of whites thought that blacks have as good a chance as whites to get any type of job for which they are qualified; only 49% of blacks held that view. Similarly, 49% of whites but 72% of blacks reported that racism against blacks is widespread in the United States (Newport 2009). Another recent survey found that 70% of whites described race relations as good, whereas only 61% of blacks felt this way. Twenty-eight percent of whites think that President Obama will go too far in efforts to aid the black community; only 5% of blacks hold this view (Gallup 2009). Indeed, for black Americans, change in race relations may not come fast enough. Fully 56% of blacks report being treated unfairly by police because of their race; only 6% of whites make this claim (PollingReport.com 2009).

Your Thoughts . . .

More than a month into the Gulf oil spill disaster, Spike Lee advised President Obama to "go off" on BP. Others argued that a black man can't risk showing anger in our society. (Recall the Gates incident in Cambridge.) What do you think?

Blacks and whites see educational opportunities in the United States very differently as well. When asked about the chances for black children to get a good education, 80% of whites but only 49% of blacks believe that there is equal educational opportunity in the United States (Saad 2007). And when polled about the chances of two equally qualified students, one black and one white, being admitted to a major U.S. college, 61% of blacks believed that the white student has the edge, while just 20% of the white respondents felt this way; 48% of whites said that the two students have the same chance for admission (Newport 2007). Indeed, when it comes to the issue of special treatment, the divide between blacks and whites is profound. A 2009 Quinnipiac poll found that 78% of blacks but only 27% of whites favor the continuation of affirmative action for racial and ethnic minorities (Quinnipiac 2009). The significance of this divide is highlighted by a 2007 Supreme Court ruling that held public school admission policies *must* be colorblind. Chief Justice Roberts wrote that "the way to stop discrimination on the basis of race is to stop discriminating on the basis of race."

The Supreme Court ruling on school admissions is in line with a popular sentiment—that is, that special treatment of minorities is no longer needed and does more harm than good. Many in the United States believe that racial inequality and discrimination are primarily things of the past. Progress has been made, and the nation is now a level playing field. Are such claims accurate? Has racial equality been achieved in the United States? Furthermore, when inequalities do arise, are they rightfully attributable to race or racism?

Before answering these questions, it is important to define the terms we will be using in this essay. **Race** is typically defined as a group of individuals who share a common genetic heritage or obvious physical characteristics that are deemed socially significant. **Racism** refers to prejudice and discrimination based on the belief that one race is superior to another. **Prejudice** refers to

Race
A group of individuals who share a common genetic heritage or obvious physical characteristics that members of a society deem socially significant

Racism
Prejudice and discrimination based on the belief that one race is superior to another

Prejudice
A prejudgment directed toward members of certain social groups

Discrimination Unfavorable treatment of individuals on the basis of their group membership

an unfavorable prejudgment of an individual based on the individual's group membership. **Discrimination** refers to unfavorable treatment of individuals on the basis of their group membership.

Public opinion polls suggest that our optimism about race relations has improved over the past decade—we have climbed back from the pessimism of the 1995 to 2000 period. And 61% of all Americans believe that race relations will get better as a result of the Obama presidency (Newport 2009). But despite these good feelings, racial divisions in America persist. And this division is most apparent in our local communities and neighborhoods.

Housing patterns in the United States clearly underscore America's racial divide. In homebuying decisions, the neighborhood's racial composition matters to white buyers. Whites are less willing than blacks to live in integrated neighborhoods. While blacks are comfortable with a 50/50 racial divide, whites are unlikely to consider communities with anything but a strong white presence (Clark 2009; Krysan 2008; Krysan and Bader 2007; Krysan et al. 2009).

Attitudes on neighborhood living arrangements reflect actual residential patterns in the United States. Despite the civil rights movement, affirmative action programs, and other equality initiatives, housing segregation is still a fact of American life: Census tract data show us that for all the neighborhoods in the top 100 metropolitan areas of the country, nearly 26% are exclusively white and nearly 43% are majority white. Twenty percent of these neighborhoods are majority nonwhite and 12% are predominantly minority. These percentages, with the exception of the predominantly minority neighborhoods, represent progress with regard to integration of neighborhoods over the past decade. (The predominantly minority **neighborhood** has grown from 9.6% in 1990.) But the exclusively white neighborhoods are located in suburban areas and are economically exclusive, whereas the predominantly minority neighborhoods are located in central cities and are low income (Turner and Rawlings 2009).

Many factors work to undermine housing integration. To be sure, integration is limited by persistent disparities in income and wealth. Whites on average have higher incomes and wealth than minorities and thereby can afford to live in more exclusive neighborhoods. There are significant levels of discriminatory practices working against minority homebuyers. The National Fair Housing Alliance estimates that the incidence of housing discrimination against minorities approaches 4 million violations each year, but most of these incidents remain unaddressed (National Fair Housing Alliance 2010). Black homebuyers, for instance, are more likely to encounter **racial steering**. In racial steering, real estate agents direct prospective buyers toward neighborhoods that largely match the buyer's race or national origin (National Fair Housing Alliance 2007; Turner and Rawlings 2009). In buying a home, minorities often receive unequal treatment when trying to finance home purchases—they are often denied essential information about the loan process and products (Turner and Rawlings 2009). Studies also show that home loan applications for blacks and Hispanics are rejected at a higher rate than those for whites, regardless of applicants' income levels (Silverman 2005; U.S. Department of Housing and Urban Development 2005). When minority loans are approved, they are more likely to come with a higher price tag. Even with equal income and credit backgrounds, blacks are more likely to receive higher rates and subprime loans than their white counterparts. Indeed, in the year preceding our recent recession, subprime lending was five times more prevalent in black neighborhoods than white neighborhoods (Morris 2010;

Racial steering Practice in which real estate agents direct prospective buyers toward neighborhoods that largely match the buyer's race or national origin

National Fair Housing Alliance 2008; Pettit and Rueben 2010). When minorities succeed in buying and financing homes, they may find it more difficult to insure their properties. **Linguistic profiling**—that is, identifying a person's race from the sound of his or her voice—occurs and negatively affects the insurance services offered to minorities (Squires and Chadwick 2006). In light of this information, it is not surprising to learn that blacks and Hispanics are less likely to own their homes than are whites. Further, since home equity is the primary source of wealth for most Americans, this gap in homeownership has a profound impact on the persistence of racial inequality (National Fair Housing Alliance 2008).

Linguistic profiling Identifying a person's race from the sound of his or her voice

But money and discriminatory practices don't tell the whole story. Equally important to housing segregation is the fact that voluntary clustering is also at work. There is evidence that racial and ethnic groups are hesitant to move to predominantly white neighborhoods because of concerns over hostilities. This is especially true for newly arrived immigrants. And whites tend to avoid mixed neighborhoods for fear of decreasing property values and/or rising crime (Turner and Rawlings 2009). In the end, we have residential patterns that preclude the development of meaningful interactions and relations between racial groups and further promote segregated living arrangements (Bonilla-Silva, Goar, and Embrick 2006).

Ironically, education may be another driving force behind segregated living arrangements. Higher-educated whites are more likely to consider race when selecting schools for their children, and they are more likely to live in "whiter" neighborhoods (Emerson and Sikkink 2006). Childhood memories also have a hand in perpetuating segregated living. When making adult housing selections, Americans tend to reproduce the neighborhoods of their childhoods, thus keeping residential segregation alive (Dawkins 2005).

Discriminatory practices are still a part of the rental process as well. Websites advertising apartments often use language that would be prohibited in newspaper classifieds (National Fair Housing Alliance 2010). (Since 2009, some state and local Fair Housing Assistance Program agencies have issued charges against people posting discriminatory ads on websites such as Craigslist.) And in e-mail exchanges, landlords can use tenant names as cues for racial/ethnic identity. In a recent experiment, white, Arab, and black names were randomly attached to e-mails inquiring about advertised apartments. Across all rental categories, landlord responses were significantly different for the three groups—blacks received fewer positive responses than either Arab or white names (Carpusor and Loges 2006).

Segregated living imposes other financial burdens on racial and ethnic minorities. Too often, jobs are located in areas where these workers can't afford to live. A recent analysis by the Urban Institute of six major metropolitan areas (Atlanta, Boston, Houston, Philadelphia, Seattle, and Washington, D.C.) found that with the exception of one location (Houston), the majority of jobs (especially low-wage jobs) were located in suburbs. But consider that low-wage workers are *more likely* to reside in central cities. And even when low-wage workers reside in the suburbs, there are still job obstacles. The fact of housing segregation and the general lack of efficient public transportation beyond urban areas means that matching workers with jobs is problematic (Pindus et al. 2007).

Some sociologists contend that race segregation ultimately translates into knowledge segregation. The 2008 high school status dropout rate for 16- to 24-year-old white students was 4.8%—a figure much lower than the 9.9% and 18.3% rates for blacks and Hispanics, respectively (Aud et al. 2010, Table A-19–1). As of 2008, approximately 30% of the 25+ white population had a college degree in contrast to 20% of blacks and 13% of Hispanics (U.S. Census Bureau 2011, Table 225). Similar discrepancies exist in other education-related practices. For example, nearly

85% of white high school seniors use computers at home at least once a week versus 63% of black and 71% of Hispanic seniors. Gaps continue at the college level. Seventy percent of white college students use the Internet at school, while only 61% of black and 57% of Hispanic college students report doing so (U.S. Department of Education 2009a).

Racial inequality in the educational sphere is an old story. America's public school system has battled the issue for decades. In 1954, the U.S. Supreme Court's *Brown v. Board of Education* decision ordered American schools to desegregate with all deliberate speed. Yet more than 50 years later, full integration still eludes public schools. In many cases, white parents effectively circumvented the desegregation ruling by relocating to suburban school districts or by enrolling their children in private schools. In Mississippi, for instance, court-ordered desegregation was met with a dramatic increase in private segregationist academies (Andrews 2002). And the migration of white and middle-class families to suburban school districts continues to hold steady today (Carr 2007; NAACP 2008). In fact in the 1990s, Supreme Court rulings put an end to many desegregation plans in school districts across the nation (Orfield and Eaton 2003). Our nation's courts, once champions of desegregation efforts, are declaring more and more school districts "**unitary**" and therefore released from desegregation plans (Baldas 2003; NAACP 2008). In June 2007, the Supreme Court delivered a blow to school diversity when it ruled against the use of race-based admission policies in public schools pursuing voluntary integration strategies. Today, many school districts look exactly at they did before the monumental *Brown* decision (Ogletree 2007).

Unitary schools
Those deemed to have eliminated a dual system of education based on racial inequality

Our public schools are becoming increasingly resegregated along racial and economic lines. Over the last 20 years, the percent of public school students who are white dropped from 68% to 55%. Over this same period, Hispanic enrollments in public school doubled, hitting 22% in 2008, while the percentage of black students in public school held steady at 16% (Aud et al. 2010, Indicator 4).

Charter schools, touted as solutions to failing public education, are as segregated, if not more so, than our public schools. To be sure, the number of charter schools is on the rise (currently there are more than 4,000 in the United States), and student enrollment in charter schools has tripled since 2000. But patterns of enrollment look familiar. At the national level, 70% of black charter school students attend intensely segregated charter schools (Frankenberg et al. 2010). A resegregation trend has also emerged in private schools. As of 2008, 75% of students enrolled in private schools are white, while only 10% are black and 10% are Hispanic (Aud et al. 2010). In sum, the typical white student attends a school that is 77% white; typical black and Hispanic students attend schools where more than half of their classmates are black and Hispanic (Orfield and Lee 2007). These developments mean that lower levels of interracial exposure—or what some are calling a "resegregation trend"—are occurring in numerous school districts across the nation (Frankenberg and Lee 2002; Kozol 2005; NAACP 2008). (See Figure 12.1.)

Charter schools
Schools that receive public money but are free of most public regulation *if* they produce performance outcomes stipulated in their charter contracts

Parents who can afford to send their children to private schools reap rewards for their financial investments. Parental satisfaction levels with teachers, academic standards, and a school's order and discipline are all higher for parents of private school children than for parents of public school students (U.S. Department of Education 2006, Indicator 38). Private schools also offer advantages over public schools in terms of student/teacher ratios and educational outcomes.

Figure 12.1 Trends in Public School Enrollments by Race and Region of Country (1988–2008)

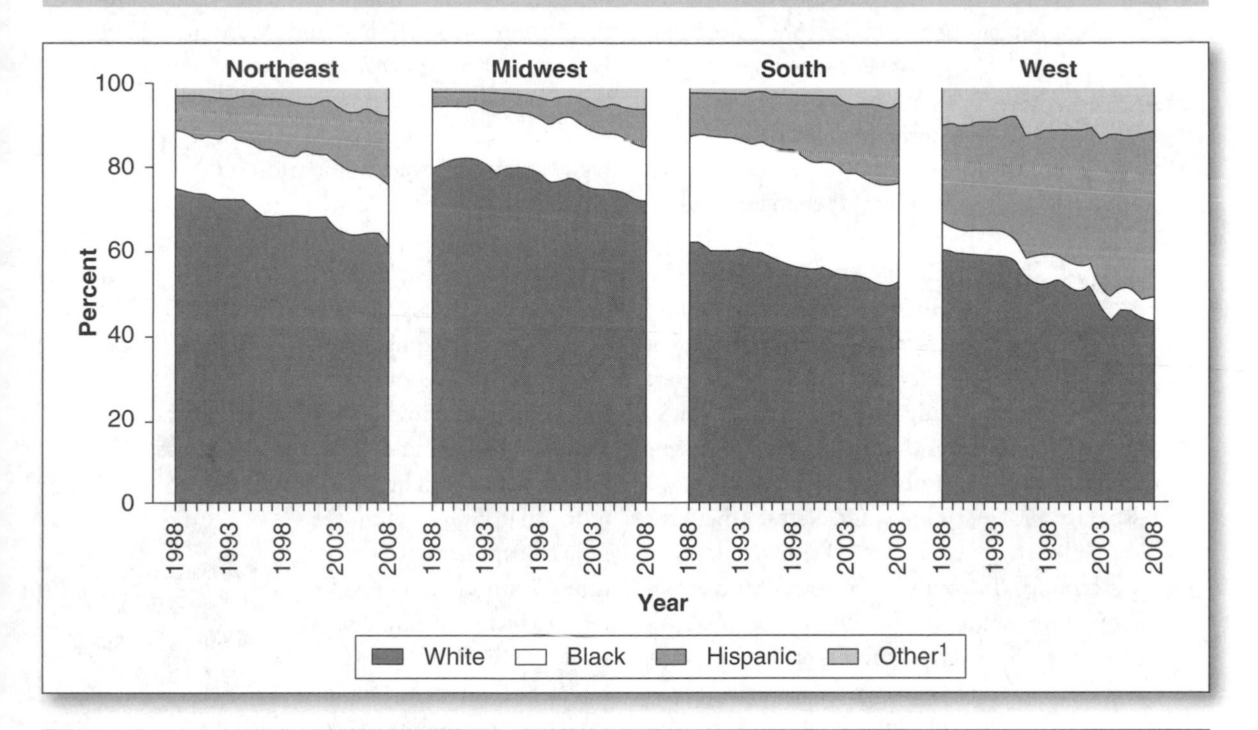

Source: U.S. Department of Commerce, Census Bureau, Current Population Survey (CPS), October Supplement, 1988–2008.

[1] "Other" includes all students who identified themselves as being Asian, Hawaiian, American Indian, or two or more races.

And private schools exhibit higher rates of teacher satisfaction across a variety of measures, including class size, availability of educational materials, colleagues, parental support, and teaching (U.S. Department of Education 2005).

Sociologist Jonathan Kozol (1991, 1995, 2001) has long studied the vast resource differences that characterize white versus nonwhite schools. In his seminal work, *Savage Inequalities* (1991), Kozol found that the poor resources of one predominantly African American Chicago Southside school forced chemistry teachers to use popcorn poppers as Bunsen burners. In contrast, students in a nearby predominantly white suburban school were enjoying a facility that housed seven gyms, an Olympic-sized pool, and separate studios for fencing, dance instruction, and wrestling. Similarly, Kozol found that PS 261 in the South Bronx housed 400 more students than permitted by local fire codes. Just a few bus stops away in the wealthy Riverdale section of the Bronx, PS 24 touted class sizes well below the city average.

Because most public school budgets are tied to local economic resources, schools in wealthy, white neighborhoods fare better than schools in poor, nonwhite neighborhoods (Cummings 2003). Furthermore, state spending on public education *decreases* as racial/ethnic diversity *increases*. In a recent analysis, the Population Reference Bureau found that those states with the highest racial/ethnic diversity spent the lowest share of their gross state product on education (3.4%), whereas those states with the lowest racial/ethnic fragmentation spent the highest share on education (4.2%). While these percentage differences may appear small, they translate into differences of billions of dollars for educational spending (Mather 2007).

The work of Kozol and others suggests that school segregation is especially destructive because of the powerful tie between highly segregated schools and concentrated poverty. Seventeen percent of all public schools are **high-poverty schools**—that is, 75% or more of their students are eligible for free or reduced-price lunches. Cities have the largest share of high-poverty schools. In 2008, 40% of all city elementary schools were high-poverty schools and 20% of all city high schools were high-poverty schools. Racial and ethnic minorities dominate the enrollments of high-poverty schools: blacks make up 43%, Hispanics 46%, Asians 4%, and Native Americans 2%, while whites constitute 34% of students in these schools. (In contrast, whites constitute 75% of low-poverty schools.) Without a doubt, students in high-poverty schools pay a high and unreasonable educational price. In high-poverty schools, a smaller percentage of teachers have master's degrees and a larger percentage have less teaching experience than their peers at low-poverty schools. In high-poverty schools, a higher percentage of students are limited in English proficiency than in low-poverty schools. At the elementary level, average reading scores, math scores, and music and visual arts scores are lower for students in high- than low-poverty schools. At the secondary level, graduation rates are lower for students in high- than in low-poverty schools. And the percentage of high school graduates who attend a four-year college is lower for graduates of high- than low-poverty schools (Aud et al. 2010).

High-poverty schools
Schools where 75% or more of students are eligible for free or reduced-price lunches

The lessons taught in predominantly white versus predominantly nonwhite schools can also differ dramatically. Students in predominantly white schools learn to be self-directed, inquisitive, and ambitious. In contrast, students in predominantly nonwhite schools are taught to obey rules and maintain the status quo and receive "signals" that discourage academic aspirations (Bowles and Gintis 1976, 2002; Kozol 1991; Langhout and Mitchell 2008; Meyers and Nidiry 2004; Oates 2009). While it has been more than 20 years since Kozol wrote his stirring indictment of our segregated public schools, the inequality he found there still exists (Books and McAninch 2006; Feldman 2003). Indeed, Kozol (2005) warns that the conditions have grown worse for inner-city youth as our schools revert to levels of segregation higher than they were in the late 1960s.

After graduation, former students find that racial boundaries are maintained in the employment sector as well. Blacks have consistently faced lower employment rates than whites. This is true during strong economic periods (Holzer 2007) and well as during weak economic times. The 2009 unemployment rate was 8.5 for whites, 14.8 for blacks, and 12.2 for Hispanics (U.S. Department of Labor 2010b). In 2009, 36% of all individual filings with the Equal Employment Opportunity Commission were for race discrimination for a total of 33,579 charges (U.S. Equal Employment Opportunity Commission 2009). Controlled **paired testing** hiring experiments—experiments in which two equally qualified candidates of different races each apply for the same job—reveal that hiring discrimination is pervasive, affecting approximately 20% of African American job applicants (Bendick and Nunes 2011; Urban Institute 1999). Other research indicates that employers' perceptions of job candidates' merits are often biased by racial stereotypes (Moss and Tilly 2001). Such findings have prompted a call for a "national report card" on discrimination as well as an expanded paired testing program to promote public understanding of the prevalence of racial discrimination (Urban Institute 1999).

Paired testing
Hiring experiments where two equally qualified candidates of different races each apply for the same job

Once on the job, discrimination continues. Just look at things on the wage front. In 2008, the real median household income for non-Hispanic whites, for blacks, and for Hispanics was $55,530, $34,281, and $37,913, respectively (DeNavas-Walt et al. 2009). Education doesn't eliminate such gaps. In 2008, the median annual earnings of full-time workers aged 25 to 34 with a bachelor's degree was $47,000 for whites, $40,000 for blacks, and $42,000 for Hispanics (Aud et al. 2010, Table A-17–1). These gaps grow still larger when minority women enter the equation. Black women working full time earn only 61% of their white male counterparts' wages. For Hispanic women, that percentage falls to 52% (National Women's Law Center 2010). Such differences may help explain why blacks and Hispanics have poverty rates that are roughly three times as high as the rates for whites (DeNavas-Walt, Proctor, and Smith 2009).

Your Thoughts . . .

Soon after his primary victory for the Senate in Kentucky, Tea Party "favorite son" Rand Paul criticized the Civil Rights Act of 1964–the act that desegregated restaurants, motels, theaters, and so on. His opposition to the law is based on his belief that the government should stay out of the affairs and choices of private businesses, even when they engage in discriminatory practices. What do you think?

Evidence of the racial divide makes its way into the entertainment sphere as well. Perhaps one of the most telling signs of racial inequality is found in children's books—a source regarded by some to be important primers for a society's culture. A recent review of Newbery and Caldecott Medal-Winning children's books found that the presence of minority characters hit a peak during the 1990s but has been waning since then (Clark 2007). Others have found the depictions of egalitarian interracial interactions in children's books to be rare (Pescosolido, Grauerholz, and Milkie 1997). Put down the books and pick up the TV remote, and racial/ethnic inequities persist. Prime-time television shows little of the diversity that characterizes our society. Whites are overrepresented both in prime-time programs and in starring roles. Note too that the eight o'clock viewing hour, the one most likely to be watched by children, is the *least racially diverse* hour of prime-time programming. And sitcoms, the most popular programming among youth, represent the most segregated programming genre (Children Now 2004). Furthermore, programming geared toward adolescent audiences is doing little to erase the racial divide: The shows watched most regularly by black teens feature black casts, but very few of these shows are also watched by white teens (Brown and Pardun 2004). Content analysis of prime-time programs that feature racial diversity of characters provided the least diversity of occupations and occupational prestige (Signorielli 2009). Network webpages use images that are predominantly male and white (Angelini et al. 2009). Television news and sport broadcasts also tend to reinforce racial stereotypes (Rada and Wulfemeyer 2005; Schaffner and Gadson 2004). A recent study of television news programming found that whites dominate news coverage both as reporters (nearly 90%) and as story "experts." When blacks and Hispanics appear in news stories, they appear as ordinary citizens and tend to be in stories about accidents, disasters, or weather events (Owens 2008).

Commercials, like prime-time programs, teach. And the lessons they teach misinform us about minorities. For example, Hispanics are seriously underrepresented and are frequently depicted in ways that reinforce stereotypes. While whites appear in ads for upscale products and home goods, minorities appear in ads for fast foods and soft drinks and for sports equipment and financial services (Harrison 2006; Jacobs Henderson and Baldasty 2003; Mastro and Stern 2003).

Clearly, many inequalities still exist in the various sectors of U.S. society. Yet many contend that such inequalities are not the product of racism. Many continue to believe that race, as a biological attribute, indicates some inherent differences in individuals' ability to achieve.

At first glance, this argument may appear valid. Biology would appear to be the unequivocal determinant of racial group distinctions. Thus, different biologies could conceivably lead to different levels of ability. Yet a biological definition of race does not produce a simple or clear racial classification scheme. In fact, identifying groups who share obvious physical characteristics proves to be a less-than-obvious task.

Using a biological definition of race, biologists and physical anthropologists can "find" as few as 3 or as many as 200+ different races. These classifications are muddied further when we note that generations of intergroup marriage and breeding ensure that no "pure" races exist. Indeed, a remarkable similarity exists across the genes of all humans: Of the DNA molecules that account for racial categories, 95% to 99% are common across all humans (Shipman 1994). Thus, if the human essence is "all in the genes," then racial similarities, not distinctions, are most noteworthy. The genome project has offered consistent evidence that there is only one race: the human race (Angier 2000; Graves 2006).

From a biological perspective, "racial differences" are best understood as beneficial, adaptive changes for our human species (Molnar 1991). For instance, the dark skin of peoples living near the equator serves as vital protection against dangerous sun rays. Similarly, the longer, narrow noses found among those living in colder northern climates help warm the air before it reaches the temperature-sensitive lungs. If the earth were to shift on its axis so that the northern hemisphere moved into direct line with the sun, we would expect an adaptive change in the skin color and nose configurations of the northern population (Rensberger 1981). Geography then is central to the variations we so readily attribute to race (Diamond 2005; Jablonski 2005; Lehrman 2003; "Race: The Power of an Illusion" 2003).

A biological approach to the race issue is really insufficient for understanding the dynamics of racial categories. Noted biological anthropologist Alan Goodman has observed that race is not about biology, but rather about an *idea* we *ascribe* to biology ("Race: The Power of an Illusion" 2003). The respected *New England Journal of Medicine* has asserted that "race is biologically meaningless" (Kristof 2003). Indeed, the task of identifying discrete racial categories has largely been abandoned by many physical anthropologists. Sociologists suggest that race is more properly understood as a *social* rather than a biological phenomenon: Race is socially constructed. The **social construction of reality** occurs when individuals create images, ideas, and beliefs about society based on their social interactions.

Social construction of reality
A process by which individuals create images, ideas, and beliefs about society based on their social interactions

The social constructionist approach suggests that racial categories emerge from social interaction, social perception, and social opinion. Historians, for instance, observe that the idea of biologically based races was advanced as a way to defend slavery in North America—that is, as a way to justify the unequal treatment of slaves in a land that promoted equality (Lee 2003; "Race: The Power of an Illusion" 2003; Smedley and Smedley 2005). Social encounters repeatedly expose individuals to specific definitions of race. If these definitions suggest clear and natural boundaries and rankings between various groups of people, the definitions can institutionalize racism as part of a society's stock of knowledge. Such definitions come to reify, or substantiate, racial distinctions that may not be supported in fact. **Reification** refers to the process by which the subjective or abstract erroneously comes to be treated as objective fact or reality. From such a perspective, we must view race as a characteristic that

Reification
Process by which the subjective or abstract erroneously comes to be treated as objective fact or reality

resides in the eye of the beholder. Change the group doing the perceiving and defining—that is, change the eye of the beholder—and you will change the racial distinctions being made.

Research by genetic anthropologists estimates that 58% of black Americans have at least 12.5% European ancestors (i.e., the equivalent of one great-grandparent; Gates 2009a). (It is also estimated that one-third of all the black players in the NBA have a white male ancestor [Gates 2009b].) Yet this biological lineage does not alter public perception. Despite evidence of white ancestry, such individuals are still classified as black. U.S. classification patterns resulted from a long-standing legal practice that mandated percentage or "one drop of blood" standards for determining racial classifications. Until 1983, for instance, the law of Louisiana dictated that individuals with 1/32 of "Negro blood" were properly classified as belonging to the black race. It was only with the 2000 census that respondents were able to identify themselves as bi- or multiracial by checking more than one race category. Were we to change the standards used for racial classification, however, very different designations of race would emerge. In Brazil, for example, any individual who has "some" white ancestry is classified as belonging to the white race. Consequently, by Brazilian standards, most black Americans would be classified as white (Denton and Massey 1989). A recent ruling by the high court in South Africa clearly shows race as a "social classification." The court authorized the reclassification of some 200,000 ethnic Chinese living in South Africa from "white" to "black" so that they could benefit from policies intended to end white domination (BBC News 2008b).

At first glance, perceptual differences of race may not seem very significant. Such differences merely underscore a major premise of the sociological perspective: Social context is an important factor in understanding, explaining, or predicting human attitudes and behaviors (see the introductory essay). **Social context** refers to the broad social and historical circumstances surrounding an act or an event. But the intriguing nature of race as a social creation becomes clearer when we view it as a significant social status or, more specifically, as a significant *ascribed* status. Some of the statuses we occupy are the result of our own personal efforts; these are achieved statuses. An **achieved status** is one earned or gained through personal effort. One's statuses as Red Cross volunteer, parent, or worker are all achieved statuses. In contrast, some of our statuses are "assigned" to us, independent of our personal efforts, desires, or preferences; these are ascribed statuses. An **ascribed status** is one assigned or given without regard to a person's efforts or desires. Age, gender, and racial status are all ascribed statuses.

Social context
The broad social and historical circumstances surrounding an act or an event

Achieved status
Status earned or gained through personal effort

Ascribed status
Status assigned, given, or imposed without regard to persons' efforts or desires

The average reader of this book will occupy many of the following statuses (try classifying each as "achieved" or "ascribed"): son or daughter, student, friend, spouse, sibling, male or female, citizen, voter, consumer, employee. Often, however, one of our many statuses will dominate the rest. This dominant status forms a master status (see Essay 6). A **master status** is a single social status that overpowers all other social positions occupied by an individual. A master status directs the way in which others see, define, and relate to an individual.

Master status
A single social status that overpowers all other social positions occupied by an individual; it directs the way in which others see, define, and relate to an individual

If we consider race in light of these status distinctions, we begin to more fully appreciate the implications of race as a social creation. For although race is a social creation, it is also an ascribed status. As such, race is imposed on the individual; one's race is beyond one's control. Race also frequently serves as a master status. As

Identity
Essential characteristics that both link us to and distinguish us from other social players and thus establish who we are

Life chances
Odds of obtaining desirable resources, positive experiences, and opportunities for a long and successful life

Social minority
A group regarded as subordinate or inferior to a majority or dominant group

a master status, race has the ability to influence the social identity and life chances of an individual. **Identity** refers to those essential characteristics that both link us to and distinguish us from other social players and thus establish who we are. **Life chances** refer to one's odds of obtaining desirable resources, positive experiences, and opportunities for a long and successful life. For races classified as social minorities, this influence is often negative. A **social minority** is a group regarded as subordinate or inferior to a majority or dominant group. Social minorities are excluded from full participation in society; they experience inferior positions of prestige, wealth, and power.

Note the irony here. Ascribed master statuses are beyond the individual's control. They are assigned, yet they have a remarkable capacity to control the individual. Indeed, certain ascribed master statuses can prove more important to one's identity than personal efforts. The irony intensifies when we acknowledge that race, an assigned status, is nonetheless a social creation. Racial designations can change as audience perceptions change.

The significance of these last few points becomes more apparent when we reconsider the real-life consequences of racial designations:

- In 2008, the poverty rate was 8.6% for whites, 24.7% for blacks, and 23.2% for Hispanics (DeNavas-Walt et al. 2009). (All groups had higher rates than in 2007.)
- In 2008, 6.7% of white children but 10.7% of black children and 17.2% of Hispanic children lacked health insurance (DeNavas-Walt et al. 2009).
- In 2008, approximately 30% of the 25+ white population had earned a college degree compared with only 20% of their African American and approximately 13% of their Hispanic counterparts (U.S. Census Bureau 2011, Table 225).
- The 2008 high school dropout rates for 16- to 24-year-olds were 4.8% for whites, 9.9% for blacks, and 18.3 for Hispanics (Aud et al. 2010).
- The average life expectancy for 2015 newborns is projected to be 77 years for white males, 71 years for black males, 82 years for white females, and 78 years for black females (U.S. Census Bureau 2011, Table 102).
- The infant mortality rate for babies of white mothers is under 6 deaths per 1,000 live births. The rate for babies of African American mothers is 13.3 deaths per 1,000 live births (U.S. Census Bureau 2011, Table 114).

Indeed, take any set of statistics regarding life chances—health and illness rates, divorce rates, crime victimization rates, death rates, and so on—and you will undoubtedly come to the conclusion that race matters.

This essay suggests that racial distinctions cannot be equated with biological or genetic differences. Race is not a simple matter of physiology. Rather, racial distinctions are more properly understood as social creations. Skin color proves to be the primary marker of racial distinctions in U.S. society; other cultures have focused on such characteristics as height or hair and eye color. No matter what a society's marker, once certain characteristics are deemed worthier than others—that is, once racial categories are created—powerful social processes such as prejudice and discrimination are set into motion.

Still, race has proven itself a highly dynamic process; the human species has shown a remarkable capacity to adapt to environmental demands. The pressing question for today and the near future is whether our social definition of race will prove equally adaptive to changes in our social and cultural environments. It is presently projected that by the year 2050, the United States will be a country where whites will be a numerical minority. One in five Americans are expected to self-identify as multiracial (Lee and Bean 2004). Given these projected demographic changes, rethinking the race issue may well be a social and cultural necessity. Perhaps by stressing the *social* nature and origins of racial distinctions, we will find that such distinctions are more amenable to change than conventional wisdom currently allows.

LEARNING MORE ABOUT IT

In *Blurring the Color Line: The New Chance for a More Integrated America* (Cambridge: Harvard University Press, 2009), Richard Alba argues that the time and conditions for significantly altering racial/ethic boundaries in the United States have arrived.

To learn more about the continued presence of racial inequality in the United States, see the most recent edition of Andrew Hacker's *Two Nations: Black and White, Separate, Hostile, Unequal* (New York: Simon and Schuster, 2003b).

W. E. B. Du Bois offers a classic treatise on the dynamics of U.S. race relations in *The Souls of Black Folk* (New York: Penguin, 1903/1982). Cornel West thoughtfully grapples with issues of race in a more contemporary book entitled *Race Matters* (New York: Random House, 1994).

David Roediger explores the social construction of race and the historical redrawing of racial lines in his book *Working Toward Whiteness: How America's Immigrants Became White* (New York: Basic Books, 2005).

In *Finding Oprah's Roots,* Henry Louis Gates Jr. provides a roadmap for using public documents and online databases for researching one's past. He also provides information on genetic testing resources currently available for tracing one's tribal roots (New York: Random House, 2007).

Devah Pager and Hana Shepherd offer a review of major research findings on racial discrimination in four areas in "The Sociology of Discrimination: Racial Discrimination in Employment, Housing, Credit and Consumer Markets" (*Annual Review of Sociology* 34: 181–209, 2008).

The transformation of the United States from a mainly biracial to a multiracial society and the "boundary" implications of these changes are considered in Jennifer Lee and Frank Bean's article, "America's Changing Color Lines: Immigration, Race/Ethnicity, and Multiracial Identification" (*Annual Review of Sociology* 30:221–242, 2004).

Shannon Harper and Barbara Reskin provide a review of the history and impact of affirmative action in the areas of education and employment in their article, "Affirmative Action at School and on the Job" (*Annual Review of Sociology* 31: 357–380, 2005).

GOING ONLINE FOR SOME SECOND THOUGHTS

You can also consult the following organizations/sites to learn more about race and ethnic relations in America and abroad:

Institute on Race and Poverty http://www.irpumn.org/website/

National Urban League http://www.nul.org

The Civil Rights Project http://civilrightsproject.ucla.edu/

NOW IT'S YOUR TURN

1. Imagine that height is a critical marker for social ranking in the United States: Shortness is valued, and tallness is devalued. Speculate on the ways in which the social structure of your hometown might change if residential, educational, and occupational patterns influenced by prejudice and discrimination were based on human height.
2. Racial categories are social creations that emerge from social interaction. Gather a sample of ads from two magazines that target different classes of readers. One magazine might be targeting an elite readership (for example, *Martha Stewart Living* or *Gourmet*), whereas the second might target a more general, less affluent readership (for example, *Family Circle* or *Good Housekeeping*). Are racial lessons delivered through these ads? Do the ads indicate any differences in life aspirations by race or ethnic group? Consider the data on life chances as presented in this essay. How does reality compare to the lifestyles projected in your sample ads?
3. Explore the impact of social structure via the Survey Documentation Analysis (SDA) web page.
   - Access the SDA Web page: http://sda.berkeley.edu/.
   - Click on the SDA archive link.
   - Click on the GSS Quick Tables link.
   - Explore the various social issues covered in the quick tables and create four to six tables that enable you to see the impact of race on Americans' attitudes and/or behaviors.
   - NOTE: Before doing your runs, look at the chart options box. Choose the option that you think offers the best visual display of the effects you are presenting. (No matter which option you choose, be sure to check the "show percents" box.)

DEVIANCE, CRIME, AND SOCIAL CONTROL

ESSAY 13

Conventional Wisdom Tells Us . . . Violence Is on the Rise in the United States—No One Is Safe

In recent decades, Americans have wrestled with a growing fear of violence. Is that fear justified? Here we review the state of violence in the United States, and we explore those instances in which the public's fears of violence are justified and those in which they are exaggerated. As such, the essay explores the many problems surrounding the detection and perception of danger and crime.

On January 8, 2011, residents of Tucson gathered at the Safeway in La Toscana Mall for a regular visit with their congressperson, Gabrielle Giffords—"Congress on your Corner," she called it. But this visit would be different from any of Giffords' previous gatherings. For on this day, some of those civic-minded Tucsonians would never return home.

Giffords, her staff, and about 30 individuals were engaged in conversation when Jared Lee Loughner approached Giffords and shot her point blank in the head. Loughner kept shooting—31 shots in all. When the barrage finally finished, 18 people were wounded and six people were dead, including Chief Judge John Roll and 9-year-old Christina Green, who was attending her first political event.

The Tucson shooting is not the most famous of the decade; it is not the most unusual or brutal of our time. Rather, this incident represents just one of many violent events—events that, some argue, have become a routine feature of modern life.

When Americans are asked to name the country's most troubling problems, violent crime always makes the list—and often, the very top of the list (Gallup 1976–2003, 2004–2010). Drive-by shootings, Amber Alerts, on-street surveillance cameras, metal detectors at the doors of our schools: In the new millennium, these images have become all too common, and they are images that can provoke a sense of fear and panic. Indeed, in surveys, many Americans report acting on these fears. Recent polls show that at least one in three of us is taking deliberate protective measures to deal with issues such as murder and violent crime: installing special locks or alarms in our homes, buying dogs or guns for protection, changing our nighttime walking patterns, and minimizing contact with strangers (Carlson 2005a; Carroll 2007b; Saad 2010).

Is the ever-present fear of violence justified? Is conventional wisdom correct in suggesting that our streets have become more dangerous than ever before? Just how likely is it that any one of us will become the victim of a violent crime?

Americans do indeed face a greater risk of violence than the inhabitants of many nations in the world. Murder, for example, as well as violent crimes such as assault or rape, occurs anywhere from 2 to 40 times more frequently in the United States than in the other developed nations of the world. The United States also has higher rates of violent crime than nations suffering from intense poverty or political turmoil—places such as Costa Rica, Croatia, Greece, India, Indonesia, Yemen, and so on (NationMaster 2010).

Rates of violence in the United States, when compared to rates found in other nations, suggest that violence is a serious problem for Americans. At first glance, such rates seem to support the conventional wisdom that violence is on the rise. However, recent statistics tell us that things in the United States may be changing. The tide of violence may be turning in Americans' favor.

Each year, the FBI provides statistics on crime in the ***Uniform Crime Reports Index for Serious Crime*** (hereafter referred to as the *UCRs*). According to the *UCRs,* violent crime in the United States reached an all-time high in 1992; that year, the nation experienced approximately 1.93 million violent crimes, including murder, rape, robbery, and aggravated assault. But recent statistics show that the number of violent crimes has dropped significantly. In 2008, the nation experienced approximately 1.38 million acts of violence. That number represents more than a 28% decrease in the violent crime rate. Many are hopeful that this trend will continue into the new century (U.S. Department of Justice 2009a).

Uniform Crime Reports Index for Serious Crime FBI-issued statistics on crime in the United States

To be sure, many will argue that official statistics such as those found in the *UCRs* grossly underestimate violence in America. **Victimization studies**—that is, statistics based on victims' self-reports and not the reports of police—present a picture of violent crime that differs significantly from the one painted by the FBI. For example, statistics from the National Crime Victimization Survey estimate that nearly 5 million Americans fell victim to violent crime in 2008 (U.S. Department of Justice 2008c). Similarly, a recent Harris poll (2006) estimates that 33 million adults have been victims of domestic violence. And the U.S. Department of Health and Human Services (2008) contends that nearly 772,000 children (about 1 in 100) are victims of abuse in the United States each year.

Victimization studies Produce statistics based on victims' self-reports and not the reports of police

Just as some researchers criticize the *UCRs* for underestimating violent crime, other researchers criticize victimization studies for overestimating the problem. Which statistics are correct? Experts in the field disagree, but the key point to remember is this: *All* statistics on violent crime—*UCRs* and victimization surveys alike—show a similar decrease in the violent crime rate from 1992 to the present. (See Figure 13.1.)

If, as statistics suggest, violent crime is waning, then what explains Americans' persistent—even increasing—fear of violence? Even with decreasing rates, does violent crime occur with staggering frequency?

Violent crime cannot be described as a frequent event. Indeed, within the world of crime, violence is quite rare. FBI statistics show that, overall, property crimes (e.g., arson, auto theft, burglary, and larceny) occur more than seven times more often than violent crimes (see Figure 13.2). (Victimization surveys suggest a similar relationship, with

Figure 13.1 Decreases in Violent Crime, 1993–2007

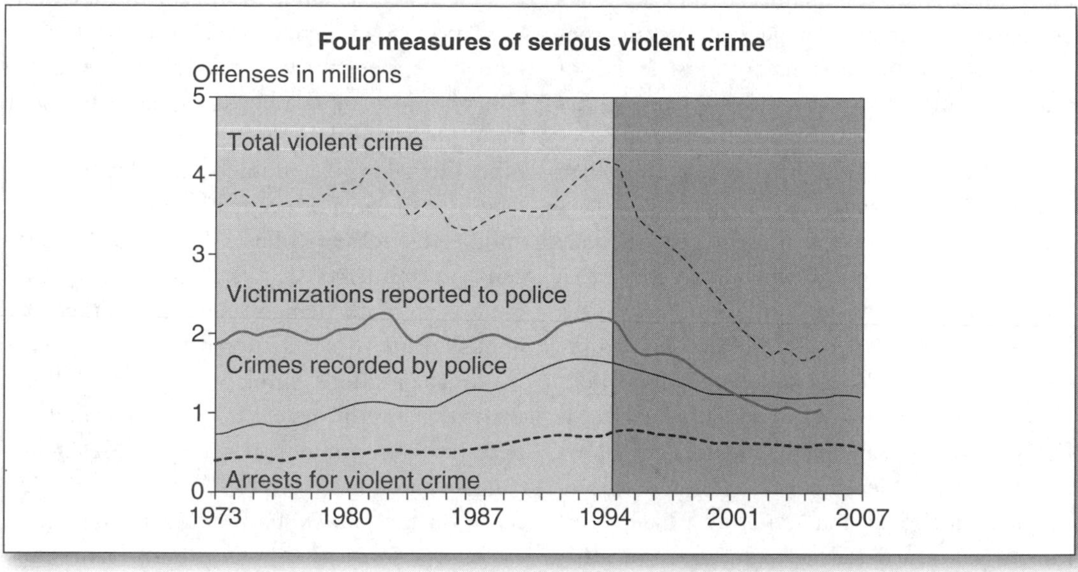

Four measures of serious violent crime

Offenses in millions

Total violent crime

Victimizations reported to police

Crimes recorded by police

Arrests for violent crime

0 1 2 3 4 5

1973 1980 1987 1994 2001 2007

Source: Bureau of Justice Statistics. Available at http://bjs.ojp.usdoj.gov/content/glance/cv2.cfm.

property crime rates 6.98 times higher than violent crime rates.) More specifically, the FBI notes that the typical American is 136 times more likely to be burglarized than murdered and more than 73 times more likely to be a victim of theft than of rape. Indeed, when it comes to fatal victimization, statistics show that Americans are more likely to take their own lives than to be killed by violent criminals (Centers for Disease Control and Prevention 2009j; U.S. Department of Justice 2009a).

Given the relative rarity of violent crime, what other factors might explain Americans' growing fear of violence? Some suggest this fear may be linked to the perceived randomness of such crimes. Americans tend to view violence as an event that can strike anyone at any time. As conventional wisdom states, "No one is safe." Crime statistics, however, do not substantiate this image (Macmillan 2001). Consider the act of murder. Most Americans picture murder as an unpredictable attack that is likely to be perpetrated by a stranger. Yet "friendly murders"—that is, murders committed by relatives, friends, or acquaintances of the victim—are more than three times more common than murders perpetrated by strangers. Furthermore, far from being random,

Figure 13.2 FBI Time Clock, 2008 

One Violent Crime Occurs Every 22.8 Seconds	
One murder	Every 32.3 minutes
One forcible rape	Every 5.9 minutes
One robbery	Every 1.2 minutes
One aggravated assault	Every 37.8 seconds
One Property Crime Occurs Every 3.2 Seconds	
One burglary	Every 14.2 seconds
One larceny-theft	Every 4.8 seconds
One motor vehicle theft	Every 33.0 seconds

Source: U.S. Department of Justice, 2009a.

murder exhibits several striking social patterns. For example, murder is a crime of the young. An individual's risk of being murdered peaks at age 24, regardless of race or gender. (Recall from Essay 6 that senior citizens are most fearful of violent crime, yet crime statistics show that seniors are least likely to become murder victims.) Murder also is a "male" crime; more than 65% of all perpetrators and 78% of all victims are male. Note, too, that murder systematically varies by race; it is an overwhelmingly intraracial crime. Whites tend to murder other whites, blacks tend to murder other blacks, and so on. The crime of murder also occurs disproportionately among the poor. In addition, socioeconomic status appears related to when and how a murder occurs. Members of the lower socioeconomic strata, for instance, are most likely to be murdered on a Saturday night, and the grisly event is likely to involve alcohol and passion. In contrast, members of the upper strata are murdered with equal frequency during all days and times of the week. In addition, murders among the privileged typically result from premeditation rather than passion (Thio 2009; U.S. Department of Justice 2009a).

Statistics on other violent crimes dispel the myth of random violence as well. Rape, for example, is rarely the product of a surprise attack. Indeed, rapes by strangers account for only about one fourth of such crimes. Similarly, simple assault more often takes place between intimates, the result of a building animosity between two individuals. Intimacy is especially characteristic of assaults involving female victims. Women are twice as likely to be attacked by an intimate or an acquaintance as they are to be attacked by a stranger (RAINN 2010; U.S. Department of Justice 2009b).

Our visions of violence seem not to match the realities of the world around us (Altheide 2002, 2009; Glassner 2000). Contrary to perceptions of violence on the rise, its high frequency of occurrence, and the randomness of violent events, violent crimes are relatively rare, highly patterned, and decreasing in recent years. Given these facts, what else might explain Americans' persistent fears and misperceptions?

One might be tempted to explain these fears by referring to the high personal cost of violence—namely, serious injury or death. However, if the risk of injury or death alone stimulated such fears, we would find similar dread surrounding other high-injury and high-mortality settings. Consider the area of occupation-related injuries and deaths. Although fewer than 20,000 Americans are murdered each year, some studies estimate that more than 60,000 U.S. workers die annually due to occupational disease or unsafe working conditions. Similarly, while the FBI estimates that approximately 1.38 million Americans become victims of violent crime, some estimates suggest that more than 4 million suffer physical harm on the job. Despite the staggering figures on occupational disease and death, Americans' fear of the work setting is negligible relative to their fear of violent crime (Centers for Disease Control 2010e; Reiman and Leighton 2009; Smith 2007). Now consider life on American roads. Americans are more than twice as likely to die in automobile accidents as they are to be murdered and nearly twice as likely to be injured in an automobile accident as they are to be injured by a violent criminal. However, few would cite a level of fear that precludes one's taking to the roads (National Highway Traffic Safety Administration 2009, 2010).

Considerations regarding the reality and cost of violent crime contribute little to our understanding of Americans' intense fear of violence. Violence in the United States is clearly a problem, but it does not appear to warrant the level of fear expressed by the American public. As a result, some sociologists contend that Americans' fear of violence may in part be socially constructed. The **social construction of reality** occurs when individuals create images, ideas, and beliefs about society based on their social interactions.

Social construction of reality
A process by which individuals create images, ideas, and beliefs about society based on their social interactions

The social constructionist approach suggests that certain social encounters expose individuals repeatedly to information on violence—information that suggests that violent crime is on the rise and that it occurs frequently, randomly, or at the hands of strangers. As a result, these data—even though they represent misinformation—come to form the public's "reality" of violence.

The mass media, especially television, are the greatest source of misinformation on violence. The National Television Violence Study (1996–1998) as well as longitudinal research conducted by scholars in both the United States and abroad (Beresin 2009; Gerbner et al. 2002; Mustonen and Pulkkinen 2009; Signorielli and Vasan 2005) provide a wealth of evidence on this point. Researchers involved in these projects have meticulously analyzed the content found in sample weeks of prime-time and daytime television. Their findings show that the rates of violent crime in "TV land" are disproportionately high compared with real-world figures.

Sixty-one percent of television programs contain some type of violence. During any weeknight, viewers see an average of three violent acts per hour. On Saturday mornings, a time period dominated by child viewers, the rate of violence increases to 20 violent acts per hour. Furthermore, some studies estimate that by the time most children leave high school, they have viewed approximately 16,000 murders on TV! The figures on TV violence are significant, for they suggest a world quite different from everyday reality. In the real world, fewer than 1% of all Americans become involved in violence. In TV land, 64% of all characters are involved in violence. Therefore, those who rely on television as their window on reality may come to view the world as a perilously dangerous place (Beresin 2009; Gerbner et al. 2002; Mustonen and Pulkkinen 2009; The National Television Violence Study 1996–1998; Signorielli and Vasan 2005).

To substantiate this claim, researchers compare both heavy and light television viewers with regard to their perceptions of violence. Respondents participating in these studies are asked a series of questions requiring them to estimate rates of murder, rape, and assault. They are generally presented with two choices in making these estimates. One choice typically reflects real rates of violence in the United States, whereas the other choice better reflects rates of violence in TV land. Results consistently show that heavy television viewers are much more likely to overestimate rates of violence than those who watch little or no television. Heavy television viewers routinely favor TV-land estimates of violence over real-world estimates. Furthermore, heavy television viewers perceive the world to be a more dangerous place than those who watch little or no TV. Thus, heavy viewers are more likely than light viewers to take the protective measures mentioned earlier: installing special locks or alarms in their homes, buying dogs or guns for protection, or changing their nighttime walking patterns (Gerbner et al. 2002; Morgan, Shanahan, and Signorielli 2009).

With more and more Americans shifting from television to computers and the Internet for entertainment, violent video games must become a part of this discussion as well. In a highly ambitious research project, Craig A. Anderson and colleagues did a **meta-analysis** of 136 research studies addressing the links between violent video games and players' perceptions and enactments of violence. A meta-analysis is a statistical technique used to assemble and summarize a large collection of quantitative studies addressing a single problem or issue. Anderson's work offers impressive support for the link between the playing of violent video games and one's perception of danger. Avid game players perceive the world as a more violent place; further, they think and behave more aggressively, show less empathy toward others,

Meta-analysis
A statistical technique used to assemble and summarize a large collection of quantitative studies addressing a single problem or issue

and engage in less prosocial behavior than occasional or non–game players (Anderson et al. 2010). Those who regularly play violent video games become more familiar and more strongly identified with these virtual worlds. As exposure to violent game worlds increases, players can lose perspective and, like heavy TV viewers, can come to see the virtual violence levels as reflective of the real world (Riddle 2010; Yao, Mahood, and Linz 2010).

Complementing the social constructionist view, some suggest that Americans' disproportionate fear of violent crime emerges from a long-standing cultural value that supports a fear of strangers. A **cultural value** is a general sentiment that people share regarding what is good or bad, right or wrong, desirable or undesirable. A **fear of strangers** refers to a dread or suspicion of those who look, behave, or speak differently from oneself. Such fears can ultimately make the world seem unfamiliar and dangerous.

Cultural value
A general sentiment regarding what is good or bad, right or wrong

Fear of strangers
A dread or suspicion of those who look, behave, or speak differently from oneself; such fears can ultimately make the world seem unfamiliar and dangerous

In the United States, cultural values instill a sense of mistrust and foreboding toward those we do not know. Couple this phenomenon with the fact that most Americans view violent crime as "stranger crime," and the misinformation that links violence to an already feared social category—strangers—serves to exacerbate and perpetuate public fears of such crimes (Dubner 2009; President's Commission on Law Enforcement and Administration of Justice 1968).

Your Thoughts . . .

What do you think accounts for Americans' long-standing fear of strangers?

The public's misplaced fears and misperceptions of violence are not without serious consequences. Such misconceptions sometimes result in an ineffective approach to crime. For example, high-profile murder cases such as the Rabbi Fred Neulander case in New Jersey, the Toni Riggs case in Detroit, or the Susan Smith case in South Carolina illustrate the danger of equating murder with strangers. In these cases, the murder victims were killed by immediate family members: Neulander and Riggs murdered their spouses, and Smith murdered her children. Yet in all of these cases, our fixation on the danger of strangers initially led to the detention of innocent people. (The false leads in the Riggs and Smith cases also involved black males; indeed, 35 black males were questioned in the Smith case.) Such mistakes substantiate the power of socially constructed scripts—scripts that depict the "typical" nature of murder and the "probable" perpetrator of the crime (Brown 1994; McDonald 2003). The same is often true for abduction cases. In 2010, the disappearance of 10-month-old Lauryn Dickens made front-page news, with initial investigations focusing on a stranger abduction. Within weeks, the investigation shifted to Lauryn's mother Shakara, who police fear may have killed the child (Associated Content 2010). Like 78% of child abductions, this event was likely not the work of a stranger but rather of a loved one (Dubner 2009).

Misplaced fears and misperceptions of violence also can detract attention from the critical sites of violence in the United States. To be sure, Americans display greater concern for violent crimes on our nation's streets than they do for violence in the home. Yet sociologist Richard Gelles notes that, aside from the police and the military, the family is the single most violent institution in our society (Gelles, Loseke, and Cavanaugh 2004; Straus, Gelles, and Steinmetz 2006).

Finally, some worry that the constant bombardment of violent media programming as well as the ways in which violent stories are told—for example, the vantage point of the viewer, the context of violence, the response of other characters to the perpetrator—will eventually desensitize readers and viewers to real-world violence. Constant media exposure may make readers and viewers more tolerant of violent acts in the real world (Cerulo 1998; Gerbner et al. 2002; Larson 2003; Morgan, Shanahan, and Signorielli 2009; National Television Violence Study 1996–1998; Signorielli and Vasan 2005).

If Americans' fear of violence is socially constructed and our perceptions of violent crime are inaccurate, should society shift its attention from the issue of violence? We suggest nothing of the kind. To be sure, any instance of violence represents one death or one injury too many. In this sense, violence may indeed be all too common in the United States. As a nation, we appear to be making strides in reducing the incidence of violence. Can levels of violence in the United States be further reduced? It is difficult to say, but any solutions to the violence problem require us to adopt a more accurate picture of the scope and patterns that characterize violent crime in America.

LEARNING MORE ABOUT IT

Barry Glassner offers an engaging look at the culture of fear in America. See *The Culture of Fear: Why Americans Are Afraid of the Wrong Things* (New York: Basic Books, 2000). In *Terror, Post 9/11 and the Media* (New York: Peter Lang Publishing, 2009), David Altheide reflects on the role of the mass media and propaganda in promoting fear and social control in our post–9/11 world.

Nancy Signorielli and Mildred Vasan offer a comprehensive review of the literature on media violence and its impact on viewers. See *Violence in the Media: A Reference Handbook* (Santa Barbara, CA: ABC-CLIO Publishers, 2005).

A classic essay by Georg Simmel, "The Stranger," offers an insightful exploration into our cultural beliefs about those we do not know; see *The Sociology of Georg Simmel,* edited by K. Wolff (pp. 402–408, New York: Free Press, 1950b).

GOING ONLINE FOR SOME SECOND THOUGHTS

To keep track of yearly increases and decreases in violent crime, visit the FBI's website and read the FBI's *Uniform Crime Reports:* http://www.fbi.gov/ucr/ucr.htm. One can also follow results from the National Crime Victimization Survey at http://bjs.ojp.usdoj.gov/index.cfm?ty=pbdetail&iid=1975.

The National Crime Prevention Council provides information on building safer, stronger communities: http://www.ncpc.org/.

NOW IT'S YOUR TURN

1. Interview from 10 to 15 people about their working "models" of crime. Be sure to obtain information on such things as the appearance of the typical criminal, the location of the typical crime, the typical criminal offense, and so on. Determine whether a general model emerges, and discuss how this model allows certain acts to escape the label "criminal."

2. This essay provides a detailed profile of the social patterns of murder: age, gender, site, and social class. Go to your local or university library and collect similar statistics for the crimes of burglary, larceny, and auto theft. Based on your data, speculate on the ways in which the "face" of murder differs from the "face" of property crimes.

3. Visit the FBI's web page (http://www.fbi.gov) and find the most recent rates for murder, rape, assault, and robbery. Then create a multiple-choice test designed to tap individuals' perceptions of violent crime rates. Your survey might include questions such as the following:

   In the United States, ________ murders occur each year:

   (a) 5,000 (b) 20,000 (c) 50,000 (d) 100,000

   Administer your survey to 10 to 15 people you know to be avid TV watchers and to 10 to 15 people you know to be only occasional TV watchers. Compare the answers of heavy and light TV viewers. Which group better estimated the actual violent crime rate?

ESSAY 14

Conventional Wisdom Tells Us . . . There's Nothing We Can Do About the Weather

Mother Nature may call lots of the shots when it comes to weather and climate. But the elements have a social and cultural side as well. In this essay, we focus the sociological eye on weather and climate, exploring the ways in which factors such as modernization, profit motives, and our entrepreneurial values can both create climate problems . . . and perhaps solve them.

"Everybody talks about the weather," wrote Mark Twain, "but nobody does anything about it." It would be hard to quibble with Twain's observations on the social aspects of weather. The weather is a conversational constant—a "go to" topic when we are at a loss for words. On the job, on a date, in line at the division of motor vehicles, we can always resort to the weather when other topics of discourse elude us: Hot enough for you? Is it ever going to stop raining?

But Twain's quote also expresses the common belief that the weather is something beyond our control. Weather professionals can attest to the strength of this belief. Despite many accurate weather forecasts, meteorologists will tell you that the public tends to focus on and remember only their big misses—the prediction of rain on a day that turns sunny or a warning of record snowfall when a mere dusting occurs. We don't think of our weather people as reliable. Yet we typically cut them some slack because the weather—well, it just happens. We have to grin and bear it.

Despite the conventional wisdom on weather, we may be more involved in manufacturing it than we realize. And the manufacturing of weather is a very social phenomenon. For this reason, weather is a topic that deserves a second look—this time, a look guided by our sociological eye.

Let's begin by reviewing some terms. **Weather** refers to the atmospheric conditions of a given moment and

Weather
The atmospheric conditions of a given moment and place

place. In contrast, **climate** refers to the average weather conditions of a region over time. Weather can change quickly, from one hour to the next. (The earth's day-to-night cycle is a powerful driver of the weather.) **Climate change** occurs over longer periods of time—that is, years and decades.

Climate
The average weather conditions of a region over time

Climate change
Any significant change in climate (e.g., temperature, precipitation, wind, etc.) that lasts for decades or longer

Now let's consider some climate "facts." While parts of the United States experienced an atypically snowy winter in 2010, the year is currently tied with 1998 as the warmest on record. For the last 25 years, the earth's average temperature has been increasing at twice the rate of the last 100 years! Globally, nine of the past 10 years have been the warmest on record. In the United States, record daily highs have exceeded record daily lows by two to one for the last decade. The oceans reached their highest recorded temperatures in the summer of 2009. Scientific evidence reveals that sea levels are rising and glaciers are shrinking. Three of the smallest Artic ice covers have occurred in the last 4 years. And since Artic ice loss feeds warming, the Artic warming trend is expected to continue for the foreseeable future. To be sure, our climate is changing and more and more scientists agree that human activity is responsible for that change (Karl et al. 2009; Richter-Menge and Overland 2010; Union of Concerned Scientists 2010a, 2010b).

To understand our hand in changing the climate, we might first start with a review of a term that has been bandied about for the last decade or so: the *greenhouse effect*. Despite some bad press, this phenomenon is, in itself, a good thing. The **greenhouse effect** refers to a natural phenomenon that makes life on earth possible. Our climate is basically driven by energy from the sun. In the normal flow of solar energy, the sun warms the earth's surface, and this energy is radiated back toward space as heat. As is the case with garden greenhouses that control heat loss, this radiated heat is trapped by an insulating blanket of gases. (We might think of "greenhouse" gases such as carbon dioxide, methane, and nitrous oxide as akin to the glass panes of greenhouses; see Figure 14.1.) The trapped heat keeps the earth's average surface temperature at approximately 60 degrees Fahrenheit. Without the greenhouse effect, the average temperature of the earth would be 0 degrees F—far too low to sustain life. So far, this sounds quite positive. So why is the greenhouse effect a bad thing? The greenhouse effect begins to be a problem when too much heat is trapped by too many gases—a situation that scientists agree is now underway. From the preindustrial era to 2005, the atmospheric concentrations of carbon dioxide, methane, and nitrous oxide have increased globally by 36%, 148%, and 18%, respectively (Environmental Protection Agency 2010a).

Greenhouse effect
Radiated heat trapped in the earth's atmosphere by an insulating blanket of certain gas molecules (e.g., carbon dioxide, methane, and nitrous oxide)

The climate changes we are witnessing are all indicative of a phenomenon known as **global warming**. Global warming refers to a rise in the earth's temperature due to the increase in greenhouse or heat-trapping gases in the atmosphere. To be sure, global warming is not a new phenomenon. The earth experienced a drastic rise in temperature after the last ice age. And over the course of the last century, the earth's temperature has increased by 1.5 degrees F (and by an average of 2 degrees F over the last 50 years in the United States; Union of Concerned Scientists 2009a). For most of the last century, sea levels have risen by .6 inch per decade, but in recent years, the sea level has increased by more than an inch each decade. By the end of the century, sea levels are expected to rise between .6 and 2 feet (Environmental Protection Agency 2010b, 2010c).

Global warming
A rise in the earth's temperature due to the increase in greenhouse or heat-trapping gases in the atmosphere.

Figure 14.1 The Greenhouse Effect

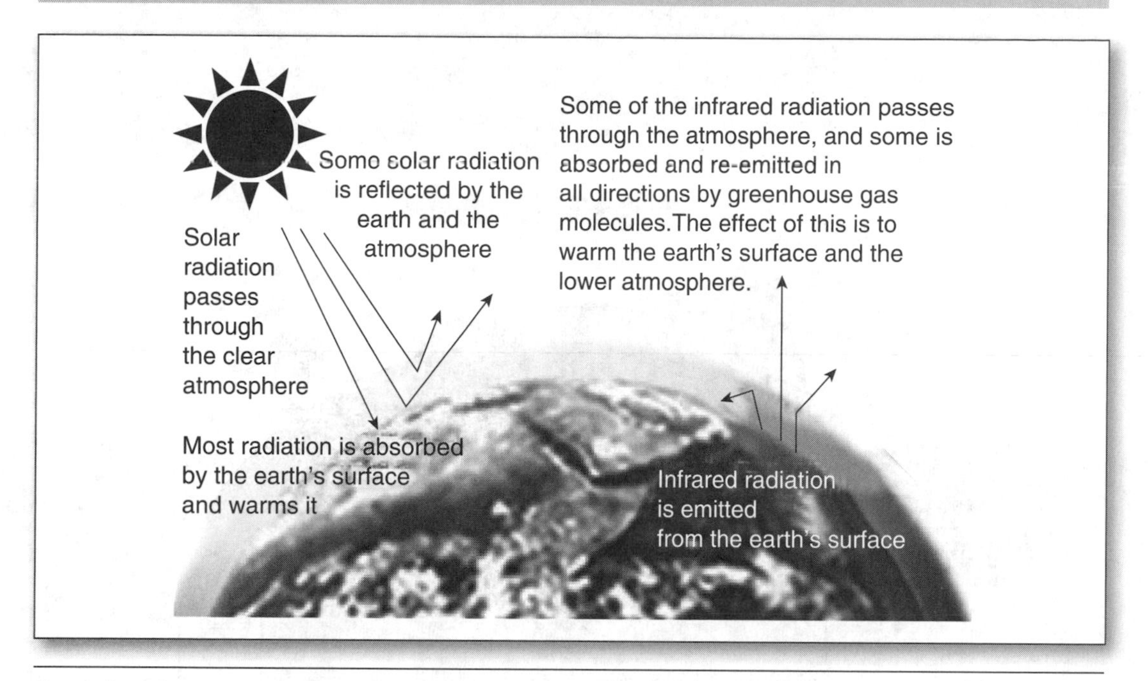

Source: http://www.epa.gov/climatechange/science/index.html.

Your Thoughts . . .

A Harris online poll found that 65% of U.S. adults believe human activities are contributing to global warming, while 77% of adults in Great Britain, 88% in France, and 92% in Germany believe this to be the case. Can you offer any reasons for the lower U.S. numbers vis-a-vis their European counterparts?

While these changes may sound modest, they are nonetheless significant. Consider that over the past 10,000 years, the earth's average temperature hasn't fluctuated by more than 1.8 degrees F. The Ice Age saw average temperatures that were only 5 to 9 degrees cooler than today. If global warming follows the predictions of either "low" (4- to 6.5-degree increases) or "high" (7- to 11-degree increases) projection models, we can expect significant changes across the United States by the end of the century. Both low and high models predict increases in the frequency of heat waves. The high projection model predicts an increase in the frequency of floods. "Twenty-year" floods will occur every 4 to 15 years. Increases in dangerous levels of air pollution (Red Ozone Alert days) and droughts are predicted. We should expect declines in crop quality and yields as well as in livestock productivity. Furthermore, increasing coastal erosion and increasing health threats from disease-carrying insects and rodents are also expected (Karl et al. 2009; Union of Concerned Scientists 2009a). Figure 14.2 shows the impact global warming has had on heavy rain events in the United States over the last 50 years. To be sure, rising temperatures demand our attention, and that attention should be focused on ourselves. The scientific community

Figure 14.2 Percentage Increases in Amounts of Very Heavy Precipitation, 1958 to 2007

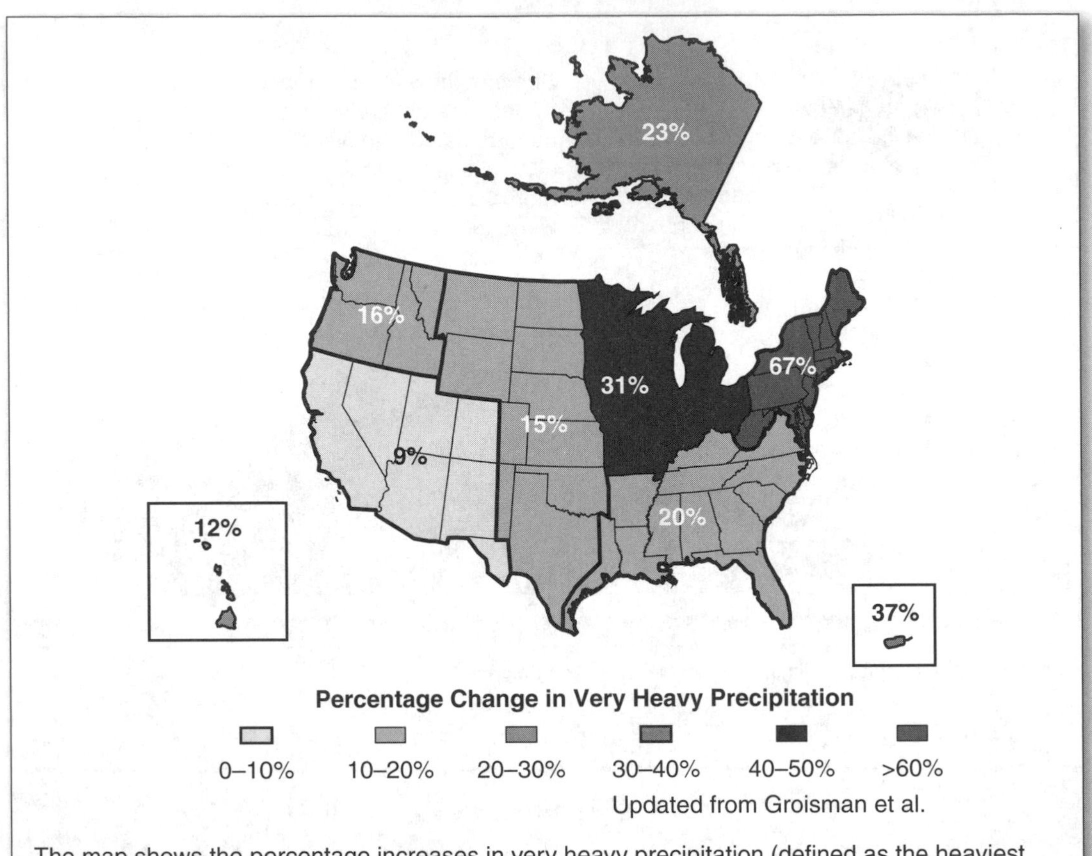

The map shows the percentage increases in very heavy precipitation (defined as the heaviest 1 percent of all events) from 1958 to 2007 for each region. There are clear trends toward more very heavy precipitation for the nation as a whole, and particularly in the Northeast and midwest.

Source: U.S. Global Change Research Program, National Climate Change. Available at http://epa.gov/climatechange/science/index.html.

is increasingly in agreement that a primary source of global warming, especially over the last 50 years, is human activity (Union of Concerned Scientists 2009a).

The human impact on the environment is really not news. Indeed, an entire field of study is devoted to this interaction—the field of **ecology**. Ecology entails the study of the interaction between organisms and the natural environment. Ecologists note that, in dealing with the environment, humans have a resource that greatly facilitates our influence—humans have **culture**. Culture refers to our design for living. It encompasses all the material and nonmaterial things and ideas we use to solve the problems of living. How does culture have an impact on the environment? Consider that humans' influence on the environment can range from low (i.e., in simple hunting and gathering societies) to high

Ecology
The study of the interaction between organisms and the natural environment

Culture
The values, beliefs, symbols, objects, customs, and conventions that characterize a group or society's way of life

(i.e., in modern, industrialized nations). In large measure, it is the material aspect of culture, in particular the technological advances of modern societies, that produces the high-impact scenario.

Technology is often used as a way of taming or bending the natural environment to better suit our current needs. Necessity and material comfort are both mothers of invention and of cultural innovation. As populations and cities and suburbs grow, fields and forests give way to housing developments and malls and parking structures. Community schools, hospitals, factories, stores, roadways, bridges, tunnels, and theme parks are built. Growing populations present growing energy and utility demands that are met with ever-expanding power grids and infrastructures. Waste management and pollution are inevitable byproducts of progress, and technology devises new ways to handle both. Our modern, technologically enhanced existence is second nature to most of us; many can't imagine life any other way. Currently, there are only a few individuals committing to "no-impact" lifestyles: living their lives so as to have no net impact on the environment. Not surprisingly, then, the high environmental impact and environmental costs of modern society often go unrecognized.

Manifest functions Obvious consequences or benefits of some social process, practice, or pattern

Latent functions Unintended benefits or consequences of a social practice or pattern

Manifest dysfunctions Obvious negative outcomes or consequences of a social practice or pattern

Merton's work on manifest and latent functions and dysfunctions helps us understand our blindness to modernity's environmental costs. **Manifest functions** are obvious consequences or benefits of some social process. **Latent functions** are unrecognized positive outcomes. **Manifest dysfunctions** are obvious negative consequences of some social process. **Latent dysfunctions** are less-than-obvious negative outcomes. Any one event or social process might give rise to each of these categories of effects. For example, a manifest function of war is the defense of a nation. A latent function is the strengthening of in-group solidarity as the nation rallies round its war effort. A manifest dysfunction of war is the loss of lives and destruction of property that accompany war. A latent dysfunction of war might be the declining status of the warring nation.

Latent dysfunctions Less-than-obvious negative outcomes or consequences of a social practice or pattern

Very often, the products and processes of modernity, while they yield immediate benefits, also present us with latent dysfunctions. So, for example, our interstate highway system obviously supports our mobility needs and helps us overcome the obstacle of distance. But our interstate highway system has also increased our dependence on the automobile and, in turn, our dependence on foreign oil. Similarly, chemicals and pesticides can make our lawn the envy of our neighbors, but they can also ruin local water tables, lakes, and streams. Unintended negative consequences are exactly what we seem to be facing with the global warming issue.

Specifically, two human activities—our increasing use of fossil fuels (coal, oil, and natural gas) and our land use practices (i.e., deforestation, landfills, agricultural activities, and development)—contribute significantly to an unexpected but nonetheless dangerous outcome: the buildup of heat-trapping gases. Our dependence on fossil-fuel-generated electricity, our use of heating fuels for our homes, and our daily reliance on cars collectively constitute the greatest source of human-caused greenhouse gas emissions. More than three-quarters of human-generated greenhouse gases in the United States result from energy-related activities—for example, the process of generating electricity accounts for 34% of our carbon emissions. The transportation sector accounts for about one-third of U.S. carbon emissions, with nearly 60% of these emissions coming from the gasoline we use in our cars (Environmental Protection Agency 2010e). (If you average 20,000 miles a year on your car, you alone are responsible for contributing 20,000 pounds of carbon dioxide to the atmosphere—and more, of course, if you're driving an SUV and still more if you're still driving a Hummer! But nobody is still driving a Hummer, right?)

Forests can help slow global warming by their sequestering function—their ability to remove carbon dioxide from the atmosphere. But when forests are cleared, carbon is released back into the atmosphere. Tropical deforestation is responsible for more than 15% of the world's heat-trapping emissions each year and accounts for more *global* warming than the total emissions from every vehicle on earth (Union of Concerned Scientists 2009b)! Farms, landfills, and coal mines are the source of another greenhouse gas—methane. Methane accounts for about 10% of the U.S. greenhouse gas emissions. (And note that methane is 20 times more potent than carbon dioxide as a heat blanket; Environmental Protection Agency 2010d.) If we continue with business as usual in our social and cultural practices, the forecast indicates significant climate change will result. Figure 14.3 provides a succinct overview of how human activity impacts greenhouse gas emissions globally.

Figure 14.3 Global Anthropogenic Greenhouse Gas Emissions in 2004

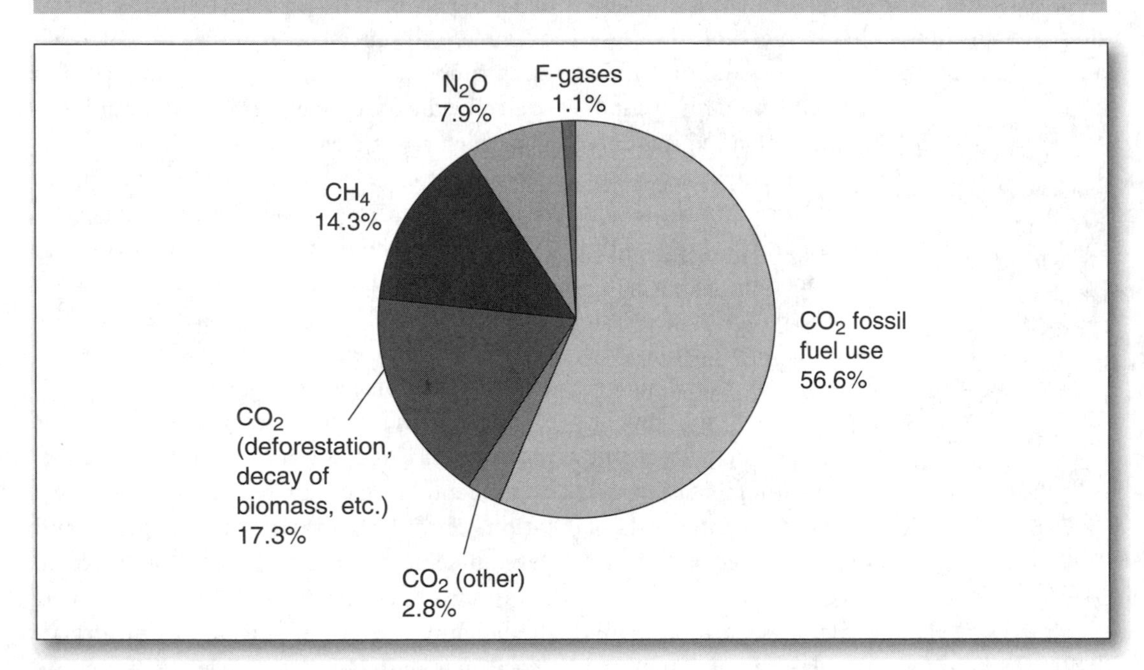

Source: Climate Change 2007: Synthesis Report. Contribution of Working Groups I, II and III to the Fourth Assessment Report of the Intergovernmental Panel on Climate Change, Figure SPM.3.(b). IPCC, Geneva, Switzerland. Available at http://www.ipcc.ch/pdf/assessment-report/ar4/syr/ar4_syr.pdf.

Note: Other CO_2 includes cement production and natural gas flaring.

Key:

CO_2—Carbon Dioxide

CH_4—Methane

N_2O—Nitrous Oxide

F-gases—Flourinated gases

Global warming has caught the attention of much of the world. The Kyoto Protocol is an international agreement to reduce greenhouse gas emissions. The protocol went into force in 2005 and sets binding targets for developed nations to reduce their greenhouse gas emissions by about 5% by 2012. As of fall 2009, 187 nations had ratified the Kyoto Protocol. On average, industrialized countries account for about one-half of global greenhouse gas emissions, so the Kyoto treaty is a step in the direction of stabilizing greenhouse gas (GHG) concentrations. While China currently leads the world in total heat-trapping emissions, the United States has the distinction of being the leader in per-capita GHG emissions and is responsible for nearly 30% of human-induced gases in the atmosphere today (Karl et al. 2009). But the United States also has the distinction of being the last major nation to refuse to ratify the Kyoto Protocol. President George W. Bush opposed the protocol on two counts: the treaty's potential for harming the U.S. economy via revenue and job losses and its exemption of developing nations. To date, the Obama administration has also argued that the agreement is too flawed for U.S. ratification.

The U.S. position on the Kyoto Protocol serves to reinforce the views of global warming skeptics. For instance, the Global Climate Coalition (GCC) was founded in 1989 by 46 corporate and trade groups representing U.S. industry. The GCC launched an aggressive campaign against the Kyoto Protocol, claiming it was a flawed agreement that would hurt U.S. business interests. More recently (November 2009), a controversy erupted over thousands of stolen personal e-mails between a few scientists who wrote dismissively about critics of climate change. Global warming skeptics used the e-mails to attack and discount climate science (they dubbed the affair Climategate). This prompted a group of leading U.S. scientists to send an open letter to Congress. The letter assured members of Congress and the public that a robust body of empirical evidence supports the warnings regarding global warming (see Box 14.1).

BOX 14.1 SECOND THOUGHTS ABOUT THE NEWS: LETTER IN RESPONSE TO "CLIMATEGATE"

DECEMBER 4, 2009

An Open Letter to Congress from U.S. Scientists on Climate Change and Recently Stolen Emails

As U.S. scientists with substantial expertise on climate change and its impacts on natural ecosystems, our built environment and human well-being, we want to assure policy makers and the public of the integrity of the underlying scientific research and the need for urgent action to reduce heat-trapping emissions. In the last few weeks, opponents of taking action on climate change have misrepresented both the content and the significance of stolen emails to obscure public understanding of climate science and the scientific process.

We would like to set the record straight.

The body of evidence that human activity is the dominant cause of global warming is overwhelming. The content of the stolen emails has no impact whatsoever on our overall understanding that human activity is driving dangerous levels of global warming. The scientific process depends on open access to methodology, data, and a rigorous peer-review process. The robust exchange of ideas in the peer-reviewed literature regarding climate science is evidence of the high degree of integrity in this process.

(Continued)

(Continued)

As the recent letter to Congress from 18 leading U.S. scientific organizations, including the American Association for the Advancement of Science, the American Geophysical Union, and the American Meteorological Society, states:

> "Observations throughout the world make it clear that climate change is occurring, and rigorous scientific research demonstrates that the greenhouse gases emitted by human activities are the primary driver. These conclusions are based on multiple independent lines of evidence, and contrary assertions are inconsistent with an objective assessment of the vast body of peer-reviewed science. . . . If we are to avoid the most severe impacts of climate change, emissions of greenhouse gases must be dramatically reduced."

These "multiple independent lines of evidence" are drawn from numerous public and private research centers all across the United States and beyond, including several independent analyses of surface temperature data. Even without including analyses from the UK research center from which the emails were stolen, the body of evidence underlying our understanding of human-caused global warming remains robust.

We urge you to take account of this as you make decisions on climate policy.

Source: An Open Letter to Congress From US Scientists on Climate Change and Recently Stolen Emails, http://www.ucsusa.org/assets/documents/global_warming/scientists-statement-on.pdf.

While the federal government has been slow to warm up to global warming, other sectors have been more responsive. Local and state governments and even big business have acted where the federal government has not. California has taken a series of steps to reduce greenhouse gas emissions: In the past few years, it has passed laws to reduce greenhouse gas emissions from cars and from ill-considered land development and urban sprawl. California is also ahead of schedule in its goal of having 20% of the state's energy come from renewable sources. The Regional Greenhouse Gas Initiative (RGGI), a coalition of 10 northeastern and mid-Atlantic states, is working to reduce carbon dioxide emission by power plants. The coalition auctions off carbon allowances to power plants and invests the proceeds in green technologies (Hopkins 2010). Some corporations are also responding. In 2003, DuPont, Ford, and other major firms started the Chicago Climate Exchange (CCX). Members of the exchange have committed to reducing their greenhouse gas emissions by 4% and to trading credits with other exchange members that are not meeting these output goals (Collier 2005). The CCX is the only operational cap-and-trade system in North America.

While culture has a hand in exacerbating the global warming problem, it may also play a role in slowing or even solving the crisis. The entrepreneurial spirit is one of the defining U.S. values. The ability to profit from global warming may be the very thing that reverses the warming pattern. As seen with the previously mentioned CCX, the trading of greenhouse gas emissions is presenting entrepreneurs with new business opportunities. A **cap-and-trade system** uses the power of the market to create financial incentives for reducing emissions. (The system was

Cap-and-trade system
A strategy for combating global warming that uses the power of the market to create financial incentives for reducing emissions

used successfully in the 1990s to reduce sulfur dioxide emissions and combat the acid rain problem.) Dairy farmers, for instance, can earn credits for reducing their methane gas emissions below cap levels by converting cow manure into fertilizer. The credits they earn by reducing methane emissions can then be sold via the Chicago Climate Exchange. And who would want to buy these credits? Other exchange participants who have volunteered to reduce their emissions but have exceeded their caps. Buying credits from low emitters can set the balance right for high emitters and make for happy participants all around. One may ask: Why, absent the United States ratifying the Kyoto agreement, would *any* U.S. businesses volunteer to lower their greenhouse gas emissions? Again, there appears to be an economic answer. Businesses are betting that consumers will agree that "green is better." As a result, they will pay more for greener industries and products (Solman 2006a).

The belief that "green is better" is one force driving research on **biofuels** as an alternative to oil. Entrepreneurs are busy turning used cooking grease into biodiesel, and corn, soy, sugar cane, wood chips, and switchgrass into ethanol. Boosters see two major advantages to developing biofuels. Biofuels are cleaner fuel sources that will reduce greenhouse gas emissions. Biofuels are also *renewable* energy sources and might very well put us on a path to energy self-sufficiency (Solman 2006b).

Biofuels
Renewable sources of energy derived from living organisms or from metabolic byproducts; they are considered carbon dioxide neutral

This last point has taken on some urgency given the recent oil disaster in the Gulf of Mexico. On April 20, 2010, a British Petroleum (BP) leased oil drilling platform exploded, killing 11 workers and starting the worst U.S. oil leak in decades and the worst environmental disaster this country has ever seen. Before a permanent seal was achieved in September 2010, it is estimated that more than 200 million gallons of oil leaked into the Gulf waters—enough to fill 40,000 tankers and 13 million car tanks. Precious wetlands, coral reefs, and the livelihoods of citizens across the Gulf region were negatively impacted. A USAToday/Gallup poll found that while 75% of Americans felt that BP deserved a great deal of the blame for the oil spill, 44% put a great deal of blame on the federal agencies that have failed to effectively regulate oil drilling (Jones 2010c). While President Obama responded to the spill by declaring a 6-month moratorium on deep-water drilling, many are hoping that the disaster will prompt major energy reform. Indeed, the spill elicited the first Oval Office address by President Obama, an address where the President argued that it is time for the United States to end its addiction to fossil fuel (see Figure 14.4).

If there is one additional harbinger of things to come on the global warming front, it may be gleaned from the insurance industry. Greener products or new gas-reducing technologies aren't the only source of big profits. The insurance industry also sees profits in global warming itself. Insurance companies regard global warming as a disaster waiting to happen and thereby as a way to sell more insurance. Smart consumers would do well to insure against global warming so that they can be in a position to recover from any economic losses (Solman 2006a). To be sure, the insurance we opt for might be in the form of a traditional policy (this would make the insurance industry happy). But the "insurance" we opt for might also be in the steps we take to reduce our **carbon "footprint."** (The term *footprint* is used by environmentalists to convey our personal contribution to carbon dioxide emissions.) Doing so would allow us to finally take up Twain's challenge—namely, doing something about the weather—or, more precisely, about the climate.

Carbon footprint
Term used by environmentalists to convey an individual's personal contribution to carbon dioxide emissions

Figure 14.4 Primary U.S. Energy Consumption by Major Source, 1949–2007

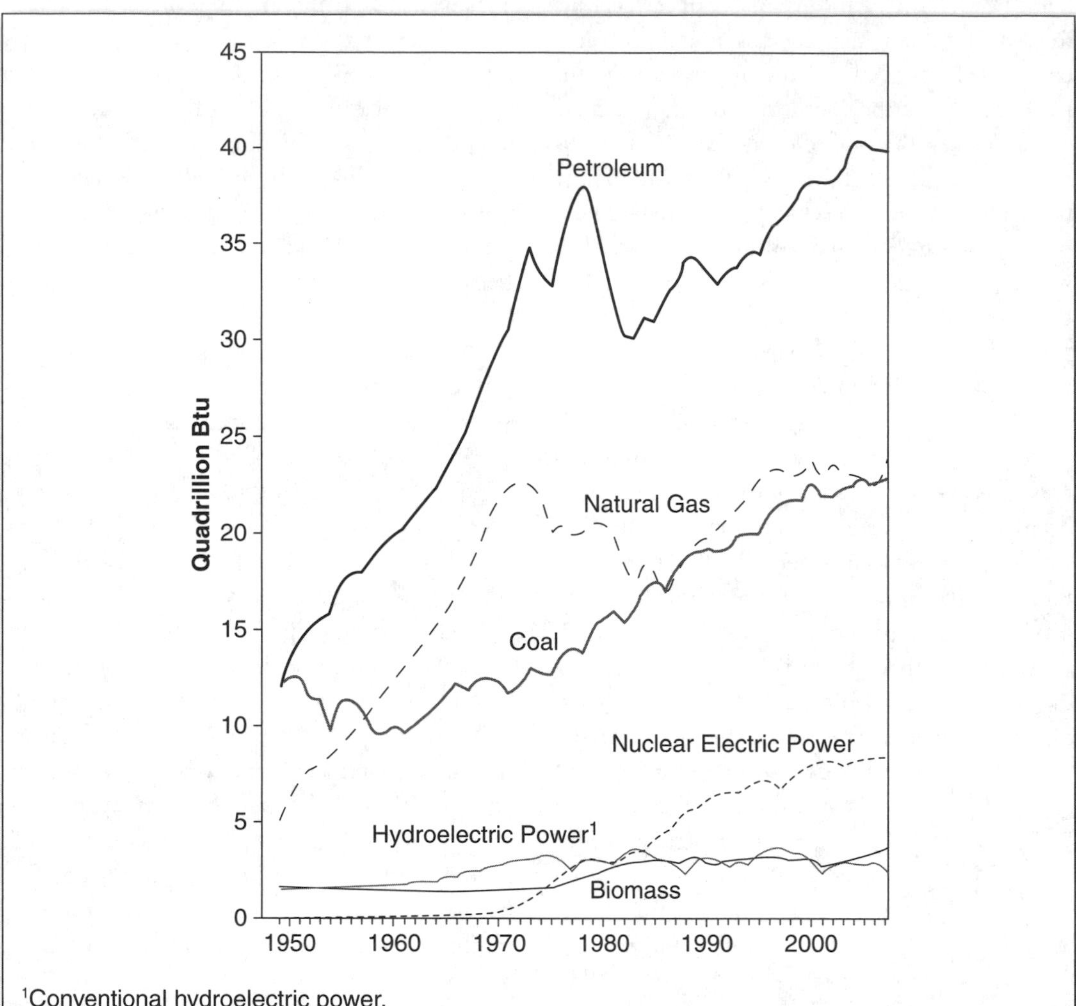

[1]Conventional hydroelectric power.

U.S. energy supply is dominated by fossil fuels. Petroleum, the top source of energy shown above, is primarily used for transportation (70 percent of oil use). Natural gas is used in roughly equal parts to generate electricity, power industrial processes, and heat water and buildings. Coal is primarily used to generate electricity (91 percent of coal use). Nuclear power is used entirely for electricity generation.

Source: U.S. Global Change Research Program, Global Climate Change Impacts in the United States. Available at http://downloads.globalchange.gov/usimpacts/pdfs/climate-impacts-report.pdf.

LEARN MORE ABOUT IT

For an in-depth discussion of the social side of environmental issues, see Michael Mayerfeld Bell's book *An Invitation to Environmental Sociology* (Thousand Oaks, CA: Sage, 2007).

Those interested in learning more about cap-and-trade systems for combating pollution will find the following 2005 article by Mathers and Manion helpful: "How It Works: Cap-and-Trade Systems." *Catalyst* 4 (1): 20–1, http:www.ucsusa.org/assets/documents/catalyst/Catalyst-Spring-2005.pdf.

GOING ONLINE FOR SOME SECOND THOUGHTS

A succinct review of the impact of global warming for the United States can be found in *Global Climate Change Impacts in the United States*: http://www.globalchange.gov/images/cir/pdf/20page-highlights-brochure.pdf.

The issue of global warming has finally made its way into the political arena, thanks in part to Al Gore's documentary *An Inconvenient Truth*. More information about the film as well as educational curriculum materials can be found at http://www.climatecrisis.net.

You can calculate your own greenhouse gas emissions (or the size of your carbon footprint) by visiting either of the following sites:

EPA: http://www.epa.gov/climatechange/emissions/ind_calculator2.html

Carbon Footprint, Ltd.: http://www.carbonfootprint.com/individuals.html

No Impact Man: The Documentary tells the story of a New York family that tried for a year to reduce to zero its impact on the environment. See a trailer for the documentary at http://www.imdb.com/title/tt1280011/.

The following Web page offers a list of the top 10 sources for environmental news: http://environment.about.com/od/activismvolunteering/tp/environews.htm.

NOW IT'S YOUR TURN

1. Talk to 20 people—10 who drive SUVs and 10 who drive small economy cars. How many in each group recognize global warming as a serious problem? Do members of each group hold many social similarities—age, gender, socioeconomic status (SES), or political affiliation? Is there a predictable social profile linked to global warming believers and nonbelievers?

2. Visit some "going green" websites (e.g., "Living with Ed—Planet Green" at http://planetgreen.discovery.com/tv/living-with-ed/) to get a good idea of green products and practices. Now try another brief survey. This time, ask 20 to 30 people about their willingness to go green. What's the social profile of those who are willing to "pay" for a green solution to global warming?

NOTE

1. Early in 2007, Americans were introduced to the "no impact man," a New Yorker vowing to live for one year in New York City without the taken-for-granted practices and products that have a negative impact on the environment. See http://noimpactman.com for more information.

ESSAY 15

Conventional Wisdom Tells Us . . . There Ought to Be a Law

There's no social ill that the law can't fix . . . or at least that is what many Americans believe. In this essay, we review various social functions of the law. We also consider whether we are overly dependent on this tool of formal social control.

"There ought to be a law" . . . it's a phrase that expresses Americans' penchant for the law. In the United States, no social realm is exempt from the rule of law. Indeed, law is a pervasive feature of American society. Family, civic duties, education, business and commerce, religion, government, and even birth and death are regulated by law. Leave this world without your legal pass (i.e., a will or a trust) and your "life" (as it were) will become the domain of probate courts and lawyers.

Americans pride themselves on being a land of laws. The U.S. Constitution, our most cherished legal document, is invoked by liberal and conservative, law-abiding and law-violating individuals alike. Americans trust in the law and symbolically display it as "blind" to status differences. The law, it is argued, renders all parties equal before the scales of justice. No one in the United States, not even the president, is considered above the law. A few years back, a federal court rejected the argument that President George W. Bush had been granted "inherent powers" to violate the laws of Congress and ruled that the Bush administration's warrantless spying on ordinary citizens was unconstitutional (American Civil Liberties Union 2006). (The Sixth Circuit Court of Appeals dismissed the case in July 2007 but refused to rule on the legality of the program.) Americans expect great things of the law and are quick to express their displeasure and disappointment when the law behaves badly. A decision by a Michigan judge to settle a holiday visitation conflict between parents by flipping a coin received national coverage in February 2002. One party to the conflict felt the judge's behavior made a mockery of the judicial process; the judge's superiors agreed that the incident hurt the judicial process (Ashenfelter 2002).

Many sociologists contend that social control represents the dominant function of the law.

Social control
The process of enforcing social norms

Norms
Social rules or guidelines that direct behavior, they are the "shoulds" and "should nots" of social action, feelings, and thought

Informal social control
Social control initiated by any party; no special status or training is required

Formal social control
Social control that is state centered

Social engineering
Using the law to satisfy social wants and to bring about desired social change

Social control refers to the process of enforcing social norms. **Norms** are social rules or guidelines that direct behavior. Social control can be informal or formal. **Informal social control** can be initiated by any party; no special status or training is required. It includes such mechanisms as socialization, gossip, ridicule, shaming, praise, or rewards. **Formal social control** is state centered, that is, enforced by the power of the state. The law is a tool of formal social control; It is legitimated and enforced by the state.

In recognizing the law's social control function, we acknowledge it as a vehicle for achieving and maintaining social order and stability. The law steps in when informal social control tactics fail to get the job done. For the most part, people readily appreciate the social control function of the law. Most would agree that certain behaviors—murder, robbery, assault, burglary, arson—are clearly social wrongs deserving the law's attention. Few would question the need to regulate such norm violations. Indeed, most would rail at the dangers posed by the absence of law in response to these behaviors.

The law, however, serves other social functions—functions about which one finds less public consensus. In addition to operating as a tool of social control, the law can function as a tool of **social engineering**. In this capacity, the law is constructed to satisfy social wants and to bring about desired social change.

Law for the sake of social engineering poses dilemmas for societies. Whose wants and desires should the law promote? Whose ideas of right and wrong should the law encode? Can we rightfully enact law that pushes forward a specific moral agenda?

Despite the dilemmas posed by socially engineered law, there is no shortage of statutes and regulations that fit into this category. Indeed, the United States boasts a substantial historical record in which certain groups have legislated right and wrong—even amid the outcries of others. The Roaring Twenties provide perhaps the most notorious example of this process. In 1919, the National Prohibition Act (a.k.a. the Volstead Act) made the sale, manufacture, and transportation of liquor, beer, and wine illegal in the United States. In part, the Volstead Act resulted from efforts of middle-class groups to counteract the vices of newly arrived immigrants (Gusfield 1963). As is the case for many instances of social engineering law, however, not everyone agreed with the act's definition of right and wrong. As a result, the passage of the Volstead Act spurred several unintended consequences. For example, outlawed saloons were quickly replaced by newly devised speakeasies. Hardware stores began to stock and sell portable stills. Libraries supplied instructional books and pamphlets needed for distilling homemade "hooch." California vintners (who actually expanded their acreage during Prohibition) introduced "Vine-glo," a grape juice that with proper care and time (60 days) could be turned into wine. Brewers followed suit, manufacturing and selling "wort," a half-brewed liquid that could be turned into beer with the simple addition of yeast. And since the Volstead Act allowed the sale of alcohol for medicinal purposes, doctors began writing alcohol prescriptions, and pharmacists began filling them in unprecedented numbers (Time-Life Books 1988). In short, Americans responded to the social-engineering nature of the Volstead Act by designing creative ways in which to avoid, evade, and ignore the law.

Americans' rebellious response to the Volstead Act was not an isolated incident. Many argue that certain drug laws have encouraged similar activities (Campos 1998). Here again, the actions of moral entrepreneurs have had unintended consequences. **Moral entrepreneurs** are individuals who seek to legally regulate behaviors that they consider morally reprehensible. Like Prohibition, efforts to criminalize certain drugs have created an extremely lucrative black market in illicit drugs. Drugs that have a negligible pharmaceutical value now command hundreds of dollars on the streets. Furthermore, the financial incentive associated with drug trafficking has provided an irresistible career option for certain individuals. Some would also argue that the legal prohibition of particular drugs ensures that increasingly potent and harmful forms of illicit drugs will continue to enter the illegal drug market. Former Baltimore Mayor Kurt Schmoke maintained that this scenario accurately described the emergence of crack cocaine in his city (Mills 1992). (See Box 15.1 for a current drug/law controversy.)

Moral entrepreneurs Individuals who seek to legally regulate behaviors that they consider morally reprehensible

BOX 15.1 SECOND THOUGHTS ABOUT THE NEWS: ARE THE STATES GOING TO POT?

As the nation struggles to get out of the grip of the great recession, many strategists are exploring new ways to cope with the financial crisis. California has come up with something that many feel could be a winning proposition: legalizing marijuana. In November of 2009, Assemblyman Tom Ammiano introduced a bill into the state legislature that would see California become the first state in the United States to legalize marijuana for recreational use by those 21 years of age and older.

Now before you think someone is just high or blowing smoke, know that this is a serious proposition that appears to have some political legs. Proponents say that efforts to outlaw marijuana use have been misguided and extremely costly for society. Pot remains a popular drug that many teenagers find easier to obtain than alcohol. Furthermore, California has already taken a step toward legalization with its medical program that permits the use of marijuana for the treatment of various illnesses. (In August of 2010, thousands showed up at a medical marijuana show in San Jose, California, and many queued up for their medical marijuana ID cards.)

But perhaps most importantly in these dire financial times, proponents argue that legalizing pot will mean huge revenues for the cash-strapped state. In addition to the monies and jobs that can be generated via the creation of a new industry, California would also collect revenues (to the tune of more than a billion dollars a year) from a user tax like we see on cigarettes and alcohol. All of this must smell sweet to a state that is suffering from declining revenues.

After the first 2009 bill overran its time limit for consideration, a second bill was introduced in February 2010. Proponents of the measures are hoping that by following a slow but steady course, they will be able to convince skeptics of the merits of the measure. So far, the legalization efforts have been endorsed by such groups as the AFL-CIO, the American Civil Liberties Union, the American Federation of State, County and Municipal Employees (AFSCME), the California Public Defenders Association, and the National Black Police Association. And while California voters recently rejected the issue (Proposition 19) in the fall of 2010, it is likely to be back on the voting block again. Keep your eye on this one. Where there's smoke . . . there may be some legal change in the air.

The examples just cited illustrate the dangers posed when we use law as a tool of social engineering. Equating the law with general public sentiment is a tricky business. In modern, complex societies where there are multiple "publics," there is no simple relationship between public opinion and the law. Thus, the ability to translate social wants and desires into law involves social power more than social consensus. Why have antismoking campaigns and legislation been so successful? Some would argue that it's all about power and status politics—smoking is inversely related to social class and status (Tuggle and Holmes 1997). And why have antigun efforts failed? Again, we would have to cite power and status politics. The NRA (National Rifle Association) is a powerful organization that has been tremendously effective in its lobbying efforts to safeguard the rights of gun owners; it spent just over $2 million in lobbying in 2009 (Center for Responsive Politics 2010). No doubt our nation's response to the latest social issue to capture the public's attention, global warming, will reflect the input of various lobbying efforts. In 2009, there were 13,746 active registered lobbyists trying to promote/secure their clients' interests. (The total number of lobbyists remains unknown because many do not register.) And if money talks, lobbyists will surely get someone's ear. In 2007, lobbyists spent an average of $17 million *each day* that Congress was in session. Amounts spent in 2008 and 2009 were still higher, and at that time of this writing, 2010 was on pace to establish a new high—showing lobbying to be immune to the current recession (Center for Responsive Politics 2010; Insurance Journal 2008). Table 15.1 lists the top 10 clients (spenders) for 2010.

When special interests successfully mobilize resources on behalf of their own causes, the law can come to contradict the general public will. In such cases, Americans cling to a belief in equality before the law, but they also understand that lawmakers consider some people more equal than others. Consider these words about U.S.

Table 15.1 Top Ten Lobbying Spenders 2010

Client	Total Expenditures
1. U.S. Chamber of Commerce	$132,067,500
2. PG&E Corp.	$45,460.000
3. General Electric	$39,290,000
4. FedEx Corporation	$25,582,074
5. American Medical Association	$22,555,000
6. AARP	$22,050,000
7. Pharmaceutical Research & Mfrs of America	$21,740,000
8. Blue Cross/Blue Shield	$21,007,141
9. ConocoPhillips	$19,626,383
10. American Hospital Association	$19,438,358

Source: OpenSecrets.org. 2010. http://www.opensecrets.org/lobby/top.php?showYear=2010&indexType=s.

lawmakers penned by noted legal scholar Lawrence Friedman (1975):

> They know that 100 wealthy, powerful constituents passionately opposed to socialized medicine outweigh thousands of poor, weak constituents, mildly in favor of it. Most people . . . remain quiet and obscure . . . This is the "silent majority." (p. 164)

While these words were written more than a quarter of a century ago, they prove timely. The recent struggles to achieve health care reform by the U.S. Congress offer compelling evidence in this regard. A single, silent majority still encounters great difficulty when attempting to make its voice heard over that of a smaller but more powerful special-interest group. Consider that from January 2005 to June 2006, drug companies spent $155 million on lobbying Congress and working to block any new legislation that would permit the federal government to negotiate lower prices for Medicare prescription drugs. Despite the fact that senior citizens, health activists, and some government officials support open pharmaceutical markets, bills permitting such changes to the Medicare program have been consistently defeated (Ismail 2007). And in the health care reform debate that dominated much of 2009, insurance companies are "credited" with changing the terms of the debate in order to benefit their industry—for example, consider the disappearance of a "public option" (Terhune and Epstein 2009). Indeed, given the final terms of the new reform bill, the insurance industry stands to gain up to 20 million new customers (Keteyian 2010).

The charge that the law hears the voices of some parties more than others extends to the highest court of the land: the U.S. Supreme Court. The Supreme Court has tremendous discretion in setting its own docket. In the opening months of its 2006 term, the Court took 40% fewer cases than it did in the previous period, a continuation of a trend started in the Rehnquist years. Within 5 years of Rehnquist becoming chief justice, the number of Supreme Court decisions dropped from 145 to 107. The number of cases with signed opinions for the 2005 term was 69, the lowest number since 1953. These declines, which are continuing in the Roberts court, are generating great debate among some legal scholars (Greenhouse 2006; Lazarus 2010; Levinson 2010). Indeed, critics argue that the decrease in cases judged worthy of plenary review is a troubling sign that the Court's docket has been captured by special interests who have the economic power to retain "expert Supreme Court advocates." (**Expert advocate** refers to an attorney who has made at least five oral arguments or a firm that has made at least 10 oral arguments before the Court.) In 1980, expert advocates were responsible for less than 6% of the petitions granted plenary review by the Court. By 2008, expert advocates were responsible for 55.5% of the cases granted plenary review in 2008 (Lazarus 2010).

Expert advocate
An attorney who has made at least five oral arguments or a firm that has made at least 10 oral arguments before the Supreme Court.

Your Thoughts . . .

In 2010, the Supreme Court eliminated a 20-year ban on corporate and labor union spending on presidential campaigns and opened the door for these bodies to make their own campaign ads to influence elections. Critics fear that corporate and labor union interests will now invariably trump the interests of the electorate. What do you think?

Is it socially harmful when law becomes disconnected from general public sentiment? Some maintain that laws contradicting public sentiment breed contempt and, more important, disregard. Consider the following examples. Unrealistic speed limits have invited many drivers to disregard traffic laws. Sunday "blue laws" or laws prohibiting the sales of cigarettes and alcohol to minors have placed many merchants on the wrong side of the law. Many middle- and upper-class families procure child care providers, housekeepers, and landscapers as a means of making their busy lives easier, but some individuals do so at the cost of violating tax and immigration laws.

Social control or social engineering: Regardless of its social function, law is a dominant force in American society. One legal scholar observes that our devotion to reason leads us to legally regulate more and more of our social interactions. We use the law to satisfy our metaphysical anxieties (Campos 1998). As these formal rules come to dominate our lives, the potential for legal work grows and grows. Even though the legal sector is expected to add the fewest new jobs in the coming decade, a 13% growth rate for legal occupations ensures this group will remain competitive with the average growth rate across all occupations. And within the legal sector, jobs for paralegals and legal assistants are expected to grow by 28% over the next decade (U.S. Department of Labor 2010c). Our nation averages one lawyer for every 525 people (Vallas 2008). And despite the high population density of lawyers, the law is still a popular career option for college graduates. Since the mid-1970s, approximately 40,000 of those graduates have entered law school each year. In the early 1960s, just about 10,000 law degrees were awarded each year, but by 2005, that figure climbed to more than 40,000 annual law degrees (American Bar Association 2009).

Against this backdrop of a legally saturated society, the United States has also acquired a reputation as a rather litigious society. Many argue that lawsuits have become Americans' favorite pastime. Have you ever lost an item at the dry cleaners? Sue! (In 2007, a D.C. administrative law judge brought a $65 million lawsuit against a local dry cleaning business because the cleaners lost a pair of the judge's pants.) Has your child been injured playing Little League? Sue! (In 2007, a Long Island mother brought suit against a Little League coach for not teaching her son how to slide properly into a base.) Have you ever been injured while you were someplace you did not belong? Sue! (In 2006, a jury awarded millions of dollars to two teens who were electrocuted while trespassing on railroad property.) Have you ever been frightened by a knock on a door? Sue! (In 2004, a Colorado woman had an anxiety attack when scared by neighborhood teens when they knocked on her door while delivering cookies. The woman sued the families of the teens for medical expenses—no word on whether or not she liked the cookies.) Have you ever sustained a burn from an excessively hot cup of coffee? Sue! (In 1992, a New Mexico woman sued McDonald's for compensatory and punitive damages after sustaining third-degree burns when a cup of McDonald's coffee spilled into her lap.) Are you an overweight fan of fast foods? Sue! (In 2002, several Americans filed lawsuits against McDonald's, Burger King, etc., arguing that the chains had not provided adequate warning about the dangers of overeating fast foods.) Indeed, each year, Americans address the wrongs (or torts) they experience in life by filing civil lawsuits. **Civil lawsuits** are generally private legal proceedings for the enforcement of a right or the redressing of a wrong.

Civil lawsuits
Private legal proceedings for the enforcement of a right or the redressing of a wrong

High-profile cases such as those just mentioned receive much popular press and prompt many to conclude that Americans are an excessively litigious group. Indeed, many conservatives in the recent health care reform debate argued that real health care reform could not happen without tort reform. They argue that tort reform is key to stopping the costly practice of "defensive medicine" by doctors fearing malpractice suits. This position does not go unchallenged—for example, it is noted that litigation accounts for only a miniscule fraction (.46 of 1%) of health care costs (Public Citizen 2010). Nonetheless, the malpractice lawsuit reform effort should prove to be a particularly interesting exchange because it pits two high-power interest groups against each other: doctors and lawyers (Stolberg 2003). Indeed, it is worth noting that the American Medical Association had a large hand in creating the American Tort Reform Association, a group dedicated to reforming the civil justice system (American Tort Reform Association 2007).

To get a better understanding about the relationship between tort action and litigiousness, we should consider the latest review of state court workloads. In 2007, more than 18 million civil cases were processed in our state courts, a nearly 5% increase over 2006. These civil caseloads encompass a wide array of cases: tort, contract, real property, mental health, probate, and civil appeals cases. Still, civil cases constitute 18% of the workload for state courts (see Table 15.2). Further, about 70% of civil cases entailed small claims and contract disputes. Tort cases represent only a small percentage of civil action suits; the number of tort cases is down by 24% since 1998 and fell 9% between 2006 and 2007 (LaFountain et al. 2009). As revealed in Table 15.3, tort cases are not the real culprit in the civil caseloads for state courts.

The charge of excessive litigation is also made with regard to **class action suits.** Class action suits are litigation in which one or more persons brings a civil lawsuit on behalf of other similarly situated individuals. To be sure, some people regard malpractice suits and class action suits as threats to health care and to big business and as negative manifestations of greedy plaintiffs and still greedier lawyers. Others, however, are less inclined to cast this litigation in such a negative light. Ever-increasing malpractice insurance premiums may be attributed to the pecuniary interests of insurance companies who want to recover stiff Wall Street losses of the past few years. A recent report by Public Citizen, a nonprofit group representing consumer interests, noted that 2009 saw a record low number of malpractice payments in the United States. Further, the value of payments has decreased by 8% between 2000 and 2009. Indeed, 2009 saw the lowest value of payments (adjusting for inflation) since 1992. Yet insurance premiums

Class action suits Litigation in which one or more persons brings a civil lawsuit on behalf of other similarly situated individuals

Table 15.2 Total Incoming Cases (in millions) for State Courts, 2007

Case Type	# of Cases	% of Total
Traffic	56.3	54.2%
Criminal	21.4	20.7%
Civil	18.1	17.5%
Domestic relations	5.7	5.5%
Juvenile	2.2	2.1%
All Cases	**103.7**	**100.0%**

Source: LaFountain et al. 2009. *Examining the Work of State Courts: A National Perspective from the Court Statistics Project* (National Center for State Courts 2009). http://www.ncsconline.org/d_research/csp/2007B_files/EWSC-2007-v21-online.pdf.

Table 15.3 Median Number of Incoming Cases (per 100,000 population) Across Sample of State Courts, 2007

Real property cases	35
Mental health cases	140
Tort cases	153
Probate cases	383
Small claims	1,298
Contract	1,349

Source: LaFountain et al. 2009. *Examining the Work of State Courts: A National Perspective from the Court Statistics Project* (National Center for State Courts 2009). http://www.ncsconline.org/d_research/csp/2007B_files/EWSC-2007-v21-online.pdf.

for doctors have continued to increase (Public Citizen 2010). Others argue that lawsuits are essential tools for achieving social justice and reform. What can't be achieved via legislation or regulation might nonetheless be accomplished via litigation (Hensler 2001). John Banzhaf, a key legal player in the recent efforts to sue the junk-food industry, believes that such action may help America come to grips with its increasing obesity problems and thereby serve the public interest (Gumbel 2002). Defenders of class action suits see them as offering solutions to the failure of government oversight of industry (RAND 2007). The idea of frivolous lawsuits is also challenged by a noteworthy statistic: Of the malpractice payments issued in 2009, nearly two-thirds were for significant or major permanent injuries (26%), quadriplegia, brain damage or lifetime care (5%), or death (33%; Public Citizen 2010).

While Americans may seem enamored with legal remedies, Americans actually have a long tradition of opting out of the legal arena and resolving conflicts via extralegal avenues. The Boston Tea Party is one of our nation's most famous extralegal tactics for fighting taxation without representation. Rather than facing the delays of legal divorce proceedings, many unhappy spouses at the turn of the century simply took their marital problems "on the road" and obtained "migratory divorces" by moving to another state (without their spouses) (Riley 1991/1997). Disgruntled workers often utilize "self-help" behaviors (that often take the form of the reallocation of office supplies or equipment) to resolve disputes with management. Today, hundreds of thousands of homeowners facing steep declines in their home values are resolving the crisis by "mailing back keys" and walking away from their mortgages (Streitfeld 2010). It is estimated that only 2% of victims of medical negligence file a claim (George 2006), and tort filings in general have been steadily declining over the last two decades (LaFountain et al. 2009). Even when legal proceedings are initiated, it is unlikely that matters will progress to trial: Either parties settle cases or judges decide them on legal motions. In 2003, only 2% of federal tort cases were decided by a trial (Bureau of Justice Statistics 2010c). Furthermore, in recent years, more and more alternative dispute resolution (ADR) processes such as mediation and arbitration are being considered as reasonable alternatives to adjudication. Consequently, there are those who argue that charges of a litigation explosion in the United States are unfounded, overstated, or public relations gimmicks (Abel 2002; Center for Justice and Democracy 2007; Cochran 2005). Indeed, a widely utilized report on tort costs is produced by a paid (and highly criticized) consultant to the *insurance* industry: Tillinghast-Towers Perin.

What explains the general legal restraint we see? One contributing factor is cost. The practice of law is expensive. The American Bar Association reports that lawyers' hourly rates can range from $60 to more than $300 an hour. Many Americans with legal problems cannot afford to resolve them via the law. Recent studies of several state courts found that low-income households need help with an average of three legal issues each year, but fewer than one in five of these problems is addressed by an attorney. On average, there is one legal aid lawyer per 6,415 people in poverty; in contrast, there is one private attorney for every 429 people above the poverty line (Legal Services Corporation 2010). Consequently, about one million potential clients get turned away from legal aid offices each year (Vallas 2008). Similarly, many middle-class individuals, ineligible for government-funded legal aid, must forgo legal redress because of prohibitive costs (George 2006). (Indeed, the high cost of legal action is one reason that some persons resort to a class action suit—it can make the cost of litigation more affordable for persons suffering moderate financial losses.) This is true despite the fact that lawyers frequently accept cases on a contingency fee basis. In contingency fee cases, attorneys receive compensation (typically 33–50% of the client's postcost financial recovery) only if they prevail in the case. While this arrangement can motivate attorneys to seek high damages, it also ensures that attorneys will refrain from taking meritless or weak cases. Pursuing no-win cases is a losing proposition. Thus, despite the rather common belief that contingency fee arrangements encourage litigation, the evidence does not support this view. A recent study by the National Center for State Courts indicates that 62% of tort cases are motor vehicle trials with median awards of less than $18,000. Punitive damages were awarded in less than 5% of tort cases (National Center for State Courts 2005). Even class action suits that many identify as contributing to our tort crisis rarely produce extraordinary monetary settlements. A recent RAND study of class action suits against insurance companies found that only 12% of cases resulted in a class settlement and the median benefit available to class members was $97. Attorneys for plaintiffs were awarded a median 30% of the common fund (RAND 2007).

The high cost of the law ensures that most of those facing social wrongs will react by adopting some informal, extralegal response. Some will avoid contact or refuse to cooperate with those who do them wrong. Others may resort to gossip or direct scorn against wrongdoers. Certain individuals elect to settle scores by seeking revenge. Indeed, many incidents of personal violence, as well as property crimes, can be attributed to such revenge seeking or "self-help" behaviors (Black 1998). Most of us, however, are likely to respond to wrongs against us by "lumping" them; we will, in effect, turn the other cheek and employ tolerance. **Tolerance** refers to inaction in the face of some offense. Indeed, tolerance is "the most common response of aggrieved people everywhere" (Black 1998).

Tolerance
Inaction in the face of some offense; an informal social control mechanism

The juxtaposition of tolerance vis-à-vis the law is a compelling reminder of the importance of sociological vision in this area of study. The law is a social phenomenon, and as such it varies greatly within its social environment. To understand its availability, distribution, usage, nonusage, and even its suspension requires us to analyze the law in light of **social context**—the broad social and historical circumstances surrounding an act or event. The post–9/11 efforts of the Bush administration provide a powerful illustration of this point. Immediately after the September 11, 2001, terrorist attacks on America, President George W. Bush signed the USA PATRIOT Act in October 2001. This act broadened the information-gathering, surveillance, and detention powers of the Justice Department; it also limited judicial oversight of these activities. Then-Attorney General John Ashcroft defended the act as an essential tool for fighting terrorism. But in the years since 9/11, critics came to view

Social context
The broad social and historical circumstances surrounding an act or an event

the act as a serious threat to fundamental constitutional rights.[1]

Another, more recent example of the importance of seeing the law in its social context is offered by Arizona's 2010 immigration law. This statute makes it a crime for illegal immigrants to be in the state. The law also requires police to verify the legal status of individuals they have stopped while enforcing other laws if those individuals "look like" they might be undocumented. Arizona officials defend the law, saying that it compensates for the federal government failing to fulfill its obligations to curtail illegal immigration, an issue of great concern for Arizona. The Obama administration sees Arizona's law as unconstitutional, arguing that the state has usurped federal powers to regulate immigration and has sought an injunction to stop the law from going into effect. In the meantime, the public remains fairly evenly divided on the issue.

The Gulf of Mexico oil spill also reveals the importance of turning a sociological eye to legal matters. The Gulf oil spill is the largest environmental disaster the United States has ever experienced. As such, President Obama issued a ban on off-shore drilling in the Gulf of Mexico in the weeks following the April 2010 disaster. Given the impact of the oil spill disaster, one might assume universal support for the drilling moratorium. In fact, however, it was met with substantial resistance. The governor of Louisiana spoke out against the ban, arguing the moratorium would hurt the state economy; oil service companies sued to have the ban lifted. In late June, a federal district court complied with that request, and in early July, a federal appeals court ruling opened the door for a return to drilling. Yet the legal maneuvering continues, revealing the dynamic nature of law. The Obama administration has promised a more narrowly crafted revised ban on deep-water drilling. The governor of Florida, in a move to protect *that* state's economy, has called for a state constitutional amendment that would prohibit all off-shore drilling in Florida.

Despite its dynamic nature and cultural importance, formal laws do not represent the whole story of social control in our society. Consider, for instance, that many individuals are using money to express their approval/disapproval of the new Arizona immigration law. Those who oppose the bill have cancelled trips to that state while others who support the new law are donating money to the state—both are powerful forms of social control in their own right. Consider, too, that while many Americans are now taking a dimmer view of off-shore drilling than before the Gulf of Mexico oil disaster, many residents of the Gulf states, even those who make their living off of the water, are reluctant to support a drilling moratorium: Nearly 80% of likely voters in Louisiana and 70% of voters in Texas still favor off-shore drilling (Rasmussen Reports 2010a). These opinions may well be controlling factors in future elections. Indeed, in advancing our understanding of law in society, we would do better to view the law as just *one* form of control among many. Pursuing such second thoughts about the law will help us see exactly how our very demand for the law and social control can contribute to much of the extralegal activity in our society.

---

[1]Perhaps in an ironic twist, some of the most vocal opposition has come from the not-so-hushed voices of librarians. In 2005, the American Library Association challenged an FBI demand—one linked to arguments of national security. The FBI sought Connecticut library use records as well as the issuance of a Justice Department gag order to preclude the library system from discussing the case. (The gag order was needed because the FBI request came via a National Security Letter [NSL]; parties receiving such letters are forbidden by law to reveal requests made within.) A federal court subsequently ruled in favor of the librarians ("American Patriots" 2006).

LEARNING MORE ABOUT IT

In *Supreme Conflict: The Inside Story of the Struggle for Control of the United States Supreme Court* (New York: Penguin, 2007), Jan Crawford Smith offers a detailed and dynamic account of conservatives' bid for control of the Supreme Court and the restriction of the Court's future role in government.

A thorough review of the theory and techniques of conflict management can be found in Donald Black's *The Social Structure of Right and Wrong* (San Diego: Academic Press, 1998).

For an analysis of how the law can be used as a tool of moral entrepreneurs, see Joseph Gusfield's classic work on prohibition, *Symbolic Crusade: Status Politics and the American Temperance Movement* (Urbana: University of Illinois Press, 1963).

Rebecca Sandefur reviews the connection between race, social class, and gender inequality with access to civil law and justice. She also considers the mechanisms that maintain and increase inequality in her article "Access to Civil Justice and Race, Class, and Gender Inequality" (*Annual Review of Sociology* 34: 339–58, 2008).

GOING ONLINE FOR SOME SECOND THOUGHTS

An interesting review of American's appetite for fast food, social change, and litigation is found in Andrew Gumbel's "Fast Food Nation: An Appetite for Litigation" (http://news.independent.co.uk/world/americas/article179042.ece).

More information about many of the most pressing (and controversial) legal issues of the day can be found at the American Civil Liberties Union's website: http://www.aclu.org.

The Center for Public Integrity provides useful information on transparency and accountability in the U.S. political process: http://www.publicintegrity.org.

The nonpartisan Rand Institute for Civil Justice conducts research and analysis to help inform public policy: http://www.rand.org/icj/.

The National Center for State Courts allows you to conduct an interactive statistics query that enables you to compare state court systems across a variety of filings (e.g., civil, criminal, juvenile, etc.): http://www.ncsconline.org/D_Research/csp/CSP_Main_Page.html.

Interviews with Chief Justice Roberts and Associate Justices Stevens, Scalia, Kennedy, Thomas, Ginsberg, Breyer, and Alito can be found online at LawProse: http://lawprose.org/interviews/supreme_court.php.

NOW IT'S YOUR TURN

1. Donald Black's (1976) theory of law asserts that social intimates (such as family members) are less likely to use the law against each other than are nonintimates. Identify and speculate as to alternate means of social control that might prove to be particularly effective in family settings. What are their implications for the legal process?

2. Consider the legal and extralegal ramifications of formally regulating our social relations in cyberspace. How have we tried to regulate cyberspace relations via the law? What do you see as some of the most likely extralegal responses to these legal rules?

3. Find out the major contributors for your U.S. Senators and Representatives by visiting the Center for Responsive Politics (http://www.opensecrets.org/about/index.php) and clicking on the "Politicians and Elections" link. Assess how they are doing on representing the interests of the people of your state.

ESSAY 16

Conventional Wisdom Tells Us . . . Honesty Is the Best Policy

. . . except, of course, when reporting your income, revealing your age, sparing the feelings of another–the list can go on and on. In this essay, we explore the conditions under which lying is viewed as normal. In so doing, we use lying as a case study that aptly demonstrates both the pervasiveness and the relative nature of deviance.

"Honesty is the best policy," wrote Ben Franklin. From an early age, parents and teachers urge us to embrace this sentiment. We learn cultural fables and tales that underscore the value of truthfulness—remember Pinocchio or George Washington and the cherry tree? Similarly, religious doctrines turn honesty into law with lessons such as "Thou shalt not lie." And in civics class, we learn that perjury—lying while under oath—is an illegal act.

Norms
Social rules or guidelines that direct behavior, they are the "shoulds" and "should nots" of social action, feelings, and thought

Prohibitions against lying are among the earliest norms to which individuals are socialized. **Norms** are social rules or guidelines that direct behavior. They are the "shoulds" and "should nots" of social action, feelings, and thought. As we grow older, we witness firsthand the ways in which dishonesty can lead to the downfall of individuals, families, careers, communities—even presidencies. Indeed, many social commentators identify the now famous Watergate incident, and the high-level lying that accompanied it, as the basis for today's widespread public distrust of the U.S. government. Of course, the Iran Contra scandal, the Clinton–Lewinsky affair, and the Bush Administration's linking of weapons of mass destruction (WMDs) to the invasion of Iraq added fuel to the fire—so much so that recent polls suggest fewer than 22% of Americans now trust the federal government (Pew Research Center 2010b).

If honesty is the best policy, how do we understand the findings of recent surveys that document widespread dishonesty in our high schools and colleges? About two-thirds of high school and college students have admitted to cheating in their student careers. Cheating includes everything from stealing tests and carrying "cheat sheets" to giving assignments and test answers to others, buying term papers, or plagiarizing other people's work (Josephson Institute 2009; Staples 2010).

If honesty is the best policy, how do we understand the deceit that is rampant in corporate America? Many Enron employees as well as countless investors had their financial futures destroyed because they believed the fraudulent bookkeeping practices tendered by J. P. Morgan, Chase, and Citigroup; the lies told by Enron's CEO, Ken Lay; and the cover-up instigated by Andersen Worldwide, the once highly respected auditing firm (Eichenwald and Atlas 2003). The Enron story is not unique. Media mogul Martha Stewart had no sooner completed her prison term issued for lying and obstruction of justice when Dennis Kozlowski, the former CEO of Tyco International, was sentenced to $8\frac{1}{3}$ to 25 years behind bars for dishonest business practices. And perhaps the biggest deception of all came in 2009, when investment consultant Bernie Madoff—now serving a 150-year prison sentence—admitted to lies that underpinned the largest Ponzi scheme in American history. Corporate lying—it seems to happen every day. One major corporation after another—AIG, Apple, BP, Bear Stearns, Citigroup, Exxon, IBM, Lehman Brothers, Microsoft, Xerox, WellPoint—has had to issue restatements of their financials to "get the numbers right." And while corporate boards of directors, the SEC, and other government regulatory boards are supposed to keep a watchful eye on CEOs, in too many cases, these bodies fail to keep anyone in check (Baker 2009).

Your Thoughts . . .

Many believe that surveillance and regulation could help minimize lying in corporations, organizations, schools, and so forth. What do you think? Will increased surveillance improve honesty? And how much surveillance would you be willing to tolerate in the name of honesty?

If honesty is the best policy, how do we understand the scandal that has rocked the Roman Catholic Church for the last few years? Hundreds of priests stand accused of sexually molesting young charges and parishioners. The scope of this scandal seems to escalate with each passing year as we slowly discover the deceptions perpetrated by the church hierarchy. In the summer of 2003, for example, the public learned of a 40-year-old Vatican document that instructed bishops to cover up sex crimes. The document was described by one defense attorney as a "blueprint for deception" (CBS News 2003). Since then, reports have shown that bishops across the United States kept the illicit behaviors of clergy under wraps by reassigning errant priests and/or buying the silence of victims. In 2009, we learned that even Pope Benedict XVI played a role in the deception (Hitchens 2010; Williams 2005).

If honesty is the best policy, how do we make sense of our transformed post–9/11 society? Dishonest claims and fraudulent documents paved the way for the terrorists' sinister activities in the months and days before the deadly attacks. Since September 11, 2001, there has been increasing interest in and demand for research and technology that will improve our ability to uncover lies and deceptive behaviors (Adelson 2004; CBS News Poll 2006; Lynch 2009). And perhaps coming full circle, Americans are learning more about the possibly deceitful practices of our own government in its efforts to manipulate public reaction and combat terrorism. An internal Environmental Protection Agency (EPA) report has acknowledged that the agency, under prompting by the White House, misled the public by declaring the "ground zero" area of New York City to be safe *before* any air testing results were obtained. Further, the White House was also accused of politically manhandling the EPA by insisting that all EPA press releases regarding air quality at ground zero be cleared by the National Security Agency (Currie 2007; Preston 2006).

Clearly, we would be lying to say that our culture firmly endorses honesty as the best policy. The conventional wisdom regarding the virtue of honesty is strong, yet we must also note that almost as early as we learn prohibitions against lying, we also learn how to rationalize the telling of lies. We learn that "little white lies" are not as serious as "real lies." We learn that context matters: Lying to strangers is not as serious as lying to friends; lying to peers is more excusable than lying to parents or authorities. We learn that lies don't count if we cross our fingers or wink while telling them. And we learn that lies told under duress are not as awful as premeditated or "barefaced" lies (Associated Press 2006; Bussey 1999; DePaulo, Kashy, and Kirkendol 1996; Ekman 2009; Elvish, James, and Milne 2010; Harrington 2009; Lawson 2000; Lee 2004; Lerer 2007; Schein 2004; Williams 2001). Thus, despite conventional wisdom to the contrary, lying stands as a ubiquitous social practice. Children lie to parents, and parents lie to grandparents. Employees lie to employers, and employers lie to regulators. Confessors lie to clergy, and clergy lie to congregations. Presidents lie to Congress and the public, and governments lie to the people. Indeed, there may be no social sphere to which lying is a stranger.

What explains the prevalence of lies when conventional wisdom so strongly supports honesty? What accounts for the discrepancies between what we say about lies and what we do? Is lying wrong? Is it deviant or not?

To start, we must first appreciate the role of cultural values in our often contradictory stance on honesty. A **cultural value** refers to a general sentiment regarding what is good or bad, right or wrong, desirable or undesirable. Cultural values can be powerful motivators that drive our behaviors. In reviewing the previous list of dishonest activities of students, corporate executives, clergy, and political leaders, it is easy to identify some overriding or motivating cultural values. Some students, for instance, committed to the value of success, may cheat in the name of getting good grades or good jobs. Some CEOs, committed to the values of wealth and power, may engage in creative bookkeeping or liability coverups to achieve corporate standing and growth. Because the Catholic Church hierarchy is committed to preserving the integrity of that institution, some bishops, in the hopes of avoiding bad press and shameful labels, seem willing to mislead parishioners about the conduct of priests. And some of our political leaders may indeed engage in hyperbole or stretch the truth. They do so in the name of national security or the passing of legislation that they believe ultimately serves the public good. In short, for some, the pursuit of high and noble values can invite expedient behavior—we do whatever we must to achieve what we think is worthy or desirable. Unfortunately, however, expedient behavior often takes us into the realm of deviance (Ekman 2009; Merton 1938).

> **Cultural value**
> A general sentiment regarding what is good or bad, right or wrong

Deviance is typically defined as any act that violates a norm. Definitions of deviance are rarely black and white. Rather, determining what is deviant is a

> **Deviance**
> Any act that violates a social norm

relative process, because norms can vary with time, setting, or public consciousness. Thus, today's deviant behaviors may be tomorrow's convention. (Think of some "deviant behaviors" that subsequently entered the realm of conformity: long hair on men, jeans on students, living together before marriage, smoking among women.) *Ideally,* norms reflect or are consistent with our cultural values. A culture that promotes the value of family should direct people to marry and have children. A culture that promotes democracy should direct people to vote and participate in community affairs. Values and norms *should* coincide, but that doesn't always happen. They can get out of balance with each other. Frequently, values dominate; we are willing to do anything to accomplish our values or goals. When values dominate, expedient behavior rules, and the normative order can be compromised. If expedient behaviors violate social norms, we enter the realm of deviance. While honesty may be the best policy, dishonesty is often the most expedient way to achieve, succeed, manage, and win. Thus, if I am truly convinced that the grade is everything, I may find it easy to lie and claim the work of others as my own. (Senator Ted Kennedy wrote of this very circumstance in his autobiography *True Compass.*) Similarly, if financial standing means everything to me, I might find it easy to boost my stock prices by lying about company practices. (Many feel that the BP oil spill and the corporate actions surrounding it fit this description to a T.) If protecting the Church from scandal is my highest priority, I might find it easy to lie about the reason for reassigning priests. (This sentiment, according to some, underpins the Church's practice of paying off abuse victims in return for their silence.) And if a war on terror is my goal, I might find it very easy to say what I must to get the job done. (Some contend that President George W. Bush did just that in linking nonexistent WMDs to Saddam Hussein prior to the U.S. invasion of Iraq.)

Because lying, like all deviant acts, is variable, sociologists distinguish between two types of lies: deviant lies and normal lies. **Deviant lies** are falsehoods always judged to be wrong by a society; they represent a socially unacceptable practice, one that can devastate the trust that enables interaction within a complex society of strangers. **Normal lies** are a socially acceptable practice linked to productive social outcomes. Individuals rationalize and legitimate normal lies as the means to a noble end: the good of one's family, colleagues, or country. A lie's relative deviancy or normalcy depends on who tells it; when, where, and why it is told; to whom it is told; and the outcome of its telling (Barnes 1994; BBC h2g2 2010; Blum 2005; Ekman 2009; Ericson and Doyle 2006; Harrington 2009; Lalwani et al. 2006; Lee 2004; Manning 1974, 1984; Ruane, Cerulo, and Gerson 1994; Ryan 1996; Seiter, Bruschke, and Bai 2002; Straughan and Lynn 2002).

Deviant lies
Falsehoods always judged to be wrong by a society; they represent a socially unacceptable practice, one that destroys social trust

Normal lies
Falsehoods that are socially acceptable practice linked to productive social outcomes; individuals rationalize and legitimate normal lies as the means to a noble end—that is, the good of one's family, colleague, or country

For example, withholding your AIDS diagnosis from your elderly mother may be viewed as an act of mercy. In contrast, withholding the diagnosis from a sex partner would probably be viewed as immoral or potentially criminal. Similarly, lying about one's age to engage a new romantic interest is likely to be defined as significantly less offensive than lying about one's age to secure Social Security benefits. As with all forms of deviance, lie classification is based on **social context**—the broad social and historical circumstances surrounding an event. We cannot classify

Social context
The broad social and historical circumstances surrounding an act or an event

a lie as deviant or normal on the basis of objectively stated criteria. In being slow to judge—be it Tiger Woods' explanation of his 2009 Thanksgiving weekend car crash, Representative Joe Wilson's public accusation that President Obama was lying about the terms of the Health Care Bill, or BP's claims that they had indeed planned for an oil drilling disaster—Americans try to carefully assess the context of any alleged misstatements.

Although deviant lies can destroy social relations, normal lies can function as a strategic tool in the maintenance of social order. Normal lies become a "lubricant" of social life; they allow both the user and receiver of lies to edit social reality. Normal lies can facilitate ongoing interaction (Associated Press 2006; BBC h2g2 2010; Blum 2005; Goffman 1974; Goleman 1985; Harrington 2009; Lalwani et al. 2006; Lee 2004; Sacks 1975; Schein 2004).

If our boss misses a lunch date with us, we tell her or him it was no big deal, even if the missed appointment led to considerable inconvenience in our day. Similarly, we tell a soldier's parents that their son or daughter died a painless death even if circumstances suggest otherwise. In both of these cases, the normal lie represents a crucial mechanism for preserving necessary social routines.

In the same way, normal lies also are important in maintaining civil social environments. Thus, when a truthful child announces someone's obesity, foul smell, or physical disability while on a shopping trip at the mall, parents are quick to instruct him or her in the polite albeit deceptive practices of less-than-honest tact. Similarly, the daily contact between neighbors inherent in most city and suburban layouts leads us to tell a rather bothersome neighbor that she or he is "really no trouble at all." In both of these cases, honesty would surely prove a socially destructive policy; the normal lie allows individuals to preserve the interaction environment.

What processes allow us to normalize an otherwise deviant behavior such as lying? In a classic article, sociologists Gresham Sykes and David Matza (1957) identify five specific techniques of neutralization that prove useful in this regard. **Techniques of neutralization** are methods of rationalizing deviant behavior. In essence, these techniques allow actors to suspend the control typically exerted by social norms. Freed from norms in this way, social actors can engage in deviance. Using the techniques of neutralization—denial of responsibility, denial of injury, denial of victim, condemning the condemner, and appealing to higher loyalties—individuals effectively explain away the deviant aspects of a behavior such as lying. Individuals convince themselves that their actions, even if norm violating, were justified given the circumstance. Once an individual has learned to use these techniques, she or he can apply them to any deviant arena, thereby facilitating an array of deviant behaviors: stealing, fraud, vandalism, personal violence, and so on.

Techniques of neutralization
Methods of rationalizing deviant behavior

Employing Sykes and Matza's techniques, then, one might *deny responsibility* for a lie, attributing the action to something beyond one's control: "My boss forced me to say he wasn't in." One might *deny injury* of the lie, arguing that the behavior caused no real harm: "Yes, I lied about my age. What's the harm?" One might *deny the victim* of the lie by arguing that the person harmed by the lie deserves such a fate: "I told her that her presentation was perfect. I can't wait for it to bomb; she deserves it." *Condemning one's condemner* allows an individual to neutralize a lie by shifting the focus to how often one's accuser lies: "Yes, I lied about where I was tonight, but how often have you lied to me about that very thing?" Finally, *appealing to higher loyalties* neutralizes lying by connecting it to some greater good: "I didn't tell my wife I was unfaithful because I didn't want to jeopardize our family."

Just as individuals learn to neutralize certain lies, they also learn appropriate reactions to normal lies. With time and experience, individuals learn that challenging normal lies can be counterproductive. Such challenges can disrupt the

Social scripts
The shared expectations that govern those interacting within a particular setting or context

social scripts that make collective existence possible. **Social scripts** document the shared expectations that govern those interacting within a particular setting or context.

If the social audience wishes to maintain smooth social exchange, then each member must learn to tolerate certain lies. By doing so, individuals downplay deviations from the social script. Like actors on a stage, individuals ignore momentary lapses and faux pas so that the "performance" can continue (Goffman 1959, 1974).

The U.S. military policy of "Don't ask, don't tell," for example, is built on such logic. Military officials look the other way to avoid the potential disruption embodied in a truthful response to the question of homosexuality. Similarly, the spouse who fails to question a partner's change in routine or habit may do so in an effort to shield the marriage from the threat posed by potential truths.

A variety of social settings require that we take someone at her or his word or accept things at face value. We learn to listen with half an ear, to take things "with a grain of salt," or to recognize that people don't always "say what they mean or mean what they say." Paul Ekman, an eminent researcher on deceit, maintains that successful liars most often depend on willfully innocent dupes (Ekman 2009). In the end, these strategies or roles offer support and tolerance for the normal lie.

This discussion of lying raises important points about deviance in general. Despite norms forbidding it, deviance happens. Studies show that nearly every member of the U.S. population engages in some deviant behaviors during his or her lifetime (BBC h2g2 2010). Indeed, Émile Durkheim (1938/1966), a central figure in sociology, suggested that deviance would occur even in a society of saints.

Theorist Edwin Lemert contends that certain types of deviance are universal. Everyone, at one time or another, engages in such acts. Lemert refers to these universal occurrences of deviance as primary deviance. **Primary deviance** refers to isolated violations of norms.

Primary deviance
Isolated violations of norms

Such acts are not viewed as deviant by those committing them and often result in no social sanctions. Deviance remains primary in nature as long as such acts "are rationalized or otherwise dealt with as functions of a socially acceptable role" (Lemert 1951, 75–76). (Note that normal lying fits easily within this category.)

Herein lies the importance of techniques of neutralization and social scripts of tolerance. The techniques and scripts can keep us and our behaviors within the confines of primary deviance. They can keep us from moving to a more significant type of deviance, which Lemert refers to as secondary deviance. **Secondary deviance** occurs when labeled individuals come to view themselves according to what they are called. The labeled individual incorporates the impression of others into her or his self-identity.

Secondary deviance
Occurs when labeled individuals come to view themselves according to what they are called. The labeled individual incorporates the impression of others into her or his self-identity

Although all of us engage in primary deviance, relatively few of us become ensnared by secondary deviance. The techniques of neutralization allow the social actor to rationalize periodic infractions of the rules. Social scripts of tolerance allow the social audience to accept such infractions as well. As such, social audiences refrain from publicly labeling the "neutralized" actor as deviant. By anchoring an individual in primary deviance, the techniques of neutralization ease the return to a conforming status. Understand, then, that when CEOs defend creative bookkeeping as a required part of their job or when bishops act in ways to spare the faithful unnecessary details about the reassignment of clergy, they are in effect working to maintain the primary status of their deviance.

When it comes to norms on lying, or any other social behavior, conformity may be, as conventional wisdom suggests, the best policy. However, when we understand the complexity involved in the workings of norm violations, we cannot help but note that deviating from the "best policy" may not be all that deviant after all.

LEARNING MORE ABOUT IT

"The Lie," by Georg Simmel, is a classic sociological essay on the topic. See *The Sociology of Georg Simmel* (New York: Free Press, 1950a: 312–316).

Brooke Harrington offers a very readable history of lying in *Deception: From Ancient Empires to Internet Dating* (Stanford: Stanford University Press, 2009).

Paul Ekman's *Telling Lies* (New York: W.W. Norton, 2009) is a highly readable and comprehensive examination of social scientific experimental research on lying. Zack Lynch offers a very readable discussion of neurological research on lying. See *The Neuro Revolution: How Brain Science Is Changing Our World* (New York: St. Martin's Press, 2009).

Carl Hausman examines lies in advertising, retail, politics, and the media and offers tips on how we might better equip ourselves to spot and stop falsehoods in *Lies We Live By: Defeating Double-Talk and Deception in Advertising, Politics and the Media* (New York: Routledge, 2000). In "Detecting the Effects of Deceptive Presidential Advertisements in the Spring of 2004," Kenneth Winneg, Kate Kenski, and Kathleen Hall Jamieson discuss the harm lies do in presidential elections (*American Behavioral Scientist* 49 (1): 114–129, 2005).

For a detailed consideration of normal lying in an occupational setting, consult Janet Ruane, Karen Cerulo, and Judith Gerson's 1994 article, "Professional Deceit: Normal Lying in an Occupational Setting" (*Sociological Focus* 27 (2): 91–109). For a look at how the normal lie operates in the world of students, see the 2007 *Washington Post* article, "At Least Half of Students Admit to Cheating."

For an explicit discussion on how corporate culture influences deceit in business practices, see Tamar Frankel's *Trust and Honesty: America's Business Culture at a Crossroad* (New York: Oxford University Press, 2005).

Gresham Sykes and David Matza's 1957 article, "Techniques of Neutralization: A Theory of Delinquency" (*American Sociological Review* 22:664–670), offers readers some insight into the dynamics that feed the legitimation process.

GOING ONLINE FOR SOME SECOND THOUGHTS

Want to test your ability to detect deception? Visit "The Lying Experiment" at Quirkology: http://www.quirkology.com/UK/Experiment_lying.shtml.

NOW IT'S YOUR TURN

1. Consider the normal lie as it exists in the world of advertising. Collect a sample of ads targeting different audiences: adults versus children, men versus women, yuppies versus the elderly. Is there any pattern in the ads' reliance on normal lying as a marketing technique? What would be the ramifications of unmasking the normal lie in advertising?
2. Consider the function of normal lying in the successful completion of the student role. Are there ways in which dishonesty has been institutionalized in the student role? What do your own experiences and the experiences of your friends suggest are the important sources of such deception?
3. Select and carefully follow the media coverage of a current scandal. Discuss the relevance of the concepts of primary and secondary deviance. Identify and document the techniques of neutralization that are being used to control or limit the public's perception of deviance.

SOCIAL INSTITUTIONS: MARRIAGE AND FAMILY

ESSAY 17

Conventional Wisdom Tells Us . . . The Nuclear Family Is the Backbone of American Society

Mom, dad, and the kids–is this the unit on which American social life is built? This essay documents the history of family in the United States, showing that the nuclear family is a relatively recent phenomenon and one that soon may be replaced by other forms of family. In addition, the stability of the nuclear family is explored in light of idyllic stereotypes.

Hearing a speech by a major political figure, surfing the offerings of the vast array of cable stations or Hulu, or listening to a Top 40 radio station all show us to be a nation hooked on nostalgia. We yearn for the "good old days," a time when life was simpler, the nation was prosperous, and nuclear families prayed and stayed together.

A return to the family—this plea rests at the heart of current rhetoric. Today's popular culture touts the world of the Andersons (*Father Knows Best*), the Cleavers (*Leave It to Beaver*), the Cunninghams (*Happy Days*), the Huxtables (*The Cosby Show*), and most recently the Barones (*Everybody Loves Raymond*) as an American ideal. (In fact, a recent survey found that Americans rated Cliff Huxtable, Ward Cleaver, and Jim Anderson as their top three favorite TV dads—that is, ones we wish we had when we were growing up; Harris Interactive 2009.) One social observer has speculated that the popularity of the television series *The Sopranos* may have been due to protagonist Tony's traditional "yearning for yesterday" views (Teachout 2002). Working dads, stay-at-home moms, and carefree yet respectful kids: Conventional wisdom promotes such units as the cornerstone of this nation.

Many believe that only a return to our nuclear "roots" can provide a cure for our ills. Only the

Nuclear family
A self-contained, self-satisfying unit composed of father, mother, and children

rebirth of nuclear family dominance can restore the backbone of our floundering society. The **nuclear family** refers to a self-contained, self-satisfying unit composed of father, mother, and children. Are conventional sentiments correct? Will a return to this nation's nuclear roots bring stability to American society?

To answer these questions, we need some facts on "family portraits" from both today and yesterday. We can start to get a feel for the current living arrangements in the United States by first considering household data. A **household** consists of all the people who occupy a housing unit. The Census Bureau recognizes two categories of households: family and nonfamily. In 2009, there were more than 117.1 million *households* in the United States. Just half (50.4%) of all *households* consisted of married-couple families. Another 12.4% consisted of female-headed families (no husband present), 4.5% were male-headed families (no wife present), and 32.7% were nonfamily households. (A **nonfamily household** consists of a householder living alone or living with nonrelative others; U.S. Census Bureau 2009a, Table H1).

Household
All the people who occupy a housing unit

Nonfamily household
A householder living alone or living with nonrelative others

Now let's narrow our focus to families rather than households. The Census Bureau defines a **family** as two or more people residing together and related by birth, marriage, or adoption. In 2009, there were 78.8 million family households in the United States and, more specifically, 59.1 million *married-couple families*. Of these married-couple families, 53.7% had children of their own while 46.3% did not have any children (U.S. Census Bureau 2009a, Table F1). From the children's point of view, however, there is little that resembles the Cleaver or the Anderson families from 1950s TV land. In 2009, of married couples with children younger than 15, only 22.6% had stay-at-home moms. (The total number of stay-at-home moms was 5.1 million; U.S. Census Bureau News 2010c). Since the 1970s, the percentage of children living with both parents has been decreasing. In 1970, just over 85% of children lived in two-parent households. In 2009, only 69% of children lived with both parents, 23.8% lived with mothers only, 3.4% lived with fathers only, and 4% lived in households with no parents present (U.S. Census Bureau 2009a, Table C2).

Family
Two or more people residing together and related by birth, marriage, or adoption

With such departures from the traditional nuclear family, are we charting new terrain? History suggests not. A macro-level analysis of family in America suggests that nonnuclear family forms are common in America's past. A **macro-level analysis** focuses on broad, large-scale social patterns as they exist across contexts or through time. The form of the "typical" American family has changed quite frequently throughout our nation's history. Indeed, historically speaking, the nuclear family is a fairly recent as well as a relatively rare phenomenon. Knowing this, it is difficult to identify the nuclear family as either the traditional family format or the rock upon which our nation was built. Furthermore, social history reveals that the nuclear family's effects on American society have often proved less positive than nostalgic images suggest.

Macro-level analysis
Social analysis that focuses on broad, large-scale social patterns as they exist across contexts or through time

Historically, American families have displayed a variety of forms. In preindustrial times, the word *family* conjured up an image quite different from the nuclear ideal. Preindustrial families were truly interdependent economic units. All members of the household—parents, children, and quite often boarders or lodgers—made some contribution to the family's economic livelihood. Work tasks overlapped gender and age groups. However, frequently, parental contributions were ended by early mortality. In the preindustrial era, average life expectancy was only 45 years (Rubin 1996). One-third to one-half of colonial children had lost at least one parent by the time they reached 21 years of age (Greven 1970).

During the early stages of industrialization, the face of the family changed. Many families of the era began to approximate the "dad-mom-and-the-kids" model. Yet early industrialization also was a time when the number of extended families in the United States reached its historical high. An **extended family** is a unit containing parent(s), children, and other blood relatives such as grandparents or aunts and uncles. Although the extended family has never been a dominant form in U.S. society, 20% of this period's families contained grandparents, aunts, uncles, or cousins (Haveven 1978).

Extended family
A unit containing parent(s), children, and other blood relatives such as grandparents or aunts and uncles

Your Thoughts . . .

The extended family appears to be making a comeback in these recessionary times. Do you think that three generations living under one roof makes for a stronger family unit . . . a stronger marital unit?

The stay-at-home mom, a mom who focused exclusively on social activities and household management, first appeared in America's middle-class Victorian families. This new role for Victorian women, however, came on the backs of working-class mothers and children. Many women and children of the working class were hired as domestics for middle-class families. In the early days of industrialization, such women and children were also frequently employed as factory workers. Only as industrialization advanced were working-class women relegated to the home sphere (Padavic and Reskin 2002). By the late 1800s, a typical family included a mother who worked exclusively at home.

Children, however, continued to work outside the home for wages. Indeed, in New England, an 1866 Massachusetts legislative report hailed child labor as a boon to society (TenBensel, Rheinberger, and Radbill 1997). Throughout the Northeast, children were regularly employed in industry or the mines (Bodnar 1976; Schneiderman 1967). In the South, children comprised nearly one-fourth of all textile workers employed at the turn of the century (Wertheimer 1977). Across the nation, approximately 20% of American children were relegated to orphanages because parents could not afford to raise them (Katz 1986).

The modern nuclear family that Americans so admire did not fully come into its own until the 1950s. A careful review of families emerging from this period suggests a unit that both confirms and contradicts conventional wisdom's idyllic models.

Historian Stephanie Coontz (2000) notes that the nuclear unit of the 1950s emerged as a product of the times. The family model of the era developed in response to a distinct set of socioeconomic factors. The post–World War II decade saw great industrial expansion in the United States. The nation enjoyed a tremendous increase in real wages. Furthermore, Americans experienced an all-time high in their personal savings, a condition that resulted largely from U.S. war efforts. The nuclear family became a salient symbol of our country's newfound prosperity.

Consumerism was a significant hallmark of 1950s nuclear families. Spending devoted to products that enhanced family life brought significant increases in our nation's GNP (gross national product). Indeed, the buying power of the 1950s family was phenomenal. During the postwar era, for example, home ownership increased dramatically. There were 6.5 million more homeowners in 1955 than in 1948 (Layman 1995). Federal subsidies enabled families to buy homes with minimal down payments and low-interest (2%–3%) guaranteed 30-year mortgages. It is estimated that half the suburban homes of the 1950s were financed in this way (Coontz 2000).

Consumerism added a new dimension to the American family. By 1950, 60% of all U.S. households owned a car (Layman 1995). Cars, cheap gas, and new highways made mobility a part of family life. The purchase of television sets converted the home into both an entertainment and consumer center. In 1950, there were 1.5 million TV sets in the United States. A year later, there were 15 million (Knauer 1998)! Indeed, 86% of the U.S. population owned TVs by the decade's end, bringing about a 50% decrease in movie theater attendance (Jones 1980). In 1954, Swanson Foods introduced a new food product to complement our fondness for the tube: the frozen TV dinner (Layman 1995). Advertisers took advantage of the new medium to sell their products to the American public like never before. By the end of the 1950s, 2 out of every 10 minutes of TV programming were devoted to advertising (Layman 1995).

Other new and improved tools and appliances (power mowers, electric floor polishers) made the home an efficient and comfortable place to live. Shopping malls appeared on the suburban landscape for ease of access by the growing suburban population (Layman 1995). The growth of the credit card industry strengthened the nuclear family's consumer patterns and encouraged a "buy now, pay later" mentality (Dizard and Gadlin 1990; Ritzer 1995). And pay later we did. Some would argue that the debt-laden 1990s and 2000s were the legacy of the spend-and-grow mentality established in the 1950s.

The buying power of the 1950s nuclear family suggested a unit with the potential to fulfill all needs. At first glance, this development may seem to be a positive aspect of the era, but some argue that it marked the beginning of a harmful trend in our society—the decline of community commitment. Such critics contend that by emphasizing the "private" values of the individual and the family, the nuclear unit intensified individualism and weakened civic altruism (Bellah et al. 1985; Collier 1991; Sennett 1977). Thus, the seeds of our later decades of self-indulgence and excess may well have been planted in the nuclear family era of the 1950s. As historian Coontz (2000) observes, "The private family . . . was a halfway house on the road to modern 'me-first' individualism" (p. 98).

Consumerism, however, is only a part of the nuclear family's legacy. There were profound changes in domestic behavior patterns as well. In contrast to the ideal lives enjoyed by the Andersons, the Cleavers, or the Cunninghams, home life of the 1950s showed some problematic developments.

For example, couples married at younger ages than earlier generations had. The formation of a nuclear family was an integral part of "making it" in the United States. As a result, many people moved rapidly toward that goal. But the trend toward early marriage was not without cost. National polls showed that 20% of the period's married couples rated their marriages as unhappy—interestingly enough, this figure is higher than any displayed in recent General Social Surveys. In the heyday of the nuclear family, millions of couples resolved their marital differences by living apart (Komarovsky 1962; May 1988; Mintz and Kellogg 1988).

The 1950s also saw couples starting families at younger ages than their predecessors. In addition, families were larger and grew more quickly than families of previous decades. More than ever before or since, married couples faced enormous pressure to have children. Indeed, *Life* magazine declared children to be a built-in recession cure (Jones 1980). But the emphasis on childbearing had some unintended consequences. The period saw a substantial increase in fertility rates, largely due to teenage pregnancies. (Kids were having kids—sound familiar?) Half of all 1950s brides were teenagers who had children within 15 months of getting married (Ahlburg and DeVita 1992). The number of out-of-wedlock babies placed for adoption increased by 80% between 1944 and 1955, and the proportion of pregnant brides doubled during this era (Coontz 2000).

City living was largely abandoned by families in the 1950s. The new young families of the era put down roots in the suburbs, which seemed ideal for breaking with old traditions. The suburbs were well suited to the newly emerging idea of the family as a self-contained unit, a unit that would supplant the community as the center of emotional investment and satisfaction. Fully 83% of the population growth in the 1950s occurred in the suburbs. Indeed, *Fortune* magazine noted that more people arrived in the suburbs each year than had ever arrived at Ellis Island (Jones 1980).

Changing family lifestyles brought changes to the workplace as well. By 1952, 2 million more women were in the workforce than had been there during World War II. This time, however, female entry into the workplace was not fueled by patriotic duty. Rather, many women were entering the workforce to cover the rising costs and debts associated with the nuclear family's consumer mentality (Rubin 1996). Others went to work to help young husbands complete educational and career goals.

The working women of the 1950s faced a workplace less receptive to them than it had been to the "Rosie the Riveters" of the World War II era. The industries that had welcomed women during the war years now preferred to keep American women at home or in dead-end jobs. Popular magazines of the day ran articles examining the social menace and private dysfunction of working women.

The pressures and rejections faced by women in the workplace had important links to the amazing increase in production and use of newly developed tranquilizers: 462,000 pounds of tranquilizers were consumed in 1958, and 1.15 million pounds of tranquilizers were consumed in 1959! Consumers were overwhelmingly female (Coontz 2000).

In considering the realities of the nuclear family, it becomes clear that neither it nor any other family form can be a remedy for social ills. Debates on the virtue of the "good old days" or attempts to identify the "perfect family" form thus seem an exercise in futility. Each historical period, with its own combination of economic, political, and social forces, redefines perfection. Necessity is the mother of invention, and major social changes of our times (delayed and reduced fertility and increases in children born to unmarried mothers or living in **blended families**, in gay families, and in single households) will no doubt force us to rethink our conceptions of families and their roles in society.

> **Blended families** Families created when those who remarry already have children from their previous marriages

Your Thoughts . . .

Which element(s)—for example, caring or enduring relationships, children, living arrangements, and so forth—do you regard as fundamental to defining family? Do you think public opinion regarding the definition of family should be used to set social policy?

These thoughts may help us better appreciate the risk entailed in trying to build current social policy around institutions whose times may have come and gone. Perhaps the best course of action today is to resist nostalgic "retrofitting" and instead assist the family in making adaptive changes to our current social circumstances. Such changes may well be the new "traditions" that future generations will yearn to restore.

LEARNING MORE ABOUT IT

Counted Out: Same-Sex Relations and Americans' Definitions of Family by Brian Powell, Catherine Bolzendahl, Claudia Geist, and Lala Steelman (New York: Russell Sage Foundation, 2010) uses survey data to provide details and insight into how Americans define family and the pivotal role children play in these views.

Stephanie Coontz offers a detailed historical review of the American family from colonial times to the present in *The Way We Never Were: American Families and the Nostalgia Trap* (New York: Basic Books, 2000). In particular, the author systematically debunks many of our most cherished family myths.

For a picture of the changing face of nuclear families, see Karen V. Hansen's *Not-So-Nuclear Families: Class, Gender, and Networks of Care* (New Brunswick, NJ: Rutgers University Press, 2005).

Explorations in the Sociology of Consumption: Fast Food, Credit Cards, and Casinos by George Ritzer (Thousand Oaks, CA: Sage, 2001) examines the irrational consequences of consumption.

Teresa Toguchi Swartz examines intergenerational family relations in her article "Intergenerational Family Relations in Adulthood: Patterns, Variations and Implications in the Contemporary United States" (*Annual Review of Sociology* 35: 191–212, 2009).

GOING ONLINE FOR SOME SECOND THOUGHTS

The Families and Work Institute offers an array of information on changes in the family and the workplace: http://www.familiesandwork.org/.

The Child Trends Data Bank offers a collection of pertinent indicators for monitoring trends in child and family welfare: http://www.childtrendsdatabank.org/?q=node.

NOW IT'S YOUR TURN

1. Use your sociological imagination to identify some key factors that prompt our present nostalgia for the past. For example, think in terms of historical and social developments that may help explain why the past looks so good to us now.

2. We noted in this essay that nuclear families were a product of specific historic and socioeconomic conditions. Consider a family type common today, the "blended family." Blended families are units that consist of previously married spouses and their children from former

marriages. What are the current historic and socioeconomic conditions that explain the blended-family phenomenon? What social policy changes might enhance the success of this family form?

3. In the essay on climate (Essay 14), we introduced the reader to functional analysis. Try identifying the various functions (manifest, latent, positive, and negative) of the following trends:
   - an increase in women electing to have babies outside of marriage
   - a continued increase in the percentage of individuals electing to live alone
   - an increase in gay couples having/raising children
   - the downward trend in percentage of children living in two-parent families

ESSAY 18

Conventional Wisdom Tells Us . . . Marriage Is a Failing Institution

High divorce rates, couples living together without being married, the need for "space," fear of commitment—have such trends doomed the institution of marriage? Here, we discuss research suggesting that the practice of marriage is alive and well despite conventional wisdom to the contrary. We also note the historical "popularity" of divorce in America and speculate on why such a trend marks our culture.

Politicians say it and social commentators lament it: Many of our current social and economic problems stem from deteriorating family values. The conventional wisdom on the matter suggests that high divorce rates are jeopardizing the future of marriage and the family. Many fear that "till death us do part" has become a promise of the past. Recent presidents (representing both political parties) and Congresses have endorsed legislation to protect and encourage marriage. Are our fears and legislative action well grounded? Is marriage really a failing institution?

Most research on the matter suggests little cause for such alarm. Indeed, a variety of indicators document that marriage remains one of society's most viable social institutions. **Marriage** refers to a socially approved economic and sexual union. A **social institution** consists of behavior patterns, social roles, and norms, all of which combine to form a system that ultimately fulfills an important need or function for a society.

Marriage
A socially approved economic and sexual union

Social institution
Set of behavior patterns, social roles, and norms that fulfill an important need or function for a society

The United States has the highest marriage rate of any modern industrial society (United Nations 2007). In 2009, only 26.1 of those

18 years of age and older had never been married (U.S. Census Bureau 2009a, Table A1). Most young adults plan to marry at some point in time, although the age of first marriage is increasing (28 for males and 25.9 for females; U.S. Census Bureau 2010b). A higher age for first marriage is itself a good sign for the institution because it is associated with increased marital stability (National Marriage Project 2009).

In the United States, the marriage institution enjoys much positive press. For Americans, marriage appears to be a key to happiness. The majority of married men (63%) and women (60%) in the United States indicate that they are *very* happy with their marriages. While these percentages are lower than those reported 30 years ago, the decline has flattened out in recent years (National Marriage Project 2009). To be sure, the link between marriage and happiness is not unique to the United States. Recent studies involving as many as 64 countries also document that married individuals report higher levels of happiness (Doyle 2002; Stack and Eshleman 1998). Furthermore, marital bliss appears to have a halo effect and creates what is known as the **happiness gap**—that is, married individuals report higher levels of general happiness than the never married (Lee and Bulanda 2005).

Happiness gap The discrepancy in reported happiness attributed to marital status: married individuals report higher levels of general happiness than the never married

In addition to raising individuals' happiness quotients, marriage also makes us healthy, wealthy, and maybe wise as well. To be sure, marriage pays well, especially for those with long unions. Married couples live more economically and save and invest more for their futures than the nonmarried. The long-term effects of these economic benefits are noteworthy: Those who never marry and those who divorce and remain unmarried suffer 75% and 73% reductions in their overall wealth, respectively. The earning power of married men is greater (by as much as 40%) than that of single men with similar education and work backgrounds (National Marriage Project 2009). In addition to boosting financial well-being, marriage is also associated with better mental and physical health benefits. Married persons have higher rates of health insurance coverage and lower prevalence of heart disease, and they live longer (Goodwin et al. 2010). Perhaps the biggest benefit of pledging "till death us do part" is reducing our chances of death. There is evidence that marriage helps keep us alive: Both married men and women show significantly lower risks of dying than the unmarried, and the effect is not simply due to married individuals selecting healthy partners. The lower mortality benefit remains even after health status is controlled (Bramlett and Mosher 2002; Lillard and Waite 1995). Having never been married is a strong predictor of premature mortality (Kaplan and Kronick 2006). (The 2007 recession has seen a subsequent drop in the divorce rate, leading some to speculate that economic hard times may encourage a new-found appreciation of the benefits of marriage [National Marriage Project 2009]).

The benefits of marriage provide important insight into another recent development: the marriage movement among gays and lesbians. Many same-sex couples want their unions to have the same protection, benefits, and recognition as are accorded to traditional marriages. Despite the 1996 Defense of Marriage Act, which defined marriage for federal purposes as being between one man and one woman, efforts are continuing for gender-neutral civil-marriage laws. Massachusetts was the first state to pass such a law, and other states (New Hampshire, Vermont, Connecticut, and Iowa) have followed suit. Such developments would appear to be in keeping with ancient practices. Newly uncovered historical evidence suggests that same-sex unions were legally sanctioned and accepted in medieval Europe (Bryner 2007).

BOX 18.1 SECOND THOUGHTS ABOUT THE NEWS: JUDGING MARRIAGE

Fans of *Mad Men* might recall the storyline about coming up with a good product pitch for Ponds face cream for women: Sell it as a way for women to achieve their ultimate dream–marriage. It was true in the 1960s and it is still true today: Marriage is a social institution with a very strong appeal. Perhaps, then, we shouldn't be surprised that groups previously excluded from entering into this state–gays and lesbians–are now demanding their right to marry.

To be sure, this is a contentious issue that is frequently in the news these days. Opponents claim that same-sex couples are an affront to the holy state of matrimony and will prove injurious to the well-being of children. Proponents insist that *all* couples deserve the chance to reap the benefits of marriage and that no harm will result to either the institution or to families.

An important showdown between the two camps occurred in California in the summer of 2010. In November 2008, 52% of California voters supported Proposition 8, a ballot measure that banned same-sex marriages. But in August of 2010, a federal district court judge overturned the ban, citing the absence of any evidence to support fears and claims that same-sex marriages pose a social harm. Indeed, the judge declared the denial of marriage licenses to same-sex couples to be unconstitutional. To be sure, the last has not been heard on this issue.

Many believe that the California ruling opened the door for the Supreme Court to step into the fray and decide whether or not gays and lesbians have a constitutional right to marriage. Stay tuned.

Marriage in the United States occurs with great frequency and generates high overall satisfaction for those who enter the union. It is also a right embraced, desired, and defended by many. What, then, explains the alarm sounded by many regarding the institution's failure? First, despite the popularity of marriage in the United States, from 1970 to 2007, we have seen a 50% decline in the annual number of marriages per 1,000 unmarried adult women (National Marriage Project 2009). Other recent changes in the American household are also a concern to some. In 2009, 50% of U.S. households were maintained by married couples—this figure is down from 55% in the 1990 census and from 71% in the 1970 census. The second most common type of household (at 27%) consists of people living alone—a figure that has steadily increased from 17% in 1970 (U.S. Census Bureau 2009a, Table H1, 2011, Table 59). Since 1960, we have also seen a 12-fold increase in the number of unmarried, cohabiting couples. It is estimated that today, more than half of all first marriages are preceded by living together—a practice that was virtually nonexistent in earlier eras (National Marriage Project 2009). Typically, though, divorce is cited as the primary threat to the health of the marriage institution. While our divorce rate has been declining since the mid-1980s (the 2008 rate of 3.5 per 1,000 population is the lowest in the last quarter-century), the United States still has one of the highest divorce rates in the world (United Nations 2007; U.S. Census Bureau 2011). Yet note that our propensity toward divorce is not a recent or modern phenomenon. (See Table 18.1 for factors that mitigate the chances of divorcing.)

Divorce is a rather old and well-established practice in America. Indeed, from the late 1800s through the turn of the new century, the United States has been a world leader with regard to the divorce rate (although for the last few years, a handful of other nations have been posting rates on par with or higher than the U.S. rate;

Table 18.1 Ways to Decrease Chances of a Marriage Ending in Divorce*

Factors	% Decrease in Risk of Divorce
Having an annual income over $50,000	−30%
Having a baby *after* marriage	−24%
Marrying *after* 25 years of age	−24%
Having parents who are not divorced	−14%
Being religious	−14%
Having some college	−13%

Source: Wilcox, W. B. (ed). 2009. *The State of Our Unions: 2009.* http://www.virginia.edu/marriageproject/pdfs/Union_11_25_09.pdf.

*During the first 10 years of marriage

United Nations 2007). The first American divorce was recorded in 1639 and occurred in Puritan Massachusetts (Riley 1991/1997). The Puritans viewed unsuccessful marriages as obstacles to the harmony deemed important to their society. Consequently, divorce was regarded as a necessary tool for the *safeguarding* of marriage, family, and community (Riley 1991/1997; Salmon 1986; Weisberg 1975).

By the Revolutionary War, divorce was a firmly established practice in the colonies, one defended via the revolutionary values and rhetoric of our newly formed nation. Thomas Jefferson, for example, in preparing notes for a divorce case in Virginia, defended the right to a divorce, using the principles of "independence" and "happiness" (Riley 1991/1997). In 1832, divorce entered the White House. Andrew Jackson was elected to the presidency despite a much-publicized marriage to a divorced woman (Owsley 1977).

By the mid-1800s, divorce was characterized as a right of those whose freedom was compromised by an unsuccessful marriage. In the years following the Civil War, the U.S. divorce rate increased faster than the rates of both general population growth and married population growth (Riley 1991/1997). By World War I, this nation witnessed one divorce for every nine marriages (Glick and Sung-Ling 1986). For a short time after World War II, there was one divorce for every three marriages of women over 30! In 1960, the U.S. divorce rate began to increase annually until the rate doubled and peaked in 1982. Today, it's thought that the lifetime probability of recent first-time marriages ending in divorce or separation is between 40 and 50% (National Marriage Project 2009). When viewed within its historical context, it is clear that divorce is an American tradition.

In considering high divorce rates in the United States, we must take care not to interpret these figures as an indictment of the marriage institution. Despite a lack of success in first-time marriages, most divorced individuals do remarry (men tend to remarry sooner than women; National Marriage Project 2009). Indeed, slightly more than a quarter of women and men aged 25 to 44 have been married two or more times (Goodwin et al. 2010). High remarriage rates indicate that divorce is clearly a rejection of a specific partner and not a rejection of marriage itself. Indeed, these remarriage patterns have prompted some observers of the family to recognize a new variation on our standard practice of monogamy: namely, serial monogamy. **Monogamy** refers to an exclusive union: for example, one man married to one woman at one time. **Serial monogamy** refers to the practice of successive, multiple marriages—that is, over the course of a lifetime, a person enters into successive monogamous unions.

Monogamy
An exclusive union (e.g., one man married to one woman at one time)

Serial monogamy
The practice of successive, multiple marriages—that is, over the course of a lifetime, a person enters into consecutive monogamous unions

The prevalence of serial monogamy in the United States suggests that the old adage "Once burned, twice shy" does not seem to apply to people who have been divorced. Indeed, the inclination to remarry—even after one has been "burned"—may provide the greatest testimony to the importance of marriage as a social institution. Consider the celebrities who keep taking trips down the aisle: James Cameron (5 times), Greg Allman (6 times), Liz Taylor (8 times), Larry King (8 times), and Zsa Zsa Gabor (9 times). Even those individuals with extremely costly failed marriages—the ultrarich with expensive prenuptial pacts—are committed to trying marriage again and again. Ronald Perelman (one of *Forbes*' richest men in America) has had four marriages to date. Donald Trump has been married three times (Zernike 2006).

Is there a positive side to divorce? Some research suggests that divorce does provide a latent function in U.S. society. A **latent function** is an unintended benefit or consequence of a social practice or pattern. Studies show that as divorce rates in the United States increase, so do rates of marital satisfaction. Considering high divorce rates in conjunction with increasing marital satisfaction suggests that, in the long run, divorce may make marriage a healthier institution (Veroff, Douvan, and Kulka 1981). There is also evidence that staying in an unhappy marriage can negatively affect one's self-esteem, life satisfaction, and overall health (Hawkins and Booth 2005). Divorce, then, can improve the well-being of some people. By dissolving unhappy unions, divorce frees individuals of distress and allows them to seek new and, it is hoped, happier unions. Within happy marriages, fear of divorce also may encourage continued dedication to maintaining a successful union. Since most divorces are followed by remarriages, divorce can also be seen as giving people opportunities for growth and second chances in life (Hetherington and Kelly 2002; Visher, Visher, and Pasley 2003).

Latent functions
Unintended benefits or consequences of a social practice or pattern

Any latent functions aside, divorce certainly carries several manifest dysfunctions as well. A **manifest dysfunction** refers to an obvious negative consequence of a social practice or pattern. For example, divorce has highly negative effects on the emotional well-being of ex-spouses. Research documents that ex-spouses face an increased incidence of psychological distress, depression, loneliness, and alcohol use (Hope, Rodgers, and Power, 1999; Johnson and Wu 2002; Williams and Dunne-Bryant 2006). Divorced individuals may also suffer a permanent alteration of their original quality of life (Lucas 2007). Financial hardship also plagues victims of divorce—this is especially true for women. In 2008, the mean income for a married individual 18 years of age and older was $46,530; the mean income for divorced females was $35,013. The mean income for divorced males 18 years of age and older was $43,028 (U.S. Census Bureau 2009c, Table PINC-02). Studies show significant decreases in the standard of living for divorced women and their children (Bartfeld 2000; Peterson 1996; Whitehead 1993). In 2008, 5.5% of married-couple families were poor in contrast to 29% of female-headed families (Institute for Research on Poverty 2009).

Manifest dysfunctions
Obvious negative outcomes or consequences of a social practice or pattern

The percentage of children living with two married parents has been declining for the last several decades (see Figure 18.1). Today, just over one-quarter of all children live in single-parent families. And each year, about one million children join the ranks of those who must deal with the aftermath of parental divorces (National Marriage Project 2009). The children of divorce often pay an undeniably high price for this breakup; in general, they experience negative life outcomes at rates two to three times higher than those for children in married-parent families (National Marriage Project 2009). They are also

Figure 18.1 Percentage of Children Under 18 in Various Living Arrangements

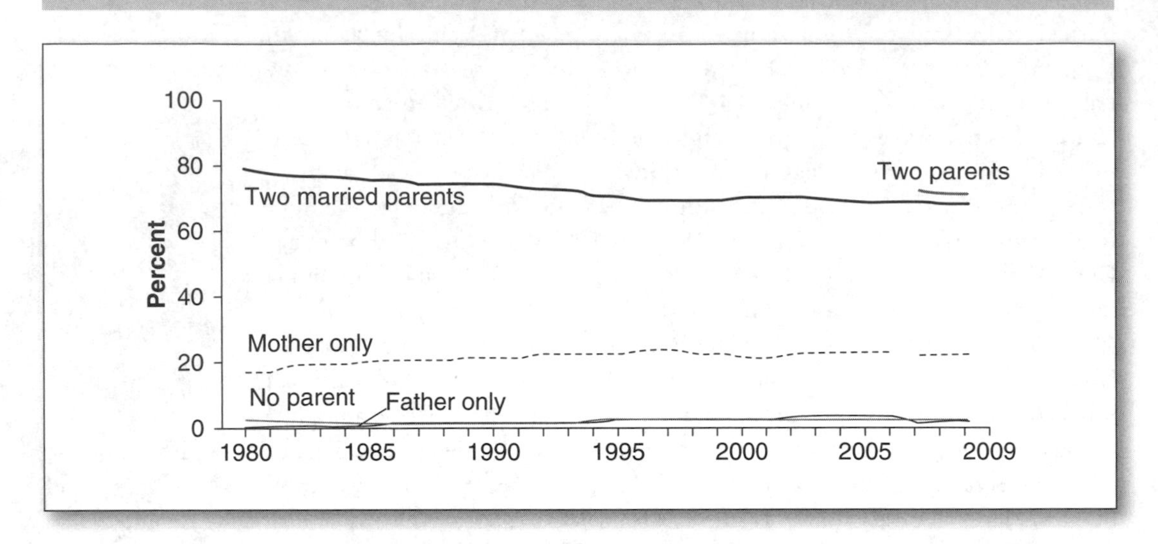

Source: U.S. Census Bureau, Current Population Survey, Annual Social and Economic Supplements. Available at http://www.childstats.gov/americaschildren/famsoc1.asp.

Note: In this figure "two parents" reflects all children who have both a mother and father identified in the household, including biological, step, and adoptive parents. Before 2007, "mother only" and "father only" included some children who lived with a parent who was living with the other parent of the child, but was not married to them. Beginning in 2007, "mother only" and "father only" refer to children for whom only one parent has been identified, whether biological, step, or adoptive.

more likely to have mental health and emotional problems in adulthood, are more likely to be divorced themselves, and are less likely to have close relationships as adults with their fathers (Nock and Einolf 2008). And while step-parents are part of many children's lives, they lack curative powers for the ills of divorce. Research suggests that children in step-families fare no better than children in single-parent families (National Marriage Project 2009).

Divorce can erode bonds between parents and children, especially those between fathers and their children. Today, roughly one-third of children live apart from their biological fathers. Children with nonresident fathers have more emotional, behavioral, and physical health problems. They are more likely to use drugs and alcohol and more likely to become delinquent. Teenage girls in single mother homes are more likely to become pregnant and teenage boys are more likely to become teen fathers. In later life, these children become adults who attain less education, earn less money, and are more likely to be divorced (Nock and Einolf 2008). As nonresidential dads' involvement with their children decreases, so does the payment of child support (U.S. Department of Health and Human Services 2008). In 2007, only 46.8% of custodial mothers received full child support payments from fathers (Grall 2009). Father-absence costs not only children but society as well. In 2006, the federal government spent close to $100 billion on programs assisting father-absent families (Nock and Einolf 2008).

Although divorce clearly poses serious problems for a society, the practice appears firmly rooted in the United States. Ironically, some suggest that the prevalence of divorce in our nation may stem from the heightened health and longevity of Americans. Historian Lawrence Stone (1989) argues that divorce is the functional substitute for death. In earlier periods of our history, marriages—good and bad alike—were typically terminated by death. The high mortality rates that characterized colonial America, for example, guaranteed that most marriages lasted less than 12 years (Fox and Quit 1980). But as life spans grew longer and longer, marriages faced new tests. In the current day, promises of "till death us do part" now constitute oaths that can traverse decades. Surely, such a profound demographic shift has had an effect on our divorce rates (Wells 1982).

Your Thoughts . . .

In the summer of 2010, Al and Tipper Gore, long touted as one of Washington, D.C.'s happiest couples, announced that they were ending their marriage of 40 years. Should marriages that last this long count as "successes" even if they end in divorce?

In addition to increased life expectancy, some contend that the ready availability of divorce in the United States helps explain the prevalence of the practice. By 1970, every state in the union permitted divorce. In that same year, no-fault divorce was introduced in California, and many other states quickly followed suit.

Yet although easy access may contribute to the high rate of divorce in the United States, history reveals that the absence of a divorce option does little to save bad marriages. For example, in states that were slow to recognize divorce as a legal option (typically, the southern states), marriages were far from indissoluble. In lieu of legal divorce, other adaptations were readily devised: annulments, desertion, migratory divorces (divorces sought after relocations to other states), premarital contracts, and separation agreements. Indeed, South Carolina's refusal to recognize divorce resulted in a curious development in the state's inheritance laws: Mistresses (another popular adaptation to troubled marriages) were legally precluded from inheriting *any more than 25%* of a husband's estate (Riley 1991/1997).

But longer life spans and ease of access do not tell the whole story regarding divorce in America. Our high divorce rates must be examined in light of some core American values. Consider, for example, American values relating to romantic love. Americans have long identified romantic love as the most important dimension of marriage. Indeed, it is the power and passion of love that convinces us to separate from our **families of orientation** (the families into which we are born) and form new unions with others—that is, **families of procreation** (the families we establish via marriage and children).

Families of orientation
The families into which we are born

Families of procreation
The families we establish via marriage and children

The principle of a love-based marriage guarantees excitement and euphoria for the individuals involved. But linking marriage to romantic love simultaneously introduces a high risk to the institution as a whole. Historian Stephanie Coontz argues that the origins of our modern divorce patterns are firmly rooted in the idea that marriage should be based on love (Coontz 2007). But love is a fickle enterprise. When the sparks of passion die, love can be fleeting. Love's fickle nature is exacerbated by the fact that people frequently fall in love with possibilities rather than realities—that is, they fall for what they *want* their future spouses to be rather than what they are (Berscheid and Hatfield 1983). Assumed similarity is common for married couples, but it is greatest for people who are dating (Schul and Vinokur 2000; Watson, Hubbard, and Wiese 2000). We also find ourselves juggling competing notions of love as we move through our lives. These competing views can make love an ambiguous basis for marriage and guarantee that individuals will have to work hard and compromise and change to keep love and marriages going (Swidler 2001). Consider also that a new "companionate" model of marriage is emerging. In this model, individuals are seeking a "soul-mate" relationship when selecting a spouse. The soul-mate standard has a romantic component. Consequently, individuals pursuing their soul mates may well be willing to delay marriage while waiting to meet their perfect mate or be willing to trade in one partner for a more satisfying other (National Marriage Project 2009). Indeed, professional matchmakers recognize the perils of matches based on romance and soul-mate searches and prefer using more rational strategies (Thernstrom 2005).

In short, love, especially passionate love, can and often does fade. And faded love can leave a marriage with no *raison d'être*, making it a union ripe for divorce. Indeed, some of the earliest U.S. divorce petitions—filed in the 1600s and 1700s—referenced lost affections and love as a justification for divorce (Riley 1991/1997). Knowing this, love may not be the strongest foundation for the institution of marriage. Indeed, many other cultures of the world do not base marriages on love; some regard a loved-based marriage as a foolish endeavor (Jankowiak and Fischer 1992). Instead, marriages are arranged for practical, social, or economic reasons. There is some indication that these alternate factors result in more stable unions with higher survival rates than those motivated by love (Dion and Dion 1996; Levine 1993; Shaw 2006).

Romantic love is only one American value that may be linked to the nation's high divorce rates. The pursuit of personal freedom, self-actualization, and self-gratification—all values on which the establishment and expansion of our country were based—are also quite compatible with our well-exercised right to divorce. Consider that the westward expansion of this country was fueled, in part, by settlers who carried the spirit of rugged individualism. Many who migrated to the West did so without the support or company of recalcitrant spouses. Under such conditions, liberal divorce laws in western states and territories proved to be a valuable mechanism for resolving such migratory conflicts.

Liberal divorce policies also proved a boon to self-actualizing business entrepreneurs. For example, when social pressure from antidivorce factions saw Nevada increase its divorce residency requirement from 6 months to a year, Nevada businesspeople vigorously protested the change. Such a change posed a threat to the valuable revenues typically generated by the divorce trade. Thus, in the interest of capturing more and more of the dollars spent in pursuing a divorce, Nevada eventually dropped its residency requirement to 6 weeks (Riley 1991/1997).

If core American values contribute to high divorce rates, then a reversal of divorce trends in the United States may demand major cultural changes—changes that would be difficult to execute. For example, increasing marital stability may require that love give way to less romantic criteria in making marriage decisions. Yet such a shift seems highly unlikely. A cursory review of television programming, movie plots, literature, or music clearly demonstrates that Americans are in love with the notion of love. Viewers of romantic programming tend to have higher idealistic expectations of love (Segrin and Nabi 2002). Our romantic delusions may also explain the fact that long-distance dating relationships are found to be more stable than geographically close dating relationships. Idealized views of love appear to fuel this paradox: Absence makes the heart grow fonder. Curiously enough, when long-distance relationships become close-distance ones, they become unstable (Stafford and Merolla 2007). Furthermore, surveys from the 1960s through the 1980s document an increase in Americans' commitment to love as the basis for marriage (Simpson, Campbell, and Berscheid 1986). More recently, an overwhelming majority of American college students indicated that they would not be willing to enter into a loveless marriage. Decidedly larger numbers of non-American students, however, were willing to entertain this possibility (Levine 1993).

Similarly, individualism's dire effects on many a happy marriage suggest that reducing divorce rates may require an exchange of individualism for concerns of community. Again, however, such a change is unlikely. Individualism is arguably the defining American value. (Consider again, for instance, that single-person households are the second most common household type in the United States today.) Proposed solutions to other social problems, such as welfare, poverty, and homelessness, recommend a recommitment to individual accomplishment and responsibility. The "me generation" of the 1980s, the decline in social connections, and the economic threats of the 1990s and today have put concerns for "Number 1" at the forefront of American attentions (Cerulo 2002, 2008; Collier 1991; Etzioni 1994, 1996; Putnam 2000).

In contrast, many feel that concerns for community have gradually come to occupy the back burner of the American agenda. In recent years, national leaders such as former Speaker of the House Newt Gingrich have been known to fight community-minded programs (i.e., the National Youth Corp), arguing that volunteerism should be a personal decision rather than an institutionalized practice. Some social scientists contend that the average U.S. citizen may now share the Gingrich mentality. Robert Putnam, a political scientist who examines trends in community participation, notes a marked decline in civic involvement. He argues that few U.S. citizens are actively engaged in solving problems such as crime, drug abuse, homelessness, and unemployment—all threats to community stability (Putnam 1995, 2000). Indeed, a recent Harris poll found that nearly half of Americans think that no one should feel *obligated* to get involved with social causes (Harris Interactive 2010a).

As community loses its place on the American agenda, some speculate that family has come to fill the void. In many ways, family has become the substitute for community. In contrast to the America of the 1930s and 1940s, where "God, home, and country" were the rule, the America of recent decades finds religion and civic concerns overpowered by issues of family. Since 1995, "spending time with family" has been in the top three favorite leisure-time activities of Americans. More than 90% of Americans report having positive relationships with their families (Harris Interactive 2008, 2010b). A growing percentage of American youth report that having a good marriage and family life is extremely important to them (National Marriage Project 2009). As we move into young adulthood, our sense of family obligation increases (Fuligni and Pedersen 2002). As we age, the emotional support we receive from our families is an important resource, one that increases our life expectancies (Ross and Mirowsky 2002). And the family is the backbone of supportive services for the aged and the disabled (American Association of Retired Persons 2003; Houser and Gibson 2008).

Ironically, Americans' heavy investments in family may make this unit increasingly vulnerable to the practice of divorce. When family experiences fail to yield the expected emotional benefits, American cultural traditions and laws make it possible to cut our losses and invest anew. Oddly enough, our restricted family focus may render the family more, rather than less, vulnerable to disruption (Giddens 1992).

In considering the reasons for high divorce rates in the United States, some additional factors require attention. Certain social structural changes also mitigate the likelihood of waning divorce rates. For example, increased labor force participation of both married and unmarried women and high geographic mobility create social contexts that increase the risk of marital dissolution (South 2001; South and Lloyd 1995). The practice of cohabiting before marriage is growing—more than half of all first marriages are preceded by cohabitation. Yet there is a body of research that suggests that marriages preceded by cohabitation tend to have higher divorce rates (National Marriage Project 2009). Furthermore, it is estimated that 40% of children spend time in a cohabiting family (Bumpass and Lu 2000).

Such arrangements as well as the reality of having divorced parents may offer children counterproductive lessons about commitment (Amato 2001). The rise in single-person households over the last few decades may make it easier for some individuals to contemplate the prospect of going it alone after a divorce. Last, there has been a dramatic change in the relationship between marriage and having children. Not only are Americans desiring and having fewer children (six out of 10 Americans think the ideal family size is two or fewer children), but children are also becoming less of a priority in marriages. By a margin of three to one, Americans see the main purpose of marriage as being personal fulfillment and happiness (68%) rather than having and raising children (21%; Pew Research Center 2007). Furthermore, more and more women are electing to have children outside of marriage. In 2006, 38.5% of all births were to unmarried women (National Center for Health Statistics 2009). These changes in attitudes and fertility patterns may mean that some unhappy spouses will perceive fewer obstacles to divorce.

Is marriage a failing institution? Hardly. But along with the health of marriage in America comes the health of divorce. Arguably, divorce is itself an established institution, one that facilitates the pursuit of some core American values. Without some changes to these values, we can well expect both marriage and divorce to be permanent features of society.

LEARNING MORE ABOUT IT

Using census and interview data, Michael Rosenfeld examines the historical and social underpinnings for the variety of alternative unions that mark today's society in his work *The Age of Independence: Interracial Unions, Same-Sex Unions and the Changing American Family* (Cambridge: Harvard University Press, 2007).

Historian Stephanie Coontz explores the place of love and emotional commitment in the history of marriage in her book, *Marriage, a History: From Obedience to Intimacy, or How Love Conquered Marriage* (New York: Penguin, Reprint Edition, 2006).

In *Alone Together: How Marriage in America Is Changing,* Paul Amato, Alan Booth, David Johnson, and Stacy Rogers examine recent adaptive changes (Cambridge, MA: Harvard University Press, 2007).

In *Same-Sex Marriage* (New York: Cambridge University Press, 2006), Kathleen Hull uses interview data from more than 70 people involved in same-sex relationships to explore the cultural practices and legal debates surrounding the issue of gay marriage.

Sara McLanahan offers a very readable review of how children fare without fathers in "Life Without Father: What Happens to the Children?" (*Contexts* 1 (1): 35–44, 2002).

GOING ONLINE FOR SOME SECOND THOUGHTS

Visit Public Agenda Online and scroll down the page and click on the *Issue Guides* link. There, you'll find an overview of major family issues, competing perspectives on future policy, and public opinions on an assortment of family topics: http://www.publicagenda.org/.

The PBS documentary *Let's Get Married* (November 2002) provides a brief introduction to the new "marriage movement," its impact on government policies, and its reception by citizens and politicians alike: http://www.pbs.org/wgbh/pages/frontline/shows/marriage/.

For opposing views on the benefits of the institutions of marriage, visit the National Marriage Project Webpage (http://www.virginia.edu/marriageproject/) and the Alternatives to Marriage Project web page (http://www.unmarried.org/).

NOW IT'S YOUR TURN

1. Trying to understand divorce in America provides a good opportunity for exercising our sociological imagination. Divorce, as it currently stands in our society, is not just a private trouble—a personal failing of the individual. Rather, it is a public issue—a phenomenon tied to myriad broader social, cultural, and historical events. Confusing public issues with private troubles results in misinformed social policy. Public issues cannot be remedied with individual-oriented solutions appropriate to private troubles. Instead, public issues require that we pay attention to changes in social forces that are larger than and transcend individuals. Recognizing divorce as a public issue, identify three appropriate targets for our reform efforts.
2. Access some recent statistics on national divorce rates, broken down by the following groups: African Americans, Asian Americans, Hispanics, and whites of Western European origin. Attempt to explain any differences you find using information on the cultural values held within each group.
3. Love is the basis on which current American marriages are built. Think of at least two other possible foundations for the institution of marriage. In each case, what are the likely ramifications of changing the "rules" of choosing a mate?
4. The Center for Marriage and Families at the Institute for American Values offers an index for tracking the health of marriage in the United States (http://familyscholars.org/). Take a look at the indicators used by the index. Are they reasonable? Are there any serious omissions?

SOCIAL INSTITUTIONS: THE ECONOMY

ESSAY 19

Conventional Wisdom Tells Us . . . Welfare Is Ruining This Country

A frequently expressed opinion when talk turns to welfare reform is that too many people are on the dole and too many recipients have other options. In this essay, we review some of the least understood dimensions of welfare and explore exactly where welfare monies are going.

In conventional wisdom, charges against the U.S. welfare system abound. Welfare recipients are thought to be lazy people. They are accused of lacking the motivation to earn an honest living, preferring instead to take handouts from the government. Furthermore, many believe that the welfare rolls are continuously growing and riddled with fraud. Welfare is often discussed as a program plagued by able-bodied con artists—people with no financial need—who nonetheless manage to collect welfare checks.

Conventional wisdom also tells us that welfare fails to help those on the dole. Rather, it creates further dependency among its recipients. As such, many argue that welfare expenditures are ruining this nation. Welfare represents too great a burden for a government facing unsettled financial times.

Is the conventional wisdom regarding welfare accurate? Is welfare simply a tax on the economy, one destroying American initiative? A fair assessment of the system requires us to take a second look.

Many of the common and persistent beliefs about welfare are largely a product of the **Aid to Families with Dependent Children** (**AFDC**) era. The AFDC was a public assistance program rooted in the Social Security Act of 1935. Concerns about AFDC helped fuel the clamor for

Aid to Families with Dependent Children (AFDC) A New Deal program created by the Social Security Act of 1935 to provide federal assistance to needy dependent children; replaced in the late 1990s by the Temporary Assistance to Needy Families (TANF) Program

Temporary Assistance to Needy Families (TANF)
A federal program created in 1996 by the Personal Responsibility and Work Opportunity Reconciliation Act to provide temporary assistance and work opportunities to needy families with children; also referred to as a "welfare-to-work" program

welfare reform. Reform materialized in 1996 with the passage of the Personal Responsibility and Work Opportunity Reconciliation Act (PRWORA), which abolished the AFDC program and replaced it with a new, time-limited program: **Temporary Assistance to Needy Families (TANF)**. TANF is a "welfare-to-work" program. It gives states much flexibility in structuring their welfare programs, but it imposes a restriction on federal funding: Families are subject to a lifetime limit of 5 years of support. TANF was implemented in all 50 states within 18 months of its 1996 passage (California was the last state to put TANF into effect). In this essay, we will look back at the AFDC era as well as at the current TANF program to see whether there is any basis for popular views about welfare.

Are welfare rolls spinning out of control? Data suggest that the answer to the question is no. A family is considered welfare dependent if more than 50% of its income is derived from welfare programs—for example, AFDC/TANF, food stamps, and SSI (Supplemental Security Income). In 1996, 5.2% of the population was welfare dependent. In 2005, 3.8% of the population was so classified. If we focus exclusively on AFDC/TANF caseloads, we see that in 2007, the average monthly number of TANF recipients was 4.7 million, a number that is 67% *lower* than the peak caseloads for AFDC in 1996. While TANF is credited with these dramatic declines, the historical record also indicates that our welfare ranks have *always* been relatively small (U.S. Department of Health and Human Services 2009b).

Do welfare recipients live "high on the hog" and have more and more kids just to keep the benefits flowing? In the pre-welfare reform period, the average monthly cash benefits to an AFDC family amounted to less than $500, an amount that was insufficient for lifting families out of poverty. Under the TANF program, cash benefits are even lower. The 2008 Congressional report shows an average monthly cash payment of $372 for a TANF family. In real dollars, the average monthly benefit has declined to just 65% of what it was in the late 1970s. Also note that in both eras, AFDC and TANF, the typical welfare family has been small. In 1996, the average size for an AFDC family was 2.8 (with 1.9 children); in 2006; the average size for a TANF family was 2.4 (with 1.8 children; U.S. Department of Health and Human Services 2009b).

Are welfare recipients simply lazy? No doubt some are. After all, laziness is a trait found in all social groups. However, it would be wrong to assume that the *typical* welfare recipient is a social loafer. The TANF program is essentially about assisting mothers with *dependent children.* (This was also true for the old AFDC program.) Today, fully 30% of adult TANF recipients are employed or in community service positions. Consider also that in 2006, 47% of the TANF caseload consisted of "child-only" units—families in which there are no eligible adults in the assistance unit (U.S. Department of Health and Human Services 2009b).

To be sure, the perceived positive value of work is the cornerstone of TANF: To receive temporary assistance, recipients *must* get job training or find work. At first glance, this would seem like a step in the right direction. TANF should provide some strong incentives that will move individuals from welfare to work. What could be wrong with this reasoning and policy? If you listen to the critics, there's plenty wrong.

Consider again the typical welfare family: mothers with dependent children. Critics of the welfare-to-work reform raise three key objections. First, TANF forces poor mothers to choose work *over* caring for their children. Eligibility requirements make full-time mothering, a choice for the economically advantaged, a nonoption for the poor (Ayers 2010; Boushey 2002; Jones-DeWeever et al. 2003). Perhaps this bind helps explain a newly emergent group: the no-work, no-welfare group. In an average month of 2005, there were 690,000 single mothers and 1.3 million children in families that received no earnings and no government cash support (Parrott 2008). Second, critics charge that TANF work requirements do little to combat the root cause of welfare: poverty. The disturbing truth is that many of the jobs currently available in our country—especially service-sector jobs earmarked for women—simply pay too little to keep a family from the grips of poverty (Edin and Lein 1997a; Padavic and Reskin 2002). Ten of the fastest-growing occupations through 2018 will be in the service sector and eight out of 10 of these jobs will offer low or very low median wages (U.S. Bureau of Labor Statistics 2009). Single mothers who leave welfare for work in one of these jobs will likely remain poor or near poor. The **poverty threshold** is the federal government's designation of the total annual income a family requires to meet its basic needs; this figure is adjusted annually for inflation. Currently, the poverty threshold for a family of three (e.g., a mother and two children) is $17,285—that is, a family whose income is below this amount is designated as "poor" (U.S. Census Bureau 2010c). In 2008, nearly 40 million people were below the poverty threshold (U.S. Census Bureau 2010d).

Poverty threshold
The federal government's designation of the annual income a family requires to meet its basic needs

Approximately 2.9% of women older than 16 are full-time workers earning hourly wages at or below the federal minimum wage (i.e., $7.25 an hour as of 2009). At best, these women will earn just over $13,000 a year (assuming a 35-hour work week for 52 weeks). This amount would fail to put *any* working parent above the poverty threshold. In general, those who transition from the TANF program to employment make only modest gains in income—gains due in part to the "earned income tax credit" (Schott 2009).

A third objection raised by critics of the welfare-to-work strategy is that the historic declines in caseloads achieved since the inception of TANF are occurring on the backs of the most vulnerable. In 2006, just 12.5% of the poor were TANF recipients. Over the course of its first decade, TANF served fewer and fewer *eligible* families. Indeed, fewer than half of families poor enough to qualify for TANF were actually receiving benefits. Since 2000, the percentage of children living below the poverty line has increased, yet the percent receiving assistance from TANF has declined (U.S. Department of Health and Human Services 2009b).

As counterintuitive as it may seem, accepting a minimum- or low-wage job can actually prove a costly proposition for the poor. Census data indicate that working actually hurts the chances of health care coverage for the poor: 46.7% of the **working poor** have health insurance while 61.1% of the nonworking poor have it (U.S. Census Bureau 2009d, Table HI03). Other follow-up research on newly employed welfare recipients indicates a similar outcome—new workers experience significant barriers to employment-based health insurance post-TANF reforms (Seccombe et al. 2007). For some, "work first" employment has come at the cost of pursuing postsecondary education (Jones-DeWeever and Gault 2008). Welfare-to-work programs can also have an adverse effect on family routines and effective child supervision (London et al. 2004; Scott et al. 2004).

Working poor
Individuals who have spent at least 27 weeks in the labor force but have incomes below the poverty level

Low-paying jobs also fail to provide a solution to a major obstacle facing poor female heads of households: child care. In 37 states, the average cost of a year of child support is *more than* the yearly tuition bill for a 4-year public college (Children's Defense Fund 2010). For families paying for child care, those below the poverty line devote a greater percentage of their incomes to meeting child care costs than do other families: 25% versus 7% (Johnson 2005). Research has shown that child care is central to improving employment outcomes and to families staying off welfare. Unfortunately, however, federal TANF child support grants peaked in 2003 and have been declining every year since (U.S. Department of Health and Human Services 2009b). To be sure, subsidies make a meaningful difference *if* they can be obtained. In a national study of low-income families, subsidy payments reduced their child care costs by more than half (Layzer and Goodson 2006). Yet in 20 states, families must have incomes that are below 175% of the poverty line to qualify for a public child care subsidy (Children's Defense Fund 2010). It is estimated that only 20% of those families most in need of help receive child care subsidies (Center on Budget and Policy Priorities 2010a). Without much-needed increases in TANF funding, hundreds of thousands of children in low-income families will lose access to child care assistance in the near future (Parrott 2005).

The harsh truth is that hard work does not necessarily save one from the need for government assistance. In 2008, 6% of those in the labor force were classified as "working poor"—that is, these individuals are employed but not making enough to climb above the poverty threshold. Furthermore, the majority of the working poor (55%) are poor *despite holding full-time jobs* (U.S. Department of Labor 2010d). Homelessness—perhaps the most visible marker of poverty—offers additional evidence that work and poverty are not mutually exclusive. Recent surveys have found that one-quarter (or more) of the urban homeless are employed. Their earnings, however, are *not* enough to pay for a place to live. In general, a worker would need to earn far more than the minimum wage in order to afford a one-bedroom (e.g., requires about a $15 hourly wage) or a two-bedroom (e.g., requires about an $18 hourly wage) apartment (National Coalition for the Homeless 2009b).

The plight of the working poor is especially stark when we consider the effects on the family. For example, a family of four, with both parents working full time at minimum-wage jobs, will just barely keep their collective heads above the poverty line (using the current poverty threshold figures). Change the picture to a single-parent family—where only one minimum-wage income pays the bills—and the family will certainly fall into the clutches of poverty. In 2008, the poverty rate for working-male-headed households with related children under 18 (no wife present) was 13.5% while the rate for working-female-headed households with related children under 18 (no husband present) was 30.6%. When the children at home are under 6 years old, the poverty rate for these working fathers climbs to 18.3%, and the rate for these working mothers climbs to 44.2%. The financial state of children in single-parent homes is quite precarious: Such children are much *more likely* to be poor than their counterparts in a two-parent family where both parents work (e.g., such families with children under 6 saw poverty rates of 3.3% in 2008; U.S. Census Bureau 2009b, Tables POV15, POV16). With the development of unforeseen hardships such as illness, layoffs, home or car repairs, and so on, a hardworking, low-income family could quickly fall into official poverty. A hardworking family could all too easily find itself in need of welfare.

Do those who enter the welfare system become hopelessly dependent on it? The recent overhaul of the U.S. welfare program was tied to this assumption. Yet a large body of research proves that the assumption is false. Those who turn to the government for financial help tend not to seek that assistance for very long. Despite the rather common belief that welfare is a chronic dependency condition—that is, a long-lasting condition assumed to destroy one's will to work—most people receive assistance for relatively short periods of time. Between 1995 and 2004, just over 65% of AFDC/TANF participants received assistance for 1 to 2 years; fewer than 3% of AFDC/TANF participants received assistance for 9 to 10 years. And between 2001 and 2003 (the most recent years for which information is available), half of TANF recipients spent only 4 months or less in the program (U.S. Department of Health and Human Services 2009b).

The welfare changes introduced by the TANF mandate relatively short assistance periods. In general, an adult-headed family is limited to 60 months of TANF-funded assistance *over its lifetime.* States are free to set shorter time limits if they so desire. As of 2008, 17 states have modified the time limits (Rowe and Murphy 2009). Given the mandated time limits, many TANF recipients are motivated to leave welfare as soon as they begin working in order to "stop the clock" on their lifetime eligibility. Such realities confirm that welfare is not an enticing lifelong choice.

It is also important to note that most welfare stays are not the result of being born to poverty. Rather, ordinary life events such as unemployment, illness, or divorce can easily push one into poverty. A decrease in earnings was the single most common event prompting single mothers to seek TANF assistance, according to a recent Congressional report (U.S. Department of Health and Human Services 2009b). Over the course of the current recession, the unemployment rate more than doubled as our economy lost millions of jobs. In February 2011, there were 13.7 million unemployed persons in the United States (U.S. Department of Labor 2011a). For the past 3 years, bankruptcy filings have seen annual increases on the order of 27% to 33% (U.S. Courts 2010). And since the start of the recession, approximately 9 million homes have been lost to foreclosures (Tracy 2010). Current estimates hold that about 16% of Americans lack health care insurance. In the past, employment was the path to health insurance benefits for workers, but this is no longer the case. Just under 60% of us have employer-sponsored health coverage. Forty percent of U.S. firms do not offer any health benefits (Henry J. Kaiser Family Foundation and Health Research & Educational Trust 2009; Robert Wood Johnson Foundation 2010). Given the current economic circumstances, it is rather easy to imagine the potentially devastating financial consequences for families facing a medical emergency today. Indeed, given the relative ease with which one can slip into poverty, it is actually quite remarkable that only a very small percentage of American families remain permanently poor and therefore permanently tied to welfare programs.

BOX 19.1 SECOND THOUGHTS ABOUT THE NEWS: SLACKING OFF IN THE RECESSION?

It is a belief that has remarkable staying power: Government "handouts" undermine the motivation to work. This sentiment was most recently floated by Newt Gingrich, former Speaker of the House and current aspirant to the White House in 2012. Gingrich created quite a stir by arguing that the problem with extending unemployment benefits in these recessionary times is that the assistance creates slackers–that is, people get used to being unproductive and are unwilling to accept job offers.

In making his comments, Gingrich referenced the case of Mike Hatchell, a former law enforcement officer and auto industry worker with 32 years of work experience. This individual has been unemployed for 59 weeks, and yes, he has turned down several job offers. In his defense, however, Hatchell said the offers were for positions offering entry-level wages that discounted his decades of work experience. Such low wages would mean Hatchell's failing to keep pace with his financial obligations: a home mortgage and car insurance, and daily living (e.g., food) expenses. These were outcomes he wasn't willing to risk.

Hatchell's final take on Gingrich's stance on "welfare" (Gingrich's characterization of unemployment benefits)? As someone who paid into unemployment insurance for 32 years, Hatchell feels totally entitled to collect on that insurance when it is needed. He doubts that "out of touch" politicians are able to appreciate this point.

Suppose we all agreed that most welfare recipients are hardworking, honest individuals who simply need some temporary help. Isn't it still the case that the welfare system represents too great a financial burden for our country? To answer this question accurately, we must carefully distinguish among the terms *poverty, public assistance programs* (a.k.a. *welfare programs*), and *social insurance programs* of the welfare state.

> **Poverty**
> An economic state or condition in which one's annual income is below that judged necessary to support a predetermined minimal standard of living

Poverty refers to an economic state in which one's annual income is below the threshold judged necessary to support a predetermined minimal standard of living. In 2008, the poverty rate was 13.2%, the highest it has been since 1997 (U.S. Census Bureau 2010d). To be sure, we have experienced double-digit poverty rates in the United States for the last quarter of a century. Yet the percentage of families receiving AFDC/TANF over this same time period has never climbed out of the single digits. A few years prior to TANF reforms, just over 5% of the population received AFDC assistance. In 2006, 1.6% of the population were TANF recipients. Similarly, since TANF reform, the number of children receiving benefits has been cut in half and is presently at an all-time low rate of less than 5%. Only a small percentage of the total population receives food stamps: just under 9% in 2006. For the past quarter-century, the percentage of all Americans who receive SSI (Supplemental Security Income for elderly, blind, or disabled individuals meeting income eligibility) has hovered around 2% (U.S. Department of Health and Human Services 2009b). Clearly, only a small percentage of Americans participate in "traditional" welfare programs. More surprising, however, is the fact that only a portion of those living in poverty actually look to the government for help via some public assistance program.

Public assistance programs are those directed exclusively at the eligible poor—that is, recipients must meet income and, most recently, behavioral requirements. One might think that once determined to be eligible, families would rush to receive their due. This is hardly the case. It is estimated that only 40% of families eligible for TANF enroll and receive benefits. In 2006, less than 27% of children in poverty were receiving TANF assistance. Similarly, only 59% of poor households participate in the food stamps program (U.S. Department of Health and Human Services 2009b). These figures would suggest that public assistance is a safety net that many Americans would rather live without.

> **Public assistance programs**
> Programs directed exclusively at the eligible poor

Your Thoughts . . .

Since the start of the latest recession, there has been a 58% increase in Americans receiving food stamps: 42 million (or 14% of the population) turned to SNAP (Supplementary Nutrition Assistance Program) in 2010 (Murray 2010). Do you think this one aspect of the public assistance safety net is easier for Americans to accept? If your parents were having trouble putting food on the table, would they be willing to use food stamps?

Underutilized or not, public assistance programs are costly. In 2006, Congress reauthorized funding for TANF with an annual cap of $16.5 billion. The 2011 federal expenditures (estimated) for the food stamp program totaled $75.1 billion, while SSI expenditures totaled nearly $53.3 billion. But as expensive as these programs may seem, their costs wane in comparison to other government expenditures. Consider, for instance, that the 2011 federal outlays for Social Security and Medicare were $738.4 and just over $497.3 billion, respectively. The 2011 budget allocates $738.7 billion for the Department of Defense base budget (Office of Management and Budget 2010). As these figures suggest, America's financial burdens lie not with antipoverty public assistance programs such as TANF and food stamps, but elsewhere. More specifically, America's financial woes are located in broad social insurance programs (e.g., Social Security and Medicare) as well as in security spending. **Social insurance programs** are those that require payroll contributions from future beneficiaries. Neither eligibility for nor benefits from these programs are linked to financial need.

Social insurance programs
Programs that require payroll contributions from the future beneficiaries; neither program eligibility nor benefits are linked to financial need

In reality, social insurance programs—not antipoverty public assistance programs—are our largest "welfare" expenditures. Indeed, these social insurance programs constitute the largest part of what we have come to define as America's "welfare state"—a state that transcends poverty per se and instead offers protection based on our more general rights of citizenship (Bowles and Gintis 1982). In 2010, only 14% of the federal budget was devoted to poverty-linked "safety net" programs while 21% was dedicated to Medicare/Medicaid/CHIP (Children's Health Insurance Program), 20% to Social Security, 20% to Defense and Homeland Security, and 19% to paying for a variety of public services. The last 6% was earmarked for paying the interest on our national debt (Center on Budget and Policy Priorities 2010b).

When we turn to facts and figures, we can quickly discredit the conventional wisdom on welfare. Yet we must also concede that year after year, conventional wisdom on this subject overpowers facts and figures. Why?

Some of our misconceptions regarding welfare no doubt are fueled by the profound changes occurring in the economic and occupational structure of our society. Americans today face a growing gap between the rich and poor—a gap that is setting all-time highs and lows. Simply put, the rich are getting richer, and the poor are getting poorer (see Table 19.1). This long-standing trend makes the United States the most unequal democracy in the world (Boshara 2003). A recent Congressional Budget Office study found that over the last quarter-century, the top 1% of families saw their *after-tax* incomes rise 281%, the top fifth saw gains of 95%, and the middle fifth experienced a 25% gain. Between 1997 and 2007, the after-tax share of the nation's total household income rose from 7.5% to 17.1% for the richest 1% of Americans while the share for the bottom four-fifths declined from 58% to 48%. The share going to the bottom fifth Americans shrank from 6.8% to 4.9%. Bush-era tax cuts have served to increase the concentration of income in the top income groups and exacerbate income inequality in the United States. In 2007, the bottom fifth of income groups received tax cuts that averaged $29 and raised after-tax income by 0.4%. The middle fifth received cuts that averaged $760 and raised after-tax income by 2.4%. The top 1 percent received tax cuts that averaged $41,077 and raised after-tax income by 5% (Sherman and Stone 2010).

Another telling way to understand the concentration of wealth in the hands of the few is by looking at the increases in millionaires and decamillionaires (individuals worth $10 million or more). In the past 25 years, the number of millionaires in the United States has more than doubled and in 2007 numbered just over 7 million. During this same period, the number of decamillionaires increased by nearly 600%, and they currently number approximately 464,000 (Wolff 2010).

In the face of present economic circumstances, intergenerational upward mobility is no longer a birthright for most Americans.

Table 19.1 Average After-Tax Income by Income Groups—1979–2007

	After Tax Income—1979	After Tax Income—2007	% Change 1979–2007
Top 1%	$346,600	$1,319,700	281%
Top one-fifth	$101,700	$198,300	95%
Next one-fifth	$57,700	$77,700	35%
Middle one-fifth	$44,100	$55,300	25%
Next one-fifth	$31,000	$38,000	23%
Bottom one-fifth	$15,300	$17,700	16%

Sources: Sherman and Stone (2010) Center on Budget and Policy Priorities. http:www.cbpp.org/cms/index.cfm?fa=view&id=3220.

Intergenerational upward mobility
Social status gains by children vis-à-vis their parents

Intergenerational upward mobility refers to social status gains by children vis-à-vis their parents. In the 1950s, the average male worker could expect a 50% increase in income over the course of his working lifetime. This expectation is no longer a safe bet for the average worker. Rather, present-day workers find themselves competing in a global economy. As a result of this turn of events, many American jobs have been lost to other countries, and wages for low-skill jobs have suffered a marked decline. Furthermore, present-day workers find themselves in an occupational landscape that is increasingly dominated by computer-related, health-related, and service-industry jobs. The occupations that are expected to see the *largest numerical growth* in job opportunities are not sure routes to great wealth. Fully seven of the projected top 10 occupations (through 2018) for future employment offer low or very low earnings (see Table 19.2). Looking ahead to 2018, it is generally the lowest-paying occupations (all of which only require on-the-job training) that will promise the greatest number of jobs (e.g., home health aide jobs, food preparation/serving jobs, and personal and home care aide jobs). Consequently, many of those entering the labor force in the next few years will be faced with a serious challenge to upward mobility, especially if they are transitioning out of TANF or are without higher education.

Shifts in the occupational and economic structure of our society have shaken the very core of American values. Americans invest heavily in the idea that they can and will work their way up the socioeconomic ladder; the economic shifts described here threaten that belief. The growing inability to achieve the American Dream has left many people frustrated and searching for someone to blame. In this regard, the poor—and welfare recipients in particular—are easy scapegoats. The logic of the welfare system is completely inconsistent with fundamental American values: individual effort, equal opportunity, success, and upward mobility. Rather than promoting hard work and achievement, welfare programs are thought to institutionalize qualities that are directly opposite. In this way, welfare recipients come to constitute an out-group in our society (Feagin 1975; Lewis 1978). An **out-group** is considered undesirable and thought to hold values and

Out-group
Group that is considered undesirable and is thought to hold values and beliefs foreign to one's own

Table 19.2 Top 10 Occupations With Largest Numerical Growth Through 2018

Occupations	Median Wages 2008	Education/training
1. Registered Nurse	$62,450	Associates degree
2. Home Health Aide	$20,460	On job-training
3. Customer Service Rep	$29,860	On-job-training
4. Food Prep/Serving (including fast foods)	$16,430	On-job-training
5. Personal and Home Care Aide	$19,180	On-job-training
6. Retail Salesperson	$20,510	On-job-training
7. Office Clerk	$25,320	On-job-training
8. Accountants/Auditors	$59,430	Bachelor's degree
9. Nursing Aides/orderlies	$23,850	Vocational degree
10. Postsecondary teachers	$58,830	Doctoral degree

Source: U.S. Department of Labor, U.S. Bureau of Labor Statistics. 2010. *Occupational Outlook Handbook, 2010–11 Edition: Overview of the 2008–18 Projections*. Available online at http://www.bls.gov/oco/oco2003.htm.

In-group
A group whose members possess a strong sense of identity and group loyalty and hold their group in high esteem

beliefs foreign to one's own. An out-group is identified as such by an **in-group**, which holds itself in high esteem and demands loyalty from its members.

Individuals who can avoid public assistance, regardless of their exact income, can count themselves as members of the hardworking in-group. Indeed, it is the negative image of the welfare out-group that keeps many poor and near poor from accessing various forms of public assistance: "I may be poor, but I'm not on welfare."

The power of American values explains why we cling to conventional wisdom regarding welfare. This power also can help us better understand why relatively few Americans denigrate society's wealthy sector, even when that wealth is gained at the expense of the working and middle class. After all, a rising tide lifts all boats. The General Social Survey (GSS) has repeatedly found that roughly three-quarters of Americans think that class differences are acceptable because they reflect what people have made of their opportunities; and about six in 10 Americans agree that large income differences are needed to provide incentives for individual effort (GSS 2010).

To be sure, those on the upper rungs of the U.S. stratification ladder are accumulating great wealth. Over the past 30 years, we have witnessed a 2,500% rise in CEO incomes (Krugman 2002). And the trend continues, even in the midst of a major economic recession. In 2008, the average salary of top corporate executives was 319 times higher than that of the average hourly worker! In 2009, CEOs of our nation's largest corporations received an average of $9.25 million in total compensation. And the accumulation of wealth continues even after the work is done—*or not done.* CEOs of the top 10 recipients of federal bailout money were paid an average of $25 million despite taking their companies to the brink of bankruptcy (Parks 2009).

When we value individual effort and opportunity, tolerance of wealth must be expected. We embrace the old adage that "what's good for GM (or more recently GE) is good for America." (In his last full year running GE, CEO Jack Welch was paid $123 million; his retirement package included perks worth at least $2 million a year [Krugman 2002]. After his retirement, Welch went on to write a memoir sharing his secrets to success; it became a best-seller.) The wealthiest have already arrived where many of us would like to go. They are proof to us that individual efforts can pay off; they are proof that the American Dream, to which we are so committed, lives on.

To sustain the power of American values, we must lay the blame for poverty on the poor themselves. There is, of course, a certain irony and destructiveness to this process. Personalizing poverty deflects our attention from the social causes of poverty, such as the changing occupational structure and a lack of education. Such a stance lessens the likelihood that we will successfully reduce poverty. Indeed, without major social changes, social reproduction theory suggests that the American Dream will continue to elude the poor. **Social reproduction theory** maintains that existing social, cultural, and economic arrangements work to reproduce in future generations the social class divisions of the present generation. Princeton economist Paul Krugman identifies corporate culture and boardroom handshakes as the invisible force behind the recent explosion in CEO pay (Krugman 2002).

Social reproduction theory
The view that existing social, cultural, and economic arrangements work to reproduce in future generations the social class divisions of the present generation

One proponent of social reproduction theory, Pierre Bourdieu (1977a, 1977b), maintains that the aspirations of lower-class children are adversely affected by their class position. The lower-class child is immersed in a social world hostile to the American Dream. The objective realities of the lower-class environment deflate hopes of success; the restricted opportunity structure inherent in a lower-class location leads to reduced life aspirations.

Social reproduction is not the only obstacle to poverty reduction. Structural functionalists remind us that the elimination of poverty is highly unlikely as long as the poor among us serve valuable social functions. **Structural functionalism** is a theoretical approach that stresses social order. Proponents contend that society is a collection of interdependent parts that function together to produce consensus and stability. **Social functions** refer to the intended and unintended social consequences of various behaviors and practices.

Structural functionalism
A theoretical approach that stresses social order and that views society as a collection of interdependent parts that function together to produce consensus and stability

Social functions
Intended and unintended social consequences of various behaviors and practices

Personalizing poverty sustains a lower class. The lower class, in turn, fulfills many needs for those in other social locations. For example, the poor provide society with a cheap labor pool. They also create countless job opportunities for others: for those wishing to help them—social workers, policymakers, and so on—as well as for those wishing to control them—police and corrections officers. The poor even provide financial opportunities for those wishing to take advantage of them—loan sharks, for example, and corporations seeking tax breaks via the food discard market, and the Wall Street players who saw the lucrative returns on selling and bundling subprime loans (Aaron 2009; Funiciello 1990; Gans 1971; Jacobs 1985). And sustaining the poverty out-group enables the social mainstream to better define and reaffirm some of its most fundamental values and beliefs.

In light of these functions, we must re-examine the notion that welfare is ruining this country. Welfare may breed dependency, but dependency for whom? Given the social functions of the poor, welfare may breed a social dependency of the masses on the few, as the poor ultimately serve as vehicles by which mainstream values are ensured.

LEARNING MORE ABOUT IT

Alison Davis-Black and Joseph Broschak consider outsourcing and examine its impact on the attitudes and behaviors of workers in their article "Outsourcing and the Changing Nature of Work" (*Annual Review of Sociology*, 35: 321–340, 2009).

A comprehensive review of the impact of the economy on the living standards of Americans is offered by the Economic Policy Institute's *The State of Working America 2006/2007* (Ithaca: Cornell University Press, 2007).

In *The Working Poor* (New York: Vintage Books, 2005), David Shipler profiles some of those caught between poverty and prosperity and in so doing challenges many common misconceptions about those living in or near poverty in the United States.

If you want to learn more about the reality and challenges of working for poverty-level wages, see Barbara Ehrenreich's best-selling *Nickel and Dimed: On (Not) Getting by in America* (New York: Owl Books, 2001).

To read more about the gender and racial base of poverty, see Kathryn Edin and Laura Lein's *Making Ends Meet: How Single Mothers Survive Welfare and Low-Wage Work* (New York: Russell Sage, 1997).

In *Rachel and Her Children: Homeless Families in America* (New York: Crown, 2006), Jonathan Kozol uses both statistics and rich interview data to provide a compelling account of homelessness in America.

The classic work on the functions of poverty is Herbert Gans's "The Uses of Poverty: The Poor Pay for All" (*Social Policy*, Summer: 20–24, 1971).

GOING ONLINE FOR SOME SECOND THOUGHTS

The following organizations and sites can also help you learn more about welfare and related topics:

The United States Conference of Catholic Bishops offers a short but insightful poverty "tour" at its website: http://www.usccb.org/cchd/povertyusa/index.htm.

The Administration for Children and Families' website offers a link for information about a wide variety of social support issues: http://www.acf.hhs.gov/.

The Center on Budget and Policy Priorities offers valuable information on public policies for low- to moderate-income families and individuals: http://www.cbpp.org/about/.

National Coalition for the Homeless provides basic information and a variety of fact sheets about homeless in the United States: http://www.nationalhomeless.org/.

NOW IT'S YOUR TURN

1. American values are one explanation for the triumph of conventional wisdom over facts. Choose another concept from the material covered thus far in your course and provide an alternate explanation of why welfare gets such a bum rap.

2. The cutoff level for official poverty is arbitrary. Identify five different consequences of setting the cutoff point higher; identify five consequences of setting the cutoff point lower. Are the consequences you identify primarily functional or dysfunctional for mainstream Americans?

3. Compare Table 19.2 in this chapter with Table **22.5** in the education chapter. Why are the occupations listed in each table different? What are some of the implications of identifying future job prospects in these two ways?

4. Take the "poverty tour" offered on the United States Conference of Catholic Bishops' website: http://www.usccb.org/cchd/povertyusa/index.htm. Next, update and customize the tour to reflect the realities of living in your geographic area. For instance, consider the local costs of housing, utilities, food, and so forth. Assess the likelihood of a family of four with poverty level income to survive in your area.

ESSAY 20

Conventional Wisdom Tells Us . . . Immigrants Are Ruining This Nation

"Why don't you go back where you came from?" This angry cry seems to be getting more and more familiar as the United States faces the highest levels of immigration in its history. Is immigration ruining this nation? This essay reviews the historical impact and future trends of immigration in the United States.

Prejudice
A prejudgment directed toward members of certain social groups

Immigrant groups
Groups containing individuals who have left their homelands in pursuit of a new life in a new country

"Why don't you go back where you came from?" This is a familiar taunt that most of us have heard. Here in the United States, it is a question often born of ethnic and racial prejudice. And, increasingly, feelings of prejudice target members of immigrant groups. **Prejudice** refers to the prejudgment of individuals on the basis of their group membership. **Immigrant groups** contain individuals who have left their homelands in pursuit of new lives in new countries.

Immigration has always been a fact of American life. The earliest European settlers were immigrants to the 3 to 8 million Native Americans who already occupied the continent. In the first census, in 1790, approximately one in five Americans was an "immigrant" slave brought from Africa (U.S. Census Bureau 2002). And the 1850 census (the first to collect data on nativity) noted that 9.7% of the U.S. population was foreign born. The Mayflower Society claims that presently there are tens of millions who are able to trace their heritage to the original 102 passengers landing near Plymouth, Massachusetts, in 1620 (Mayflower Society 2007). Our immigration past has not been lost on our leaders. Franklin Roosevelt observed (and the point was later reiterated by the great communicator Ronald Reagan), "All of our people all over the country—except the pure-blooded Indians—are immigrants, or descendants of immigrants, including even those who came over on the *Mayflower*."

The U.S. immigration experience can be divided into four major waves. The earliest wave consisted mostly of English immigrants who arrived in the United States long before official entry records started to be recorded in 1820. This first group also consisted of immigrants from Scotland, Ireland, Germany, and other northern and western European nations. The second major wave occurred between 1820 and 1860 as many people who were being pushed out of Europe by forces of industrialization relocated to the United States and joined the westward expansion of the nation. About 40% of the immigrants who arrived during this period were from Ireland alone. The third wave occurred between 1880 and the outbreak of World War I. During this period, more than 20 million immigrants from southern and eastern European nations arrived in the United States. Most of these immigrants moved to east-coast and midwest cities. By 1910, more than half of the workers in New York City, Chicago, and Detroit were immigrants. These three waves of activity were followed by an immigration lull (between 1915 and 1965) brought about by two world wars, the Great Depression in the United States, and the appearance of U.S. immigration quotas. We are currently in the midst of a fourth wave of immigration that started after 1965. This wave has a decidedly different look than previous waves and reflects a change in U.S. immigration policies that gives preference to family members of those already residing in the United States and to skilled workers in demand by U.S. employers. With this fourth wave, the predominant origins of immigrants shifted from European to Latin American and Asian countries (see Figure 20.1). The current wave also is distinguished by an increase in the number of unauthorized immigrants entering the United States (Martin and Midgley 2010; Martin and Zurcher 2008). (See Table 20.1.)

The United States has the most foreign-born residents of any country, and immigrants account for at least one-third of recent U.S. population growth. While the current *percentage* of foreign-born residents is around 13% (a figure lower than that for the early 1900s), the *number* of foreign-born residents reached 39 million in 2009. About 30% of this group are unauthorized residents—that is, **illegal aliens**. (The United States also has the most unauthorized residents of any country; Martin and Midgley 2010.) While more than half of these illegal immigrants evaded border controls when entering the United States, about 45% entered legally but overstayed the terms of their visas (Martin and Midgley 2010; Martin and Zurcher 2008). Between 1990 and 2009, approximately 19 million immigrants became **legal permanent residents (LPRs or "green card recipients**"; see Box 21.1 for other immigration terms) of the United States (U.S. Department of Homeland Security 2010). Approximately two-thirds of legal permanent immigrants are family sponsored—that is, they are relatives of U.S. citizens or of permanent immigrants (Monger 2010). Despite our nation's immigration background, public opinion polls between the 1960s and 1990s indicated that a majority of Americans were in favor of reducing immigration levels in the United States. This restrictive sentiment abated somewhat

Table 20.1 Daily Foreign Visitors/Settlers per day in the United States

Approximately 100,000 foreigners arrive in the United States each day:

- 3,100 foreigners arrive with immigration visas (and on path to possible naturalized citizenship)
- 99,200 foreigners arrive as tourists and business and student visitors

Approximately 2,000 unauthorized foreigners settle in the United States each day:

- More than half enter illegally.
- The rest are visitors who violate their visas.

Source: Martin, P. and E. Midgley. 2010. *Population Bulletin Update: Immigration in America 2010.* http://www.prb.org/Publications/PopulationBulletins/2010/immigrationupdate1.aspx.

Illegal aliens
Foreigners in the United States without valid visas

LPRs
Legal Permanent Residents

during the economic good times of the late 1990s, but after 2001 (and 9/11), Americans once again began expressing greater concern about immigration issues and asking political leaders for immigration reform. By the spring of 2010, at a time when the United States was still faced with high unemployment and a challenging economy, a Gallup poll found that 10% of Americans (16% of those living in the West; 14% of Republicans) identified immigration as the *most important* problem facing the nation. These numbers were no doubt fueled in part by the controversy sparked by an Arizona law designed to fight illegal immigration by making it a crime to be in the United States without proper documentation. The law also allows state and local police to detain those suspected of being illegals unless they can produce evidence to the contrary. (A Gallup poll found that 51% of Americans who knew about the Arizona law were in support of it; Jones 2010a).

Your Thoughts . . .

In the summer of 2010, a debate arose over the issue of "anchor babies"—that is, children born in the United States to illegal immigrant mothers. These children, by virtue of their U.S. births, are citizens of the United States and thus provide permanent U.S. ties for other family members. Some proponents of immigration reform question whether anchor babies are entitled to citizenship given the illegal status of their mothers. What do you think?

BOX 20.1 IMMIGRATION TERMS

Aliens: citizens of a foreign country. The United States distinguishes four legal statuses for aliens: legal immigrants, temporary legal migrants, refugees, and unauthorized migrants.

Legal immigrants (a.k.a. legal permanent residents [LPRs] or green card holders): foreigners granted visas that allow them to live and work permanently in the United States. After 5 years, legal immigrants can apply to become naturalized citizens. There were more than 1.1 million legal immigrants admitted to the United States in 2009.

Naturalized citizens: legal immigrants (at least 18 years of age) who have lived in the United States for at least 5 years, paid application fees ($675), undergone a background check, and passed English and civics tests. In 2008, we saw a record high in naturalized citizens: 1,046,539.

Temporary legal migrants: foreigners in the United States for specific purposes (e.g., visiting, studying, or working). The United States issues 25 types of nonimmigrant visas, including B-visas for tourists, F-visas for foreign students, and H-visas for foreign workers. There were 33.6 million temporary legal migrants in the United States in 2006.

Refugees and asylees: foreigners allowed to stay in the United States because of fear of persecution in their home countries. Refugees may become legal permanent residents after living in the United States for a year. There were 177,368 refugees/asylees in the United States in 2009.

Unauthorized migrants (a.k.a. illegal aliens): foreigners in the United States without valid visas (estimated at 11.5 million in 2006).

Source: Lee 2010; Martin and Midgley 2006; Monger 2010.

Given America's immigration history, the current calls for reform seem somewhat ironic. We find ourselves casting doubt on the value of immigrants in a nation long considered the "land of immigrants." To be sure, our immigrant roots are well established (see Table 20.2). As evidenced by annual parades and festivals, a great many "hyphenated-Americans" take pride in their diverse ancestral roots. At the same time, our daily newspapers, television newscasts, and Internet blogs document a growing chorus of anti-immigration sentiments. Commentaries filled with fear, distrust, and hate are becoming a staple of talk-radio broadcasts. The Federation for American Immigration Reform (FAIR) has called for severe reductions in the number of U.S. immigrants to combat excessive population growth, environmental degradation, job loss, and low wages. FAIR works closely with talk radio to get its immigration reform message out to the American people. The tragic events of September 11, 2001, have prompted many Americans to rethink the wisdom of freely embracing foreigners. Building a 700-mile wall along the Mexican border, increasing the number of border patrol agents, assigning National Guard troops to border patrols, making illegal crossings a felony, and issuing ID cards are all part of the current immigration debate. Not surprisingly, immigration reform is a pledge heard from many political candidates, especially aspiring presidential candidates. But the reform-minded are not waiting for the federal government to act. Many critics maintain that immigrants place an unfair burden on state budgets. In the first quarter of 2010, 45 states introduced 1,180 immigration-related bills (National Conference of State Legislatures 2010).

Table 20.2 Total and Percentage Foreign Born for U.S. Population: 1900–2009 (Numbers in Thousands)

	Foreign-Born	
Year	**Total**	**Percentage**
1900	10,341	13.6
1910	13,516	14.7
1920	13,921	13.2
1930	14,204	11.6
1940	11,595	8.8
1950	10,347	6.9
1960	9,738	5.4
1970	9,619	4.7
1980	14,080	6.2
1990	19,767	7.9
2000	31,108	11.1
2008	37,961	12.5
2009	38,517	12.5

Source: Migration Policy Institute Data Hub: Foreign-Born Population and Foreign Born as a Percentage of the Total U.S. Population, 1850–2009. Available online at http://www.migrationinformation.org/datahub/charts/MPIDataHub-Number-Pct-FB-1850.xls.

Is the current conventional wisdom on immigration justified? Do these sentiments reflect a new anti-immigration trend? Or are these anti-immigration sentiments more common and longstanding than we realize?

Your Thoughts . . .

During the most recent debate over immigration reform, a black-eyed "Dora the Explorer" showed up in a fake mug shot for illegally crossing the border and resisting arrest. (Find the picture at http://www.huffingtonpost.com/2010/05/21/dora-the-explorer-illegal_n_584541.html.) Do you think this spoof on the very popular child cartoon character offers any valuable insight into age differences regarding the immigration debate in the United States?

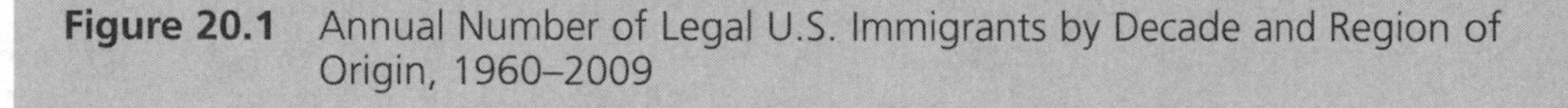

Figure 20.1 Annual Number of Legal U.S. Immigrants by Decade and Region of Origin, 1960–2009

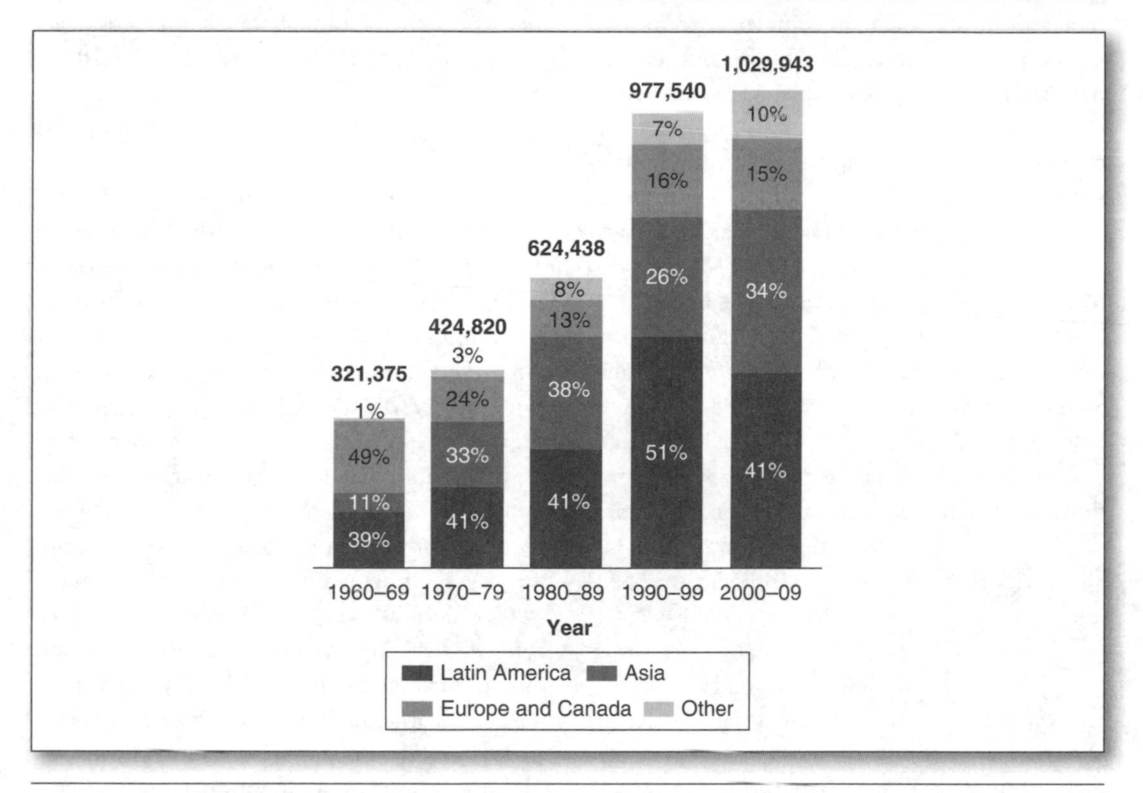

Source: Population Bulletin Update: Immigration in America 2010. Available at http://www.prb.org/Publications/PopulationBulletins/2010/immigrationupdate1.aspx.

The immigration history of the United States is nothing if not complex. Despite the message delivered by the Lady in the Harbor, the United States has seldom greeted immigrants with totally open arms. Descendants of the first immigrant settlers, white Anglo-Saxon Protestants from England, were slow to welcome other newcomers. Rather, they expressed concern about "new" and undesirable immigrants and organized against those arriving from Germany, Ireland, Poland, Italy, and other white ethnic countries (Fallows 1983). The 1850s saw the rise of a political party—the Know-Nothings—whose unifying theme was decidedly anti-immigration. In the 1860s, James Blaine of Maine sought to curb Catholic immigration by seeking an amendment to the U.S. Constitution banning states from providing aid to schools controlled by religious groups. His efforts established the groundwork for restrictions on government aid to religious schools and the current school voucher debate (Cohen and Gray 2003). In the same decade that the Statue of Liberty first beckoned immigrants to our shores, a group of U.S. residents founded the first all-WASP (white Anglo-Saxon Protestant) country club; these residents also established the Social Register, a list identifying the exclusive "founding" families of the United States (Baltzell 1987). In the 1920s, President Hoover freely expressed clear anti-immigration sentiments when he encouraged New York City's mayor, Fiorello La Guardia, the son of immigrants, to go back where he belonged. The mass migration period of 1880 to 1924 saw nativists and politicians working hard to restrict immigration. Campaigns to impose literacy tests in order to hold the tide on immigration were repeatedly mounted in the late 1800s and early 1900s. And from 1921 to the mid-1960s, the government used a quota system to regulate and limit immigration. In 1965, Congress passed a new law that replaced quotas with a complex system that grants priorities to three categories: foreigners with relatives living in the United States, people needed to fill vacant jobs, and refugees. These changes produced a major shift in immigration patterns (more and more immigrants originating in Latin America and Asia) and renewed calls for immigration reform (Martin and Midgley 2006). These examples suggest that, although the United States proudly touts its immigration history, immigration in the United States has always been characterized by a love–hate relationship.

Americans' love–hate stance toward immigration may be the product of certain core cultural values. A **cultural value** is a general sentiment that people share regarding what is good or bad, right or wrong, desirable or undesirable.

Cultural value
A general sentiment regarding what is good or bad, right or wrong

We are a nation strongly committed to economic opportunity and advancement. At various times and to various parties, the labor of immigrants has provided one sure route to economic betterment. For example, estimates suggest that nearly half of colonial-era European immigrants came to America as indentured servants who were willing to work off their debts for a chance at a better life in the new land. Similarly, the forced immigration of African slaves provided cheap labor for the South's labor-intensive agricultural development (Daniels 1990). The construction of the transcontinental railroad and the economic development of the West depended on the willing and able labor of Chinese immigrants, and Japanese immigrants were welcomed as cheap, reliable labor for Hawaiian sugar plantations. By 1910, immigrants constituted 14% of our national population, yet they made up more than one-half of the industrial labor force (National Park Service 1998). During World Wars I and II, young Mexican men were invited into the United States as guest workers (via the Bracero program) to help fill the farm labor shortages that developed as American soldiers were shipped overseas for combat duty (Martin and Midgley 2006). In short, immigration has benefited many U.S. enterprises, industries, corporations, and war campaigns.

The ties that link immigrants to traditional American cultural values have not only benefited big business but also advanced the lives of countless immigrants. Indeed, the crush of immigrants in the mid-19th century was prompted by the immigrants' hopes that they could escape their own poverty via the economic expansion that was taking place in the United States. Such promises of economic betterment continue to attract immigrants to our shores, even amid current trends toward economic globalization. The promise proves a potent one. Throughout the early 1990s, even the lowest-paying jobs in the United States were an improvement over those most new immigrants left behind. A day's wage for an unskilled worker in Mexico comes to 47.60 pesos (or $3.65 in U.S dollars). Compare this with $58 a day one can earn at a minimum-wage job in the United States. A minimum-wage job in China can bring in 270 yuan a month (or roughly $39). A minimum-wage job in the United States yields about $1,160 a month (International Labour Organization 2010). In short, work in America is a winning proposition and a pull for many workers of the world. See Table 20.3 for some compensation figures for production workers of various countries.

The cultural emphasis on economic advancement and opportunity helps explain the affinity between the United States and immigrants, but such emphasis also helps explain our long history of resisting immigrants. American tolerance for immigrants decreases whenever immigrants prove a threat to the economic well-being of "traditional" American workers. Indeed, the strongest support for immigration restrictions has often come from organized labor (Schuman and Olufs 1995).

Recall that Chinese immigrants played a critical role in the construction of the transcontinental railroad. However, the Chinese Exclusion Act of 1882 was passed when Chinese immigrants began to be viewed as a threat to the white labor force. Similarly, Mexicans were welcome immigrants to the United States during the labor shortages imposed by World Wars I and II and again during the farm labor shortages of the 1950s. However, in the 1960s and the 1980s, when traditionally white labor jobs were in jeopardy, attempts were made to stem the flow of Mexican immigrants (Martin and Midgley 2006; Schuman and Olufs 1995).

In the 1990s, economic changes created a double bind for the traditional American workforce. Specifically, many low-wage jobs left the United States for more profitable locations abroad. (Again, see Table 20.3 to understand the financial incentives that tempted employers to move their jobs out of the United States.) At the same time, more and more foreign workers entered America and offered direct competition for the low-wage jobs that remained. Add to the mix one other development emerging from the economic boom of the 1980s and the limited number of American students in the science and engineering fields: a growing demand for highly skilled temporary workers. College-educated immigrants are eligible for tens of thousands of work permits (H-1B visas) each year. These immigrants are very much in demand by American businesses, especially high-tech industries. The number of H-1B visas doubled during the 1990s and totaled just under 410,000 in 2008 before decreasing (by 17%) to 339,000 in 2009 (Monger and Barr 2010). The demand for the skilled H-1B visa holder remains strong among high-tech firms and these visa holders can stay in the United States as long as employers are willing to sponsor them. But opponents to this influx of highly skilled immigrants argue that the group only serves to undermine the job opportunities, wages, and educational pursuits of their American counterparts (Martin and Midgley 2006; Martin and Zurcher 2008; Monger and Barr 2010). Thus, the immigration squeeze on American workers is felt by those at both ends of the skills spectrum. (In 2009, about 15% of U.S. workers were foreign born; Martin and Midgley 2010.) Clearly, these economic realities play a significant role in fueling current anti-immigration sentiments.

Table 20.3 Hourly Compensation Costs for Manufacturing Production Workers in Select Locations, 2011 (in U.S. Dollars)

Country	Dollars per Hour
Philippines	1.50
Mexico	5.38
Poland	7.50
Taiwan	7.76
Brazil	8.32
Hungary	8.62
Czech Republic	11.21
Singapore	17.50
United States	33.53

Source: U.S. Department of Labor, Bureau of Labor Statistics. 2011b. "International Comparisons of Hourly Compensation Costs in Manufacturing, 2009." http://www.bls.gov/news.release/pdf/ichcc.pdf

Americans' anti-immigration attitude is further explained by some basic processes of group dynamics. We refer specifically to conventional patterns by which in-groups and out-groups develop. The people who constitute the group to which one belongs form an in-group. An **in-group** holds itself in high esteem and demands loyalty from its members. In-groups

In-group
A group whose members possess a strong sense of identity and group loyalty and hold their group in high esteem

then define others as members of an out-group. An **out-group** is considered undesirable and thought to hold values and beliefs foreign to one's own. American society consists of a variety of ethnic groups. These groups are frequently ranked relative to their tenure in the country, but each wave of immigration to a country establishes new population configurations. In general, the most established immigrant groups cast themselves in the role of the in-group. Such groups define themselves as the "senior" and thus most valid representatives of a nation. These in-groups cast those that follow them in the role of out-groups. Recent arrivals are stigmatized as elements foreign to an established mold (Spain 1999).

Out-group
Group that is considered undesirable and is thought to hold values and beliefs foreign to one's own

Research demonstrates that members of an in-group carry unrealistically positive views of their group. At the same time, in-group members share unrealistically negative views of the out-group (Hewstone, Rubin, and Willis 2002; Tajfel 1982). As groups improve their status, they are more likely to display in-group bias (Guimond, Dif, and Aupy 2002). Threats to in-groups can increase the derogation of lower-status out-groups (Cadinu and Reggiori 2002; Hopkins and Rae 2001). Low-status minorities are the most susceptible to in-group devaluation (Rudman, Feinberg, and Fairchild 2002). Exposure to prejudice about out-groups can increase the negative evaluations of those groups and hinder social ties with them, thus reinforcing destructive social dynamics (Levin, VanLaar, and Sidanius 2003; Tropp 2003). Because newcomers are viewed relative to those with earlier claims, the very process of immigration perpetuates social conflict. Indeed, the mechanics of immigration seem to guarantee a hostile boundary between the old and the new, between established ethnic groups versus recent arrivals.

In light of Americans' historical relationship with immigrants, should we simply dismiss current anti-immigration sentiments as "business as usual"? Perhaps not.

As the United States entered the 20th century (1900–1909), 92% of immigrants were of European origins. Indeed, these European origins are frequently credited with facilitating past immigrants' transition to U.S. culture. The many shared customs and characteristics and the spatial dispersion of various European ethnic groups facilitated the assimilation of each new European immigrant wave. **Assimilation** is the process by which immigrant groups come to adopt the dominant culture of their new homeland as their own. But the last 100 years have seen a dramatic change in regional background of immigrants to the United States: In 2009, only 10% of immigrants had European roots (U.S. Department of Homeland Security 2010).

Assimilation
The process by which immigrant groups come to adopt the dominant culture of their new homeland as their own

BOX 20.2 SECOND THOUGHTS ABOUT THE NEWS: BUILD IT AND THEY WON'T COME

The topic of immigration reform has been in the news for the last several years, providing fodder for both national and local politicians. A proposed solution to illegal crossings from Mexico that gained some traction a few years back was the building of a fence along the United States–Mexico border. Congress approved the plan in 2006 and to date 3 billion taxpayer dollars have been spent on the yet-to-be-completed project—the barrier currently covers only a third of the 2,000-mile border. (*The Fence* is a short documentary film about the project.)

More recently, states have been taking matters into their own hands and pursuing other options for curbing the flow of illegal immigrants. In the summer of 2010, Arizona passed a law that would permit law enforcement officers to ask for proof of legal residency from those they suspect are in the country illegally. Those unable or unwilling to comply with the request will face deportation. Public sentiment is with the new law and several other states are considering similar legislation. Local, national, and even international critics of the Arizona law argue that it amounts to racial or ethnic profiling and will encourage discrimination against people with Hispanic backgrounds. President Obama and the Department of Justice maintain that immigration reform is a federal, not a state, issue. Defenders of state action counter that the federal government has neglected its responsibility.

The states are on point here. The late Senator Edward Kennedy issued a call for comprehensive immigration reform in the 1960s. Comprehensive reform entails more than fences and deportation. Employer demands for cheap labor, global outsourcing of jobs, educational reform, guest worker programs, and even humanitarian amnesty programs all need to be on the table. To date, however, Congress has failed to answer the call for such comprehensive reform. Still, the American public wants Congress to act—securing the borders is ranked just behind job creation as a top priority for the federal government (Jones 2010b). The prospects for federal action are somewhat dubious, though, given the current state of political partisanship and obstructionism. For now, it appears that immigration reform will remain in the news for some time to come.

Today, the vast majority of immigrants are from Asian or Latin American nations. Table 20.4 lists the top 10 countries of origin for legally permanent immigrant residents in 2009. By the year 2050, most immigrants will hail from Latin America, Asia, Africa, the Middle East, or the Pacific Islands (Schmidley 2003). By 2050, Hispanics are expected to make up 25% and Asians 10% of the U.S. population (Martin and Midgley 2006). On the one hand, these shifts in immigration and population patterns should hardly be cause for concern. The immigration history of these new groups, like the history of previous immigrant groups, has largely been a story of success. Asian men and women, for example, have the highest median income of *any* single racial group. In 2007, the median income for Asian families was nearly $13,000 higher than the median income for white families (U.S. Census Bureau 2011, Table 696). In 2008, the median value of Asian owner-occupied homes was approximately $400,000 (about twice the median value for non-Hispanic white owner-occupied homes). In these challenging recessionary times, Asian Americans are experiencing a lower unemployment rate than whites (7.5% vs. 8.7%; Jacobsen and Mather 2010). For those 25 years of age and older, 50% of Asian Americans versus 31% of whites have bachelor's degrees or higher (Jacobsen and Mather 2010). Approximately 49% of Asians aged 16 years and older work in management or professional occupations; 38% of whites 16 and older are so employed (U.S. Department of Labor 2010e: Table 10).

Hispanic immigrants to the United States can point to similar triumphs. Although it is true that the economic conditions of Hispanic Americans tend to lag behind national averages, the median income of Hispanic households was just over $40,000 in 2007 (U.S. Census Bureau 2011: Table 696). The percentage of Hispanic households with incomes more than $100,000 has more than doubled since 1980 (U.S. Census Bureau 2011, Table 689). There were 2.3 million Hispanic-owned businesses in 2007 and they generated more than $345 billion in receipts (U.S. Census Bureau 2010e). The educational profile of Hispanics has improved greatly in recent years. The status dropout rates (i.e., the percentage of 16- to 24-year-olds who are not enrolled in high

Table 20.4 Top 10 Countries of Origin for Legal Permanent Immigrants to the United States, 2009

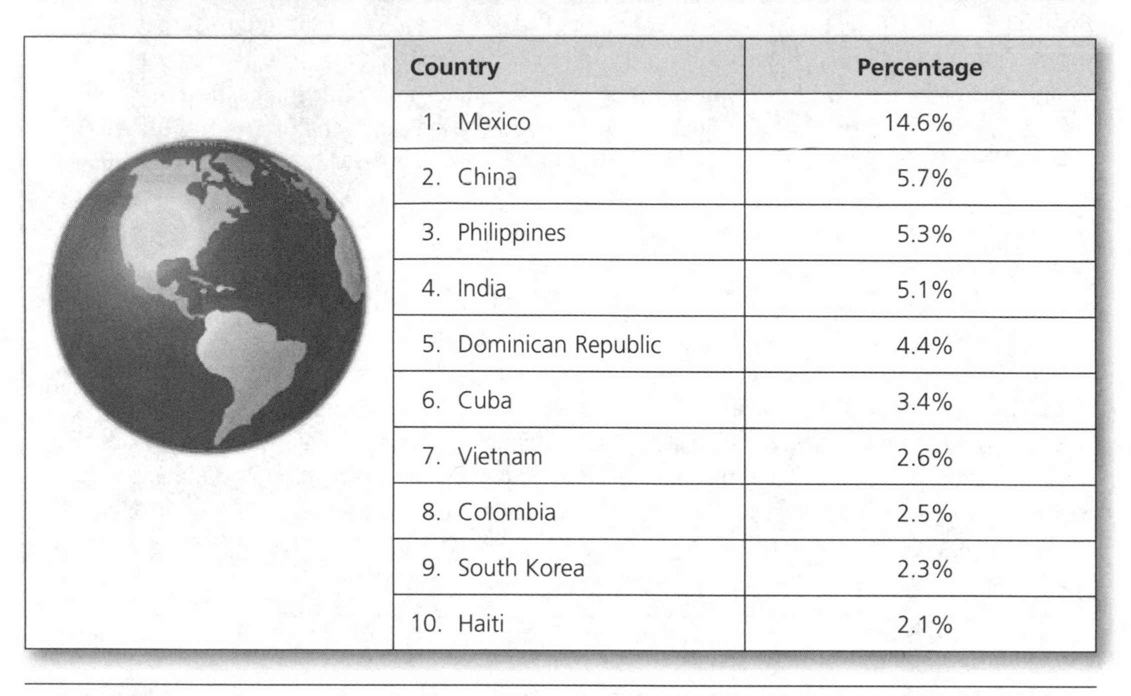

Country	Percentage
1. Mexico	14.6%
2. China	5.7%
3. Philippines	5.3%
4. India	5.1%
5. Dominican Republic	4.4%
6. Cuba	3.4%
7. Vietnam	2.6%
8. Colombia	2.5%
9. South Korea	2.3%
10. Haiti	2.1%

Source: Monger, R. 2010. *U.S. Legal Permanent Residents: 2009*. Annual Flow Report, Department of Homeland Security, Office of Immigration Statistics. Available online at http://www.dhs.gov/xlibrary/assets/statistics/publications/lpr_fr_2009.pdf.

school nor have earned a high school degree) have decreased over the last 30 years (from 35.2 in 1980 to 21.4 in 2007). The percent of Hispanic high school graduates who immediately transition to college has been increasing for the last 30 years: In 2008, 61% of such graduates immediately enrolled in college. Between 2000 and 2007, Hispanics experienced a 48% increase in graduate and professional school enrollments (Planty et al. 2009). Hispanic households also reaffirm important core American values. The family, for instance, is greatly respected in Hispanic culture. Seventy-eight percent of Hispanic households are family households (compared with 67% of white households). (A **family household** is one where one or more persons are related to the householder by marriage, birth, or adoption; U.S. Census Bureau 2009e). Hispanic immigrants are one of the driving forces behind the return of multigenerational family households (Pew Research Center 2010). A smaller proportion of Hispanics than whites are divorced (8% vs. 11%; U.S. Census Bureau 2011, Table 56). Hispanic American youth also report the strongest sense of family duty during young adulthood (Fuligni and Pedersen 2002). And in terms of politics, the Hispanic vote will surely receive more and more attention in local, state, and national elections as the number of Hispanic voters continues to grow. (In the 2008 presidential election, 9.7 million Hispanic Americans voted, two million more than in 2004. Perhaps the turnout was motivated in part by the fact that the 2008 presidential election saw the first Hispanic American

Family household
Household where one or more persons are related to the householder by marriage, birth, or adoption

vie for the nomination of the Democratic party: Bill Richardson.)

While the faces of immigrants have changed in the past 50 years, intergenerational assimilation and upward mobility are still the norm. There is a very clear association between length of time in the United States and both income and home ownership rates. While earnings of immigrants are initially lower than those of natives, the gap disappears over time. For the foreign-born arriving after 2000, about 24% own their own homes, but for those arriving before 1990, nearly 68% own their own homes (Passel and Cohn 2009; Pew Hispanic Center 2010). Immigrants also keep pace with native-born Americans in terms of entrepreneurial activity (Camarota 2007). And the eventual mastery of the English language is another key to immigrants' economic success. Historically, the shift from speaking another language to speaking English has occurred over three generations. Among recent immigrants, however, the shift seems to be occurring within two generations (Martin and Midgley 2006; Portes 2002). Age at arrival, length of time in the United States, and level of education are all factors that hasten the acquisition of fluency in English (Rumbaut 2009). This faster turnaround time may also reflect the current reform mood of the country. A 2010 national survey found that 87% of Americans think English should be the official language of the United States (Rasmussen Reports 2010b).

Your Thoughts . . .

It is often noted that language is the glue that holds a group together and the bedrock of national identity. Is there merit to the argument for making English the official language of the United States? Or should we go so far as to mandate "English only" as a way of fostering full integration of immigrants? What do you think?

Given the current mood of the nation, it is perhaps no surprise that more and more immigrants are seeking U.S. citizenship (naturalization). During the 1950s and 1960s, the average number of immigrants naturalizing each year was fewer than 120,000. Between 2000 and 2009, that yearly average jumped to approximately 680,000. In 2008, an all-time record was set for the number of immigrants who became naturalized citizens: 1.04 million. For those who naturalized in 2009, the average time lapse between becoming a legal permanent resident and becoming a citizen was 7 years (Lee 2010). (See Box 20.1 for conditions for naturalizing.)

In light of the general success rates posted by recent immigrants in general and by many Asian and Hispanic Americans, we must consider that current anti-immigration sentiments may be based on issues of race. Visible physical differences, as well as a lack of familiar cultural practices, make the assimilation of "new" immigrant groups more difficult than it was for earlier immigrants. New, non-European immigrants may lack the physical and cultural similarities necessary for eventual acceptance as part of society's in-group.

If immigration and population trends develop as predicted, anti-immigration sentiments fueled by issues of race may get worse before they get better. Demographers project that by 2050, the U.S. population will be 438 million; 82% of the population increase from now until 2050 will be due to immigrants and their children. By 2025, the percent of foreign born in the United States will surpass the highs set a century ago during our last great immigration period. By 2050, Hispanics will account for 29% of the U.S. population while non-Hispanic whites will constitute a numerical minority (47%; Jacobsen and Mather 2010). If these projections are accurate, future immigration, in a very profound sense, will change the status quo. The practice of assimilation may necessarily give way to multiculturalism. **Multiculturalism** accentuates rather than dilutes ethnic and racial differences. Such an environment might strip in-groups of dominance and power. In contrast to the adversarial stance of the in-group/out-group design, a multicultural structure demands that all groups be viewed as equally valued contributors to the mainstream culture.

Multiculturalism
A perspective that accentuates rather than dilutes ethnic and racial differences

Immigration projections suggest that the United States is moving closer to being a microcosm of the world. Our nation will experience an increase in the diversity that already characterizes the younger generations of Americans. Such changes could bring us closer to fully realizing the motto that appears on all U.S. currency: *E Pluribus Unum*, one formed from many. Thus, current and future attitudes toward immigrants in America will hinge on our readiness to deal with fundamental population changes.

Certainly, some Americans will resist this development, arguing that it threatens our national identity and changes our national face. Contemporary movements against bilingual education and for a national language offer evidence of such resistance. Nevertheless, others will view our changing population as a positive economic opportunity. Consider the fact that economists forecast a very different world for coming generations of Americans. More and more of us will be earning our livings in the service sector of an increasingly global, postindustrial economy. Postindustrial economies place a high premium on knowledge and information (Drucker 1993). Occupations requiring postsecondary degrees are expected to see the fastest growth over the next decade (Ramey 2010). Immigration bodes well for educational advancement. There is a growing body of research that suggests that strong pro-education values as well as high valuation of close family ties found in some immigrant groups produce higher academic achievement (Harris et al. 2008; Keller and Tillman 2008; Plunkett et al. 2009; Suarez-Orozco et al. 2009). In 2008, 11% of the foreign born had completed a graduate or professional degree (compared with 10% of native-born Americans; Jacobsen and Mather 2010). Some studies also indicate that bilingualism is associated with positive educational outcomes (Lutz and Crist 2009; Portes 2002; Proctor et al. 2010). Indeed, globalization makes multilingualism a way of life and a pathway to empowerment (UNESCO 2003). Multilingual workers will clearly occupy a position of advantage in an increasingly global economy and workplace (Portes 2002). To be sure, bilingualism is being touted as a valuable resource in our current tight job market.

Framed in this way, new immigration patterns may help supply us with a new source of cultural capital (Archdeacon 1992). **Cultural capital** refers to attributes, knowledge, or ways of thinking that can be converted or used for economic advantage. Interestingly enough, as we witness various indicators attesting to the force of globalization, we also see more and more ethnic groups asserting their identities and pushing for political recognition and autonomy (Guillen 2001). By their familiarity with cultures now central to the world market, immigrants to the United States may well give our nation a competitive edge in a global playing field. Once again, immigrants to the United States may be the national resource that makes the United States a significant player in a new world economy.

Cultural capital
Attributes, knowledge, or ways of thinking that can be converted or used for economic advantage

LEARNING MORE ABOUT IT

In *Learning a New Land: Immigrant Students in American Society,* Carola and Marcelo Suarez-Orozco and Irina Todorova report on the results of a 5-year study that examines the challenges faced by immigrant children in our public schools (Cambridge, MA: Belknap Press, 2008).

In *Migration Miracle: Faith, Hope and Meaning on the Undocumented Journey* (Cambridge, MA: Harvard University Press, 2008), Jacqueline Hagan uses interview data from immigrants from Mexico and Central America to explore the intersection between religion and migration journeys.

The New Americans: A Guide to Immigration Since 1965 (Waters, Reed and Marrow, eds.) is a collection of articles that provide a wealth of information about the diverse groups making up America today (Cambridge, MA: Harvard University Press, 2007).

Aviva Chomsky explores the many misconceptions that distort our thinking and views on immigration in *"They Take Our Jobs!" and 20 Other Myths About Immigration* (Boston: Beacon Press, 2007).

In *Immigrant America: A Portrait,* 3rd edition (Berkeley: University of California Press, 2006), Alejandro Portes and Ruben G. Rumbaut present a highly readable account of the complexities of immigration in America and mobilize data in the service of dismantling myths and misconceptions.

Richard Alba and Victor Nee offer an insightful review of assimilation of current immigrants in *Remaking the American Mainstream: Assimilation and Contemporary Immigration* (Cambridge, MA: Harvard University Press, 2005).

Bloemraad, Korteweg, and Yurdakul look at the rise in international migration and its implication for citizenship in their 2008 article "Citizenship and Immigration: Multiculturalism, Assimilation and Challenges to the Nation State" (*Annual Review of Sociology* 34: 153–179).

Mary Waters and Tomas Jimenez look at the degree to which recent immigrants are hitting the benchmarks of assimilation in their article "Assessing Immigrant Assimilation: New Empirical and Theoretical Challenges" (*Annual Review of Sociology,* 31:105–125, 2005).

GOING ONLINE FOR SOME SECOND THOUGHTS

The U.S. Citizenship and Immigration Services (formerly the INS) offers a wealth of information on immigration. Visit its website at http://www.uscis.gov/portal/site/uscis.

Separate the fact from fiction about the 11 million-plus undocumented immigrants in the United States by accessing the link "Undocumented Immigrants: Myths and Reality" at the Urban Institute's website: www.urban.org/publications/900898.html.

A discussion by Philip Martin of the various "doors" used by foreigners to enter the United States can be found at http://www.prb.org/Publications/PopulationBulletins/2010/immigrationupdate1.aspx.

You can listen to an interview with sociologists Jennifer Van Hook, Pennsylvania State University, and Jennifer Glick, Arizona State University, where they discuss the social and cultural challenges facing immigrant children at http://www.prb.org/Journalists/Webcasts/2010/usimmigrant children.aspx.

The Migration Policy Institute offers a wealth of information on immigration. Visit it at http://migrationpolicy.org/.

NOW IT'S YOUR TURN

1. Visit a library that has back copies of local telephone directories. Examine the entries in the yellow pages for a variety of categories—beauty salons, physicians, restaurants, and so on. What insights about immigration patterns can be gleaned from your data as you move from year to year?
2. Conduct several in-depth interviews with people who were born in other countries and immigrated to the United States. Find out about the conditions of their immigration, the reception they received in their new communities, and, if appropriate, the reception they received at their new workplaces or their new schools. Try to vary the immigration background of your interview subjects; that is, choose individuals who came from different foreign countries. Consider whether one's status as an immigrant functions as a master status (see Essay 6 or 12).
3. Use your own college community to locate children of recent immigrants. Prepare an interview guide that will allow you to explore whether these individuals exist in two social worlds or cultures. (For example, how do language, food, fashion patterns, and so on vary from school to home?)

SOCIAL INSTITUTIONS: MEDIA AND TECHNOLOGY

ESSAY 21

Conventional Wisdom Tells Us . . . Technology Is Taking Over Our Lives

This essay examines new communication technologies and explores their role in contemporary social life. We begin by considering the ways in which technology has altered social inequality. We move on to examine technology's impact on the development of community and intimacy. And we conclude by exploring the impact of new technologies on our definitions of social relations, social actors, and the public and private spheres.

Look around. No matter where you are, you will see the signs. Technology rules the day. On the road? Electronic message boards and Amber Alert systems are sending immediate updates on road conditions, emergencies, or criminal matters. Now look at the drivers around you. Drive time has become talk time (or, dangerously, text time) as more and more of us keep in touch with family and work via our cell phones. Check the scene at work or school. Everyone seems to have cell phones in their ears, iPods in their hands, and laptops on their shoulders. See something extraordinary in your travels? Grab the cell and send a picture! Hear a funny joke or a juicy piece of gossip? Share the wealth—text it to your best friend!

These days, it is easy to stay connected to family and friends. But the information boom is about much more than our social lives. Increasingly, governments, corporations, and law enforcement agencies are privy to our comings and goings. Surveillance cameras or electronic toll devices can record our physical whereabouts. "Cookies" can track our travels and tastes as we browse through cyberspace. Crime-fighting computers and medical data banks can verify our identities; they know us right down to the chromosomes in our DNA. Even the mundane trip to the grocery store carries a technological flair. Computerized registers are able to scan our items and immediately deliver customized coupons that fit our tastes.

New communication technologies Developments such as fiber optics, the Internet, rapid satellite transmissions, and virtual reality imaging

The **new communication technologies**—developments such as fiber optics, broadband, the Internet, rapid satellite transmissions, robotics, and virtual reality imaging—can take us to places that we've never been before. While sitting in front of our computer screens or while walking with our iPhones, we might participate in a virtual march on Washington; we might support our favorite cause via online volunteer work; we might participate in an online support group or tour an art gallery, a historic site, or a home we are thinking about purchasing. Movies can be delivered on demand to our computers; the biggest concert tour of the year may be simulcast on the Internet. Opportunity is knocking . . . although for some it may not always be an easy door to open. Computer users really must become computer "techies" who can troubleshoot their way through various hardware and software problems. Luckily, this knowledge will serve the techie well in the future. Appliances, cars, and homes of tomorrow will all be "smart" and require consumers to be programming savvy. Downtime may be gone time. The Internet is never closed; technology whirls on, 24/7.

Conventional wisdom suggests that technological advancements are taking over our lives. Critics charge that we are slaves to our hi-tech toys, drowning in a sea of points, clicks, and wireless connections. Indeed, many believe we spend so much time with the new products of technology that we have forgotten or withdrawn from the "real" world. Is conventional wisdom correct in its antitechnology stance? Have we lost control of our lives . . . or merely learned a different way to live them? A number of sociologists are examining these issues in detail. Their research suggests that new communication technologies may be less "dangerous" than critics suggest.

Your Thoughts . . .

How many times a week do you tweet or post to Facebook? How many times a day? Do these activities enhance or interfere with your strongest relationships?

Digital divide The formation of an information underclass–a portion of the population that cannot afford to access and capitalize on the things that new technologies have to offer

What is it that people fear about the new communication technologies? One common concern involves the issue of inequality. Some researchers argue that new communication technologies are creating a digital divide. A **digital divide** refers to the formation of an information underclass—a portion of the population that cannot afford to access and capitalize on the things that new technologies have to offer. To be sure, *all* technological innovations result in a digital divide. Radios and televisions, telephones and automobiles—all seemed to be a luxury of the wealthy when first introduced. But as these technologies were diffused within society at large, some of the access inequalities tended to disappear. New communication technologies are following this developmental pattern. For example, the racial, gender, and urban–rural divides that once characterized computer access and usage have diminished considerably (although men and women use the Internet for different things; see Colley and Maltby 2008). But in the areas of age, income, and education, much work remains. It is still the case that the young, the wealthy, and the well educated have better access to new technologies and are more technologically proficient

(Pew Research Center 2007, 2010c). It is important to note, however, that technology per se does not sustain or create inequalities between social groups. Rather, the structure of the contexts in which technology is used can either exacerbate or alleviate social inequality. This point holds true not just in the United States but also globally (Hampton 2010; Pew Foundation 2007; Pew Research Center 2010c; Robison and Crenshaw 2010).

Conventional wisdom also argues that new communication technologies are destroying people's involvement in civic and community life. Critics fear that individuals are substituting online activity for more traditional offline affiliations—an experience that critics contend is inferior to traditional modes of civic activity. Some fear that Internet users may be overwhelmed with available information. As a result, users may participate in selected interest groups that do little more than reinforce their existing beliefs. Others fear that users may encounter online activists who are much more extreme than their offline counterparts. Such encounters could be intimidating, resulting in users' complete withdrawal from the civic arena. Still others contend that users may decide that navigating the dense terrain of the Internet simply is not worth their effort (Calhoun 1998; Fernback 2007; Katz and Rice 2002). While such fears abound, recent studies on the Internet and community fail to lend them much credence. Current research suggests that online interactions tend to supplement, not replace, offline community involvements. In addition, heavy Internet use is actually associated with high levels of participation in offline voluntary organizations and politics. Thus, far from being overwhelmed or exceptionally selective, Internet users tend to be highly interested in a wide array of current events and highly involved in civic and political activities (Collins and Wellman 2010; Jensen, Danziger, and Venkatesh 2007; Mesch and Talmud 2010; Smith 2009, 2010; Smith et al. 2009; Wang and Wellman 2010; Wellman 2004).

Your Thoughts . . .

Has the Internet changed your political activity? If so, in what ways?

One cue to the link between community building and the Internet rests with the remarkable growth in **e-philanthropy**—that is, online giving or "cyber giving." For example, Networkforgood.org has built an army of volunteers, linking them to 200,000-plus local, international, and virtual volunteer organizations. Via the site, more than $400 million has been raised and distributed to more than 50,000 different nonprofit organizations. Further, the group was able to take fundraising viral, raising an additional $3.2 million via fundraising widgets (Networkforgood.org 2010). The increasing popularity of e-philanthropy may be due to its complementary tie to our daily work patterns: 71% of all online donations are made at work. Since many of us are online during a normal working day, e-giving has become a quick and satisfying way to support our charitable causes (Andresen 2007; *Business Wire* 2002).

e-philanthropy
Online or "cyber giving"

E-philanthropy has seen successful in the area of online volunteering as well. The largest website for online volunteering is VolunteerMatch.org. This site connects potential volunteers with nonprofit groups. An individual simply submits a list of volunteering contingencies (i.e., the distance one is willing to travel, the time frame for one's volunteer work, and the type of cause or issue with which one wants to be associated). The site then searches for an organization that matches an individual's criteria. Since its launch in 1998, VolunteerMatch has generated well over 4.9 million volunteer referrals—with the greatest growth in numbers occurring in the past 5 years (VolunteerMatch.org 2010).

E-philanthropy received a significant boost in the days following the September 11, 2001, terrorist attacks on America. On September 10, there were but a few dozen donations to AOL's Networkforgood.org. On September 14, there were 12,000 donations totaling more than $1.26 million. Indeed, in the first 6 months following 9/11, it is estimated that 10% of relief donations (or $150 million) came via the Internet. The Red Cross alone raised more than $63 million online after 9/11 (Larose and Wallace 2003). The giving trends initiated by the 9/11 disaster have grown with each passing year. In 2004, for example, a full 39% of donations for victims of the South Asian tsunami came from online donations (Bhagat 2007). In 2009, more than $560 million was raised online for victims of the Haiti earthquake. Of course, the most successful online campaign is linked to Hurricane Katrina. In 2005, the Bush-Clinton Katrina Fund, established to benefit the survivors of Hurricane Katrina, took *less than 24 hours* to raise its first million dollars in online donations, and within 8 days more than $580 million had been raised for relief efforts (Clinton Foundation 2005; Preston and Wallace 2010).

Many believed that online donations and the formation of communities around the various candidates would be the fuel of the 2008 presidential elections. They were right. In the 2008 campaign, candidates learned to use the fundraising power of the Internet in new and innovative ways.

> John Edwards convinced more than 10,600 supporters to give from $6.10 to $54 in honor of his 54th birthday, in exchange for an emailed copy of his mother's pecan pie recipe. It took just two days for former Tennessee Senator Fred Thompson, a Republican, to collect more than $350,000 in donations on his Internet site. Mr. Obama, meanwhile, promised to have dinner with four randomly selected donors, who could give as little as $5 (Schatz 2007).

Indeed, the Obama campaign perfected the formula for the creating an online workforce. Not only did two-thirds of his contributions come from online giving (Federal Election Commission 2009; Vargas 2008), his supporters were most likely to share online campaign content with others, sign up for regular election updates, set up political alerts and gatherings for those they knew, and sign up for volunteer activities (Smith 2009).

Beyond civic activity, critics of new communication technologies voice other concerns. Fears regarding the Internet and the destruction of community are matched by concerns for technology's effect on intimacy. In many ways, the intimacy debate centers on the difference between direct and mediated communication. According to conventional wisdom, intimacy demands **direct communication**, face-to-face or physically copresent exchange. For intimacy to grow, we must be with others, see, hear, and touch them. Only then is a relationship "real." But online communication is **mediated communication**—an indirect connection funneled through a mechanical medium. Many feel that mediation makes communication impersonal, fleeting, and ingenuous. As such, it is an inappropriate means by which to establish deep and lasting connections.

Direct communication Face-to-face or physically copresent exchange

Mediated communication An indirect connection funneled through a mechanical medium

The bias against mediated communication as a vehicle of intimacy is well established. Indeed, many would argue that those who build technological ties are really isolates and loners living fantasy lives and creating anonymous, meaningless worlds. To be sure, there are some well-executed studies that support these concerns. Certain works suggest that some Internet users become unduly drawn into cyberspace and thus neglect their offline relationships (see, e.g., Nie and Erbring 2002; Schroeder and Ledger 1998; Shapiro and Leone 1999). Other studies contend that the anonymity of Internet communication creates identity conflict and confusion (Turkle 1996, 1997) or the free use of deception (Epstein 2007). But in the final analysis, a greater number of studies forward a positive picture of technology's role in intimate relationships. Research shows that the Internet complements and enhances existing relationships—that is, people use e-mailing, texting, and social networking sites to enhance face-to-face relationships (Katz and Rice 2009; Tang 2010). The Internet also facilitates new friendships and bonds that might not otherwise have been possible (Boase et al. 2006; Briggle 2008; Horrigan and Rainie 2006; Jensen et al. 2007; Wellman 2004; West, Lewis, and Currie 2009). Others have demonstrated that the Internet has become a central tool in coping with life's most serious problems. For example, about one-half of Americans say that the Internet played a central role in helping them cope with a family member's major illness or their own medical problems. Similar numbers reported reliance on the Internet for important career-related decisions (Boase et al. 2006; Hoover, Hastings, and Musambira 2009; Horrigan and Rainie 2006). Thus, as Katz and Rice conclude, "The Internet is quite a social environment, inhabited by quite social folks" (Katz and Rice 2002; also see Katz and Rice 2009).

Sociologists Karen A. Cerulo and Janet M. Ruane suggest that an accurate assessment of new communication technologies will require a more flexible way of conceptualizing social relations. **Social relations** refer to the types of connections and the patterns of interaction that structure the broader society. Traditionally, sociologists have argued that physical copresence is integral to important relationships—relationships that enable communities, friendships, and intimacy. Thus, physical copresence has become the standard by which to judge the quality and importance of interaction. Cerulo and Ruane suggest that bodies may not be the most important part of the intimacy equation. Rather, the cognitive context in which communication occurs may make certain forms of direct and mediated exchange equally valuable and real. Frequent and balanced interaction among individuals with overlapping backgrounds, strong and long-term bonds, and the development of trust—these are the qualities that make us feel connected to others. When these things are present, in online or offline settings, individuals perceive the interaction to be central to their lives (Cerulo 1997; Cerulo and Ruane 1998; Cerulo, Ruane, and Chayko 1992; Chayko 2002).

Social relations
The types of connections and patterns of interaction that structure the broader society

The American courts have taken such ideas to heart. Indeed, over the past decade, the courts have substantially broadened their definition of intimacy. A New Jersey court, for example, entertained a divorce and adultery case in which the third party and the offending spouse never physically met. The courts considered an online romance, void of any physical contact, as sufficient grounds for divorce by reason of adultery. Similarly, in a Massachusetts custody battle, a probate judge granted a mother custody of her children and permitted her to leave the state and move to New York. The judge granted the children's father two weekend visits per month and two weekly visits—*virtual* visits made possible through Internet conferencing technology. The father's lawyers protested, arguing that "you can't hug a computer." But in the judge's estimation, the children's physical presence was not vital to a meaningful visit. Outside the United States, court systems have gone much further. In Japan, for example, a man was arrested for using software-generated robots to virtually assault and burglarize characters located in the online space of a game called Lineage II. In another case, a Japanese women faces up to 5 years in prison for "murdering" her husband's avatar. Across the globe, a

Dutch court sentenced teens to 360 hours of community service because they virtually assaulted a classmate and stole some of his virtual goods. And in Great Britain, a woman was granted a decree of divorce after she discovered her husband was having an affair with another man in the online community Second Life (Morris 2008; Power 2010).

BOX 21.1 SECOND THOUGHTS ABOUT THE NEWS: CYBERBULLYING

In 2010, a case of *cyberbullying* and suicide captured the U.S. media markets for days. **Cyberbullying** refers to the *repeated harassment, mistreatment, or humiliation executed by one or more persons toward another with the specific use of cell phones or cameras and the Internet.*

Cyberbullying
The repeated harassment, mistreatment, or humiliation executed by one or more persons toward another with the specific use of cell phones or cameras and the Internet.

Allegedly, college student Tyler Clementi's schoolmates secretly recorded Clementi in an intimate homosexual encounter. The footage was then immediately streamed on the Internet, thus allowing millions to view the most private details of Clementi's life and discuss his actions with a familiarity formerly reserved for one's neighbors, classmates, or personal acquaintances. When Clementi discovered the postings, he was overcome and distraught. Soon afterward, he jumped to his death from the George Washington Bridge.

To be sure, many aspects of the Clementi case are not new. Bullying has long been a part of social life. But the wide availability of online venues allows bullies to greatly expand their reach. Within seconds, one can not only be captured by a webcam or cell phone, but that information can also be made available to millions worldwide.

It is worth noting that approximately 20% of high school and college students report being the victim of cyberbullying, with women more likely to be victims than men. Note too that as perpetrators, men and women engage in different forms of cyberbullying. Women are more likely to spread vicious rumors while men are more likely to post embarrassing pictures or videos.

Nationwide, cyberbullying is not yet legally defined as a severe problem. While many states have bullying laws, only six states include cyberbullying in those laws. Among those six states, only one—Nevada—allows for the criminal sanctioning of cyberbullying.

Whatever its form and wherever it occurs, cyberbullying offers another example of the ways in which technology necessarily broadens our definition of intimacy. What was once your information to share can now become the personal knowledge of millions you do not know (Cyberbullying Research Center 2010; Hinduja and Patchin 2010a, 2010b).

Embedded in the many events discussed in this essay is a central sociological question. In light of new technologies, who is a social actor? Is the title restricted to another living, breathing human being? Or can it be a television, a computer, a robot, an avatar, or a website? Cliff Nass stands at the forefront of research addressing social interaction between humans and nonhuman objects, specifically computers, robots, and avatars (Nass and Brave 2005; Reeves and Nass 1996). Nass finds that the increased communicative capacities of these techno-objects have had notable effects on people's perceptions of the "other." Nass argues that certain types of objects evoke a sense of intersubjectivity in humans, encouraging individuals to respond to such entities in fundamentally social ways. As a result, nonhuman objects become an active part of social interaction as opposed to mere props used by humans to enhance or steer social interaction.

Over the years, Nass and associates have revisited a number of classic social psychological experiments—experiments that were designed to test person-to-person responses in social interaction. In updating the experiments, the researchers made one critical change. Now, the experiments tested person-to-computer, -robot, or -avatar responses. Results showed that people—even the most technologically sophisticated people—interacted with these entities just as they interacted with humans. Subjects were polite to computers, robots, and avatars; they responded to praise from them, and they viewed them as teammates. Subjects liked computers, robots, and avatars with personalities or social characteristics similar to their own. They trusted computers, robots, and avatars that manifested caring orientations more than those that did not. They found masculine-sounding computers, robots, and avatars extroverted, driven, and intelligent while they judged feminine-sounding computers, robots, and avatars knowledgeable about love and relationships. They even altered their body posture and mood according to the size and perspective of the screen images before them. In essence, new technologies endowed these objects with critical interactive and communicative capacities, encouraging humans to perceive and react to these entities as legitimate partners in social interaction. (For a good review of this literature, see Cerulo 2009.)

Sherry Turkle (2007), the founder of MIT's Initiative on Technology and the Self, reports similar results. Turkle observed and interviewed both the elderly and children as they interacted with robots such as Sony's robotic dog AIBO, Hasbro's My Real Baby and Furby, Leonardo, a social robot created at the MIT Media lab, and the Japanese robots Paro and Healing Partners. She chose these robots because they can recognize their owners, obey their commands, and adjust their personalities in accord with owners' utterances and actions. When faced with the robots, human users respond in social ways, according to Turkle. Among the elderly, interacting with robots diminished anxiety (especially in those suffering from dementia) and brought a sense of real companionship to mentally stable individuals. (For example, users of Healing Partners reported that "the robots are like having grandchildren around all the time"; Litke 2006.) Similarly, children perceived and related to the robots as "an autonomous and almost alive self," who, while clearly not human, were fully capable of a meaningful relationship.

Both Nass's and Turkle's work suggests that new technologies have blurred the boundaries between people and objects, allowing both to operate as viable actors in the social terrain. We are witnessing similar changes with reference to the boundaries that distinguish people from animals or the living from the dead. Via virtual imaging and advanced robotics, new communication technologies create a reality in which seemingly anything can happen—be it talking to your dog and having him or her answer back, experiencing a passionate romance with the man or woman of your dreams, or engaging in a conversation with a spouse, friend, or parent who is no longer alive. Just make-believe? On the one hand, yes. However, technology's projection of animals, fantasies, and spirits as viable social actors has had tangible consequences. In the past 10 years, we have witnessed increases in behaviors that confirm the reality of such images. Americans are, for example, spending more time with their pets—buying them clothes and furniture; taking them to spas, to psychotherapists, and to psychics; bringing them to work; and sending them to vacation resorts. Similarly, Americans are spending more time and money on mediums and other activities designed to contact spirits and angels. Mediums such as John Edwards and James Van Praugh have become media stars! (Indeed, the waiting time for a private reading with John Edwards is nearly 3 years!) Furthermore, Americans are gobbling up technological products that merge sounds and images of past and present. CDs and videos that bring us duets between the living and the dead are now hot properties. So—who is a social actor? Is the title reserved for a living, breathing human being? New technologies make the question a complicated one to answer (Cerulo 2009; Cerulo and Ruane 1997).

Technology does offer a potentially frightening "Big Brother" scenario when we consider the issue of privacy. Indeed, some argue that the potential for increased anonymity demands alternate ways of facilitating trust and maintaining social order. This, of course, is the very idea behind electronic satellite "boxes" used to keep track of an increasing number of parolees. These devices allow the government to monitor convicted felons as they move about in local communities (Lee 2002). But surveillance has also entered the world of ordinary, law-abiding citizens. The never-blinking security camera has become a mainstay of social control in our stranger-based and now terrorist-threatened society. Technology experts tell us that, each year, we are witnessing a 25% growth in video surveillance. By the end of 2010, most major cities will support on-street surveillance cameras and one-third of American households will contain security systems (Holtzman 2006; Lee 2006).

A recent survey done by the New York Civil Liberties Union found some 8,000 video cameras across the city; 3,000 additional surveillance cameras can be found in New York City subways. And in September 2010, New York City launched a "ring of steel" in mid-town that will ultimately see the addition of 3,000 integrated video cameras. Yet the distinction of the most extensive surveillance system may well belong to the Windy City. Former Homeland Security secretary Michael Chertoff says that Chicago has a sophisticated network of private and public cameras (estimated at 10,000+) that make it the most watched city in the United States (Babwin 2010).

Surveillance tapes, of course, know no distinction between the private and public realms. Behaviors that social actors may intend as private backstage exchanges (a furtive kiss in a parking deck) may very well play as front-stage performances on security video screens. Surveillance also occurs on less obvious levels. Our movements over the course of a day can be traced via electronic toll and parking passes, phone records, and credit card activity. And as many people have haplessly discovered, our travels on the Internet can be far too public an affair. Computer cookies can reveal our interests and modes of decision making.

While opportunities for the misuse of surveillance tools abound, it is nonetheless true that technology can also enhance our privacy. Our ability to live a totally anonymous or secluded existence is supported by technology that enables us to work, play, and conduct all of our social and business affairs without ever going public (Nock 1993). With this view, electronic surveillance is not an ogre but rather a response to the social changes of modern, urban living. Clearly, technology offers a solution to the social control dilemma posed by the fact that we desire and lead increasingly private lives.

Since 9/11, Americans appear to be more tolerant of governmental surveillance and seemingly more willing to regard any opposition to the government's watchful eye as unpatriotic. Indeed, with the passing of the USA PATRIOT Act (2001), the Terrorist Surveillance Act (2006), and the Protect America Act (2007), the government, in the name of fighting terrorism, is able to collect a staggering array of private information about individuals. And a recent Harris poll found that Americans are solidly behind video surveillance, especially as it relates to fighting terrorism: 96% support such measures in public places (BRS Labs 2009). On the other hand, Americans are not so complacent about *all* electronic intrusions. Consider, for instance, that in the first 4 days of its availability, more than 10 million Americans registered their phone numbers with the national Do Not Call list in order to stop telemarketing calls. As of 2009, the number of registered cell and land phones had grown to 191 million (Federal Trade Commission 2010)! Similarly, we consider spam to be the bane of the Internet; globally, it is responsible for 14.5 billion messages per day, and it is estimated to make up 45% of all e-mails (Spamlaws.com 2009). On an average day, Americans use the Internet to forward more than 300,000 examples of spam to the FTC, all in a concerted effort to help that agency fight these unwanted electronic intrusions (FTC 2007b). Clearly, we value our privacy, but we also, for now, appear willing to draw distinctions between governmental and commercial intrusions and between what we think are invasions by law-abiding versus law-violating individuals (Liptak 2002). Our discerning response suggests that high tech itself isn't viewed as the culprit. Rather, we are willing to judge technology by the cause it serves.

Technology—new-fangled, oppressive devices or new, essential tools of social life? If history proves informative, we will likely find some truth in both views. We can also be confident about a few other things. Just as we get comfortable with the latest innovations, still more advanced bells and whistles will arrive to push our buttons and sound new alarms. New technologies will prompt new debates about the pros and cons of high-tech living. And society will take it all in its technological stride.

LEARNING MORE ABOUT IT

For a very thorough and balanced treatment of new communication technologies, both the pros and cons, see James E. Katz and Ronald E. Rice's *Social Consequences of Internet Use: Access, Involvement and Interaction* (Cambridge, MA: MIT Press, 2002) and their follow-up article, "Technical Opinion: Falling Into the Net: Main Street America Playing Games and Making Friends Online (*Communications of the ACM* 52 (9): 149–150).

For some interesting ideas on the digital divide, see Patricia Randolph Leigh's book *International Exploration of Technology Equity and the Digital Divide: Critical, Historical and Social Perspectives* (Hershey, PA: Information Science Publishing, 2010).

To read Clifford Nass's fascinating work on the humanization of technology, see either *The Media Equation* (with Byron Reeves, New York: Cambridge University Press, 1996) or *Wired for Speech: How Voice Activates and Advances the Human–Computer Relationship* (Cambridge, MA: MIT Press, 2007). Sherry Turkle's work on robots as social actors can be found in *Evocative Objects: Things We Think With* (Cambridge, MA: MIT Press, 2007). Karen A. Cerulo reviews technology and its impact on the definition of the social actor in her article "Non-humans in Social Interaction" (*Annual Review of Sociology* 35: 531–552, 2009).

For a fascinating excursion into privacy and the Internet, see David Holtzman's book *Privacy Lost: How Technology Is Endangering Your Privacy* (San Francisco: Jossey-Bass, 2006).

GOING ONLINE FOR SOME SECOND THOUGHTS

For more information on cyberbullying, visit the Cyberbullying Research Center at http://www.cyberbullying.us/.

Are your parents using Facebook and Twitter? Could be. Take a look at usage patterns by age for social networking sites. The statistics are collected by the Pew Research

Center, and an interesting graph can be found at http://pewresearch.org/pubs/1711/older-adults-social-networking-facebook-twitter.

Meet Leonardo, one of many "social robots" created by the MIT Personal Robots Group. You'll find him at http://robotic.media.mit.edu/projects/robots/leonardo/sociallearning/sociallearning.html.

NOW IT'S YOUR TURN

1. Visit an online chat group or Facebook group you've never been to before. You can participate or just watch. However,
   - visit the same group three times, and
   - spend at least 30 minutes during each visit.

   How did your online visits compare to face-to-face interactions? Is intimacy possible in these forums? Did you feel more vulnerable to other interactants in the online forums? Less vulnerable? Were you less honest in your online discussions? More honest? Is online communication more useful for some tasks than others?

2. Select five charity mailings (from different organizations) that you have received in the recent past. Access each organization's website. Do a systematic comparison of the "snail mail" pitch versus online presentations and appeals. What similarities and differences do you find, especially with regard to each organization's ability to personalize its message and requests?

SOCIAL INSTITUTIONS: EDUCATION

ESSAY 22

Conventional Wisdom Tells Us . . . Education Is the Great Equalizer

Conventional wisdom tells us that educating the masses will bring equal opportunities to people of all races, ethnicities, and genders. In this essay, we explore the truth of this claim and review the progress we have made in bringing high-quality education to all.

The United States has earned a reputation as the land of opportunity, and the opportunity that so many of us desire is the improvement of our socioeconomic lot. **Intergenerational upward mobility**—the social status gains by children vis-à-vis their parents—is a key dimension of the American Dream.

Intergenerational upward mobility
Social status gains by children vis-à-vis their parents

Historically, education has been offered as the route by which such mobility can best be realized. Our free common public school system, established just prior to the Civil War, was founded on the principle that everyone, regardless of social background, should be educated. Lester Ward, a prominent sociologist of the late 1800s, thought that universal education would eliminate the inequalities associated with social class, race, and gender. Similarly, educational reformer Horace Mann promoted an expanded educational system as the antidote to poverty (Katz 1971).

Such sentiments have survived the test of time. We are a nation fueled by the belief that education will lead to equal opportunity for individual achievement and success. Former "education president" George Bush (the elder) aptly captured this cultural value, characterizing education as the "great lifting mechanism of an egalitarian society. It represents our most proven pathway to a better life."

The conventional wisdom on education reflects a structural functionalist view of society. **Structural functionalism** is a theoretical

Structural functionalism
A theoretical approach that stresses social order and that views society as a collection of interdependent parts that function together to produce consensus and stability

approach that stresses social order. Proponents contend that society is a collection of interdependent parts that function together to produce consensus and stability. This perspective links education to social stability in two ways. First, in taking their place in the education system, students learn the key norms, values, and beliefs of American culture. Second, by affording all students a chance to develop their skills and talents, education can channel the "best and the brightest" to key social positions.

Does the conventional view of education paint an accurate portrait? Is America's education system really the great equalizer?

To be sure, the American education system has grown dramatically over the past century. At the turn of the last century, only 10% of U.S. youth earned high school degrees, and only 2% earned college degrees (Vinorskis 1992). By World War I, primary education became compulsory in every state; by World War II, the same was true for secondary education. Now fast forward to today. In 2009, nearly 89% of Americans (aged 25–29) had earned a high school degree, and about 31% had earned a bachelor's degree (Aud et al. 2010, Table A-22–1). Higher education is also seen as a possibility for more and more Americans. The percentage of high school graduates who immediately enroll in college is taken as an indicator of the *accessibility* of higher education. In 1972, only 49% of high school graduates immediately enrolled in college. But in 2008, that figure increased to nearly 69% (Aud et al. 2010, Table A-20–1).

The face of our nation's college population has become more socially diverse as well. At the turn of the 20th century, college students were primarily the sons of white upper-class professionals. In contrast, today's college population includes the sons and daughters of all social classes and all racial and ethnic groups. Currently, minority students account for one-third of those enrolled in degree-granting institutions (U.S. Department of Education 2009b). In 2009, 53% of blacks 25 to 29 years old and [illegible]% of Hispanics in this age group had completed some college (Aud et al. 2010, Table A-22–1).

In support of conventional wisdom, one must note a strong and positive association between income and education. In 2008, for example, the median yearly income for those with advanced degrees was $50,000. Contrast that figure with the median yearly income of those lacking a high school degree: only $23,500 (Aud et al. 2010, Table A-17–1). As educational levels *increase*, the poverty rate *decreases*—24% of those without a high school degree are below poverty level compared to 4% of those with a college degree or higher (U.S. Census Bureau 2009b, POV29). Table 22.1 illustrates this connection; the table shows the poverty rates for the **working poor**—that is, individuals who have spent at least 27 weeks in the labor force but have incomes below the poverty level. Again, the positive benefit of education is quite apparent (U.S. Department of Labor 2010d).

Working poor Individuals who have spent at least 27 weeks in the labor force but have incomes below the poverty level

There is also a strong positive relationship between education and health, one that holds across income levels and age (Baum and Ma 2007). As education levels increase, so too does

Table 22.1 Poverty Rates of Working Poor by Education Levels, 2008

Education Level	Poverty Rates
Less than high school diploma	18.3%
High school degree, no college	7.2%
Some college, no degree	5.9%
Associate's degree	3.5%
College degree	1.7%

Source: U.S. Department of Labor, U.S. Bureau of Labor Statistics. 2010. *A Profile of the Working Poor, 2008.* http://www.bls.gov/cps/cpswp2008.pdf.

the percentage of people meeting recommended physical activity levels (U.S. Census Bureau 2011, Table 208). The world over, increased education translates into better nutrition, higher immunization rates, and lower mortality rates (World Bank 2010). And as educational achievement increases, so does individuals' civic involvement: Voter registration and turnout increase with education, as does volunteer work. And being open to the opinion of others increases with education (Baum and Ma 2007; U.S. Census Bureau 2011, Table 406). The benefits of education also extend to those who detour from life's conventional path: Prisoners' exposure to educational programs while incarcerated is associated with lower rates of recidivism (Mercer 2009; Winterfield et al. 2009).

The statistics just quoted suggest that education's links to the good life are right on target. However, on closer examination, one finds several situations that can weaken the strength of that bond. Education's "lifting mechanism" may not be fully functional for all social groups. Research verifies that the economically disadvantaged don't share equally in the benefits of higher education. The pattern is very clear: As income decreases, so does the percentage of students who both expect to attend and who enroll in college. Only 29% of students with the lowest socioeconomic status (SES) background expect to get a college degree, and only 22% expect to earn a graduate or professional degree. But among students with the highest SES background, 33% expect to earn a college degree, and 53% expect to earn a graduate or professional degree (U.S. Department of Education 2006, Table 29.1). While about 56% of students from low-income families immediately transition to college, a full 82% of students from high-income families do so (Aud et al. 2010, Table A-20–1). In short, family income remains the best predictor of both who will go to college and what college students will attend (Callan 2007). The pattern is in part explained by the preparation of students. Lower-income high school graduates are less academically qualified for college than are their wealthier counterparts. Yet even when we look only at those students who are qualified for admission to a 4-year college, the relationship between income and enrollment stubbornly persists: The higher the income, the higher the enrollment rate (Baum and Ma 2007; McSwain and Davis 2007).

Over the last decade, tuitions and fees of public 4-year colleges and universities have risen about 5% per year beyond general inflation. Indeed, these costs have outpaced those for medical care, food, and housing. During the same 10-year period, family incomes have decreased or remained stagnant. Our latest recession also means that currently, families are facing high unemployment rates. Consequently, today many families are finding the prospect of a college education to be beyond their reach (College Board 2009a). According to our national report card on higher education, 49 out of 50 states received flunking grades with regard to various indicators of affordability (California is the only state with a passing grade on affordability). In all but two states (New York and Tennessee), the percentage of family income (after financial aid) needed to meet college costs has *increased.* And this financial burden hits lower-income families hardest. Students from low-income families pay 40% of their family income to enroll in public 4-year colleges, while those from middle- and upper–income families pay 25% and 13%, respectively (National Center for Public Policy and Higher Education 2008).

For most students, financial aid is a necessary route to a college degree. In 2009 through 2010, full-time undergraduates received financial aid packages of just over $11,000—roughly $6,000 in grant aid (85% from college/universities and federal government) and nearly $5,000 in federal loans (College Board 2010a). From 1998 through 2008, the amount of student loan dollars doubled (National Center for Public Policy and Higher Education 2008). Indeed, the high cost of higher education presents two troubling developments: (1) Many qualified but financially strapped students forego college and (2) most students graduate from college with debt. The decadewide college degree loss due to financial barriers is estimated to be between 1.7 and

3.2 million (Advisory Committee on Student Financial Assistance 2008). Today, more than two-thirds of college graduates have federal loan debt, with the average debt being more than $23,000 (PIRG 2010). And even after receiving financial assistance via grants and loans, the 2008 low-income family still faces a cost of $7,800 a year for a child to attend a 2-year public college, $10,300 for access to a public 4-year college, and approximately $18,000 for access to a private 4-year college (Aud et al. 2010, Table A-47–2). Unfortunately, federal and state financial aid has not allowed middle- and lower-income families to keep pace with the increasing costs of higher education. In the early 1990s, the average Pell Grant (the largest source of federal aid for financially needy students) covered 76% of tuition costs at public 4-year colleges. In 2009 to 2010, the maximum Pell Grant (awarded to only 25% of Pell recipients) covered only 35% of those expenses and only 15% of the cost of a private 4-year college (College Board 2010a).

Your Thoughts . . .

Estimates show that it takes 14 years for typical college students to earn enough money to compensate for their college borrowing and the lost earnings incurred during their college years. Do you think this is a reasonable timeline?

Today, colleges and universities provide the largest amount of student financial aid. And while one might expect that this institutional aid might be greatest for those in the most financial need, this is *not* the case (Ehrenberg 2007). While it is true that between 2000 and 2007, the proportion of institutional grant dollars going to need-based students increased from 35% to 44%, it is still the case that 56% of these grants are awarded on a non-need basis (College Board 2010a). This anomaly reflects colleges' and universities' concern with their competitive ranking—a standing that benefits from admitting students with the best college prep backgrounds (and in turn students from higher-income families). And even at public community colleges, institutions with a tradition of making college affordable for all, half of the total grants dollars in 2007 to 2008 were distributed without regard to financial need (College Board 2010b).

For some students, the answer to the high cost of a college education is the community college. Approximately 43% of undergraduates attend 2-year institutions. The enrollment rate for 2-year colleges grew significantly during the 1970s, and total enrollments are expected to reach new highs through 2019 (Aud et al. 2010). Two-year community colleges have been described as the safety net for a state's educational system. They have played a particularly important role for first-generation college students as well as for low-income and minority students: 52% of American Indian, 52% of Hispanic, 45% of Asian, and 42% of black undergraduates attend a community college (Ayers 2010). They are seen by many financially strapped students as a practical and reasonable alternative to the escalating costs of 4-year colleges: The yearly average cost at a community college is $2,544 versus $7,020 for 4-year public colleges (Wilson 2010).

As valuable a role as community colleges play, however, it is important to note some caveats of pursuing a college career via this route. Like those of 4-year public colleges, community college costs have been increasing above inflation rates (Provasnik and Planty 2008). At the same time, budgetary support for these institutions has grown more and more problematic. State and local support has decreased in recent years, forcing more community colleges to engage in private fundraising as well as to look to Washington for federal grants. These budgetary constraints have meant that many 2-year colleges have had to turn away hundreds of thousands of students in recent years (Evelyn 2004). And despite the popular view that community colleges are stepping stones in the higher-education landscape, they do not always function this way for enrollees. A higher percentage of first-time freshmen at community colleges than those at public 4-year colleges leave school without completing a degree (45% vs. 17%, respectively, in 2006). This pattern holds true even for those community college students who originally intended to transfer to a 4-year college (Provasnik and Planty 2008). Furthermore, community college degrees often don't yield the same educational dividends of 4-year institutions. Consider, for instance, that less than 1% of transfer students to elite private 4-year colleges and 8% of transfers to elite public 4-year colleges are from community colleges (Chronicle of Higher Education 2006). Programs of study at community colleges are often criticized for being narrow and thus limiting students' expectations (Ayers 2010; Beaky 2010). The implications of terminating college with an associate's versus a bachelor's degree are striking when we focus on the financial payoff of higher education. In 2008, median annual earnings for an associate's versus a bachelor's degree were $36,000 and $46,000, respectively (Aud et al. 2010).

The greatest returns (educational, occupational, and financial) on a college degree are reserved for graduates of elite or selective private colleges (Black and Smith 2006; Brand and Halaby 2006; Long and Evan 2008). Harvard economist Caroline Hoxby has determined that students attending the top-ranked selective colleges earn back their educational investments many times over during their working careers (Hoxby 2001). "Selective" colleges, however, are very selective about their student bodies. High-ability but low-income students are underrepresented at selective colleges (Hill and Winston 2010). Gatekeeping practices (used to control students' educational progress), escalating costs, increasing use of merit- rather than need-based financial aid packages by elite universities, and **"legacy" admission** policies all ensure that access to such institutions is restricted largely to members of the most privileged social classes (Ehrenberg 2007; Glenn, 2009; Schmidt 2008).

> **"Legacy" admission**
> The practice by some colleges and universities that gives preference to applicants who are relatives of alumni

Education's equalizing mechanism often seems to fail ethnic and racial minorities as well. The 2008 high school dropout rate for 16- to 24-year-old white students was 4.8%—a figure much lower than the 9.9% and 18.3% rates for blacks and Hispanics, respectively. In 2005, for the 25- to 29-year-old age group, 93% of whites, 87% of blacks, and 63% of Hispanics had completed high school. And in 2009 for the same age group, 37% of whites but only 19% of blacks and 12% of Hispanics had completed a bachelor's degree. In terms of the perceived accessibility of higher education, minority students face a more difficult transition path than white students: In 2008, 72% of white high school graduates immediately went on to college compared with 56% of black high school graduates and 64% of Hispanic high school graduates (Aud et al. 2010, Tables A-19–2, A-22–1, A-20–3). When minority students go on to college, they frequently enter 2-year colleges: 42% of black and 45% of Hispanic undergraduates attend community colleges (Ayers 2010). And while minority enrollments in graduate programs have at least tripled (and quadrupled for some groups) since 1976, in 2008, the percentage of graduate students who were black or Hispanic was just 12% and 6%, respectively (Aud et al. 2010). The burden of financing higher education is also heavy on minority students, especially black students: 27% of black versus 16% of white college graduates have student loan debt of $30,500 of more (Baum and Steele 2010).

Low educational achievement does more than deprive people of degrees and earnings. Low educational achievement imposes heavy costs on literacy. **Literacy** refers to the ability to use printed and written information to function in society and is assessed in three areas: prose, document, and quantitative. Across all three areas, as educational attainment increases, so too do literacy scores. The percentage judged proficient—that is, possessing the skills necessary for complex and challenging literacy activities—also increases with educational attainment (see Table 22.2; U.S. Department of Education 2007, Indicator 18). Our adult reading habits are also fueled by education—the higher the education level, the more likely we are to report reading newspapers, magazines, and books on a daily basis (U.S. Department of Education 2006, Indicator 20).

Literacy
The ability to use printed and written information to function in society

The cycle of low educational achievement is hard to break: Parental involvement in the educational lives of their children—attending school meetings or teacher conferences or volunteering at school—is inversely related to parental education (U.S. Department of Education 2009b, Table 23). Dropout rates of students from low-income families are 10 times higher than those of students from high-income families students (Cataldi et al. 2009). While 82% of high school graduates whose parents have college degrees or more immediately transition to college, only 54% of graduates whose parents have a high school degree or less do so (Aud et al. 2010, Table A-20–1). College completion rates for students whose parents lack college degrees are 62% versus 76% for students with a least one parent with a bachelor's degree (Baum and Ma 2007).

Perhaps the most dramatic failure of education's equalizing powers is witnessed in the area of gender. Although women are clearly present in the classroom today—they constitute 57% of college enrollments and 59% of graduate enrollments (Aud et al. 2010, TableA-7–1)—this is a noteworthy break with the past. The educational history of women in the United States bespeaks little in the way of equal opportunity or achievement. In the 1900s, the doors to high schools and subsequently colleges were opened to women. Indeed, in 1907, there were 110 women's colleges

Table 22.2 Percentage of Adults (16+) Proficient in Prose, Document, and Quantitative Literacy by Education Level: 2003

	Adults Proficient in Various Types of Literacy		
Education Level	**Prose**	**Document**	**Quantitative**
Less than high school	1%	2%	1%
High school graduate	4%	5%	5%
Some college	11%	10%	11%
Associate's degree	19%	16%	18%
College degree	31%	25%	31%
Graduate study/degree	41%	31%	36%

Source: U.S. Department of Education, 2007a, Table 18.2. Available online at http://nces.ed.gov/programs/coe/2007/section2/table.asp?tableID=693.

in the United States. However, only 32% of these women's colleges met even the most basic standards of a true higher education program. Rather, most women's colleges were engaged in the task of preparing women for their place in society—that is, as homemakers. Government aid policies of the era reinforced this traditional tracking. Vocational training such as cooking, sewing, and home economics qualified for federal subsidies; commercial training did not (Stock 1978).

Women who wanted a "real" higher education were limited by restrictive college admission policies. Most elite schools of the East simply refused to accept women. Western institutions had more liberal policies, but such policies were generally driven by financial motives: They admitted women in an effort to ward off financial disaster. The most blatant example of this practice, however, occurred at the University of Chicago. Faced with bankruptcy in 1873, the university decided to admit women. When the financial situation of the university improved, the institution's stance toward female admissions changed dramatically. Women were immediately relegated to a separate junior college (Stock 1978).

The post–World War II era further compromised women's access to education. Prior to World War II, the percentage of women attending college had been increasing steadily. However, postwar college admissions gave absolute priority to war veterans. Such policies forced women back into the home, despite the work they had done during the war to keep America productive (Stock 1978). Furthermore, the postwar policy signaled the beginning of a long-term trend. Even today, despite their heavy investment in higher education, women do not enjoy returns equal to those of men; the financial benefits of education are significantly lower for women than for men at every level of educational achievement (see Essay 11).

Instances of class, racial, ethnic, and gender inequality lead many to doubt the conventional wisdom on education. Indeed, conflict theorists question education's ability to equalize. **Conflict theorists** analyze social organization and social interactions by attending to the differential resources controlled by different sectors of a society. Conflict theorists suggest that the U.S. education system actually transmits inequality from one generation to the next (Bidwell and Friedkin 1988; Bowles and Gintis 1976, 2002; Collins 2001; Kozol 1991, 1995, 2005; Swartz 2003).

Conflict theorists
Sociologists who analyze social organization and social interactions by attending to the differential resources controlled by different sectors of a society

The conflict view of education parallels structural functionalism in acknowledging the role of education within the socialization process. **Socialization** refers to the process by which we learn the norms, values, and beliefs of a social group, as well as our place within that social group. But in contrast to structural functionalists, conflict theorists argue that the goals of socialization vary according to the social class of students.

Socialization
The process by which we learn the norms, values, beliefs, and symbols of a social group and our own place in that social group

In Essay 12, we noted Jonathan Kozol's observations regarding racial inequalities in U.S. schools. Kozol argued that whites and nonwhites often learn different lessons within American schools. Kozol notes similar inequalities in comparisons of various social classes. (Indeed, Kozol argues that much racial inequality in the United States is fueled by factors that link minority racial status to low economic status.)

Through their elementary and high school education, lower- and working-class students are taught attitudes and skills that best prepare them for supervised or labor-intensive occupations. These include respect for authority, passivity, the willingness to obey orders, and so on (Kozol 2005; Solomon, Battistich, and Hom 1996).

They also frequently receive the message that school is not the place for them (Langhout and Mitchell 2008; Oats 2009). Low-income students who elect to go the more financially feasible route of community college often find themselves in vocationally oriented programs that squelch aspirations and function instead as job training centers (Beaky 2010; Wilson 2010). In contrast, middle- and upper-class children are taught critical thinking and skills essential to management-level jobs and professional careers—that is, responsibility and dependability. Higher-quality schools and positive interpersonal reinforcements better prepare these students for their college and postcollege careers, where they will continue their training in management and professional skills (Kozol 1991, 2005; Oates 2009).

BOX 22.1 SECOND THOUGHTS ABOUT THE NEWS: SCHOOL DAZE

With the arrival of the 2010 school year, some students in Los Angeles were anticipating more than just reading, writing, and arithmetic–more than 4,000 Los Angeles, California, students were waiting for the doors to open on the Robert F. Kennedy K–12 Community Schools. The $100 million complex is located at the site of the Ambassador Hotel in L.A. where Kennedy was assassinated in June 1968 as he sought the Democratic nomination for president. In acquiring the site nearly 20 years ago, the district had to do court battle (at a cost of $9 million) with Donald Trump, who wanted to build the world's tallest skyscraper on the land. The district also had to battle preservationists who wanted to stop the school's construction in order to preserve key parts of the Ambassador Hotel. (A compromise was reached with the preservations.)

The complex is L.A.'s (and the nation's) most expensive public school ever, costing the district roughly $580 million. (Additionally, the district sought and received another $6 million in the summer of 2010 to assure a fall opening.) The cost per student is estimated to be around $135,000! Still, the RFK complex is not an anomaly. In recent years, there have been dozens of $100 million schools built across the nation, and several of them are located in low-performing districts–for instance, New York City and New Brunswick, New Jersey (Hoag 2010).

The RFK Center was funded many years before our current economic recession through a $20 billion voter-approved bond project. Unlike other schools in high-poverty areas, the RFK complex will be a gem of a facility that will include a public park, a pool, a state-of-the-art auditorium, a public art installation, and preservations from the original Ambassador Hotel. Given today's hard economic times, some are questioning the wisdom of such an expensive school, and the school board seems to be listening to the concerns. When the board approved the additional funding needed for the fall opening, it resolved to do some serious cost cutting. For example, it proposed saving some $74,000 by changing district policies for car wash expenses!

Are "Taj Mahal" schools an extravagance we can't really afford? Or is this a long-overdue step in providing disadvantaged students a "rich" environment for learning? Give it some second thoughts.

More recently, the legislative mandates of No Child Left Behind have produced negative changes in classroom practices for low-income and minority children. The program's focus on high-stakes testing has resulted in questionable pedagogical practices, more "teaching to the test" and a neglect of other learning outcomes for students in high-poverty and high-minority schools (Cummins 2007; Jehlen 2009).

Beyond socialization, the funding and delivery of education can also maintain and reinforce class divisions. At every level of government, policy makers reward those who have resources with more resources (Carey and Roza 2008). For instance, federal funds awarded through Title 1 depend on the wealth of a state: States that spend more money on education get more Title 1 funds. Consider also that local revenues provide more than 40% of the total funding for public school budgets. Many states use local property taxes to augment these funds. This arrangement tends to put wealthier suburban school districts at a clear advantage because they can generate more monies via higher property values. Thus we wind up with the tragic flaw of school funding: Those with more resources get more resources (Carey and Roza 2008).

The size of schools and the number of students in classrooms can also make for very unequal educational experiences. Mid-sized schools offer the most positive environments for learning. Large schools (900+ students), however, dominate central city school districts. Student–teacher ratios are highest in our largest elementary and secondary public schools (U.S. Department of Education 2007, Indicator 30). As the size of schools increases, so too does the level of school crime (Aud et al. 2010).

Overcrowding is another problem that diminishes the educational experience for students. Approximately 8% of our public schools are severely overcrowded—that is, operating at 25% above their student capacity. Another 10% of public schools are overenrolled at 6 to 25% above capacity. Overenrollment is more likely in schools with more than 50% minority enrollments, in large schools, and in urban schools. A common response to overcrowding is to increase class size (reported by 44% of public school principals). In 2005, 30% of students in public schools attended schools that were overenrolled (Chaney and Lewis 2007).

Another way of assessing the educational experiences of students is by considering the impact of **high-poverty schools**—that is, schools where 75% or more of students are eligible for free or reduced-price meals. In 2008, 17% of all public schools were high poverty, and overall, 20% of elementary and 6% of secondary school students attended these schools. A snapshot of student enrollments, however, would show that high-poverty schools are minority schools (see Table 22.3). Students in high-poverty schools are at an educational disadvantage on several fronts: in terms of teacher qualifications, faculty teaching out of field, national assessments of reading and math scores, high school graduation rates, and rates of college attendance (Aud et al. 2010; National Science Board 2008; see Table 22.4).

High-poverty schools
Schools where 75% or more of students are eligible for free or reduced-price lunches

Remarkable progress has been made with regard to computer technology in our schools. In 2005, public schools averaged 154 instructional computers as compared to 90 in 1998. And virtually all public schools have Internet access. Still, the percentage of *instructional rooms* with Internet access as well as the average number of instructional computers is lower in high-poverty than low-poverty schools (U.S. Department of

Table 22.3 Racial/Ethnic Characteristics of Students Attending High-poverty Schools 2007–2008

Elementary High-poverty Schools	
% Hispanic	46%
% Black	34%
% White	14%
% Asian/Pacific Islander	4%
% American Indian/Alaska Native	2%
Secondary High Poverty Schools	
% Hispanic	44%
% Black	38%
% White	11%
% Asian/Pacific Islander	4%
% American Indian/Alaska Native	3%

Source: Aud, S., W. Hussar, M. Planty, T. Snyder, K. Bianco, M. A. Fox, L. Frohlich, et al. 2010. *The Condition of Education 2010* (NCES 2010-028). Washington, DC: National Center for Education Statistics, Institute of Education Sciences, U.S. Department of Education. http://nces.ed.gov/pubs2010/2010028_1.pdf.

Table 22.4 Educational Outcomes for High- vs. Low-poverty Schools (2007–2009)

	High-poverty Schools	Low-poverty Schools
% of 4th graders at or above basic reading level	45%	83%
% of 4th graders at or above basic math level	64%	93%
% of 8th graders at or above basic math level	53%	87%
% of 8th graders at or above basic math level	49%	87%
High school graduation rates	68%	91%
High School graduates enrolled in 4-year colleges	28%	91%
% of schools reporting 20 or more incidents of violence	38%	15%

Source: Aud, S., W. Hussar, M. Planty, T. Snyder, K. Bianco, M. A. Fox, L. Frohlich, et al. 2010. *The Condition of Education 2010* (NCES 2010–028). Washington, DC: National Center for Education Statistics, Institute of Education Sciences, U.S. Department of Education. http://nces.ed.gov/pubs2010/2010028_1.pdf.

Education 2009b). And when we look at student home use of computers, a sizeable gap still remains between white and nonwhite students (U.S. Department of Education 2007). Indeed, teens with the highest access to the Internet are those from white, college-educated, and higher-income families (Purcell 2010).

Teachers themselves may also play a role in transmitting educational inequality (Brophy 1983). Public schools with high minority enrollments are more likely to employ teachers with less experience or lacking regular certification. Students in high-minority and high-poverty high schools are more likely to have faculty who are teaching "out of field"—that is, teachers who did not major or are not certified in the subject they teach (Aud et al. **2010**; Education Trust 2008; National Science Board 2008). Research has consistently found that social characteristics of students (e.g., low socioeconomic and minority status) negatively affect teacher perceptions and expectations of student performance (Anderson-Clark et al. 2008; Auwarter and Aruguete **2008**; Langhout and Mitchell et al. **2008**; Tenenbaum and Ruck 2007; Wood et al. 2007). Still, there is also research that indicates that teachers who hold high academic expectations can offset the negative effects too frequently associated with class, race, ethnicity, and even low parental expectations (Bamburg 1994; Jacobs and Harvey 2010; Mehan, Hubbard, Lintz, and Villanueva 1994; Omotani and Omotani 1996; Wood et al. 2007).

Finally, conflict theorists cite tracking as an important source of educational inequality. **Tracking** is a practice by which students are divided into groups or classes based on perceived ability. Although tracking is meant to group students in terms of academic ability, in reality, it tends to create economic, racial, and ethnic clusters. Since the 1980s, the *form* of tracking has been modified from total program (e.g., college prep tracks vs. vocational tracks) to course-level distinctions (e.g., honors courses in English, history, math, etc.), yet the inequality of the system remains. Studies that compare the performance of low-, medium-, and high-ability tracks show that tracking benefits only the high-ability groups (Condron 2008; Huang 2009; Kao and Thompson 2003; Lleras and

Tracking
A practice by which students are divided into groups or classes based on their perceived ability

***De facto* segregation** Segregation due to actual behavioral practices

***De jure* segregation** Segregation that is imposed by the law

Self-fulfilling prophecy A phenomenon whereby that which we believe to be true, in some sense, becomes true for us

Rangel 2009). Thus, critics of tracking argue that the practice does little to equalize opportunity but rather creates ***de facto*** and ***de jure*** (legally sanctioned) **segregation** in our schools (Nelson 2001; VanderHart 2006). It is also argued that tracking fosters a self-fulfilling prophecy. A **self-fulfilling prophecy** is a phenomenon whereby that which we believe to be true, in some sense, becomes true for us. Within the tracking system, students do as well or as poorly as they are expected (or given the opportunity) to do (Lewis and Cheng 2006; Tach and Farkas 2006; Torff 2008).

The inequalities found in the U.S. education system are likely to grow worse in the near future. Our current recession is taking a hard toll on education. The prospects of continuing school layoffs, increased class sizes, shortened school years, and school budgets decimated by steep funding cuts have prompted Secretary of Education Duncan to warn of an impending educational catastrophe. The call for a federal bailout (of some $23 billion) is undermined by our historic federal budget deficits. The recession also delivers a serious blow to the growing problem of college affordability. If recent trends in spiraling costs are left unchecked, college will be beyond the reach of far too many families. Indeed, our international leadership in college access is waning; the United States currently ranks 15th out of 29 nations with regard to the percentage of 18- to 24-year-olds who have completed college (Callan 2008). These developments are especially worrisome given the short- and long-range forecasts for higher education and economic recovery. Both prospects are tied to the United States successfully competing in an increasingly global and information- and technology-driven economy. Our edge will depend on the educational preparation of American youth for this new economic order.

Forecasted changes in the job market indicate that occupations requiring some postsecondary education (e.g., associate's degrees, bachelor's degrees, and beyond) will experience greater growth than occupations requiring on-the-job training. As indicated in Table 22.5, seven out of the top 10 fastest-growing occupations today require a bachelor's degree or higher (U.S. Department of Labor 2010f). Clearly, an education system that fails to offer high-quality public education and restricts access to college and post-college training assures a bleak future for the undereducated. Students who are absent from these settings will also be chronically absent from the jobs of tomorrow and from upward mobility. But there are larger implications as well. Consider the much-anticipated aging of the Baby Boomers. Most of the projected job growth across major occupations will be due to the need for replacement workers (U.S. Department of Labor 2010f). The retirement of the Baby Boomers—the most educated generation in U.S. history—is anticipated to cause a decrease in the average education level in the United States. Such a change is predicted to result in decreasing personal incomes and decreasing tax revenues. The chances for such declines will certainly increase if we fail to improve the educational prospects of the fastest-growing segments of the youth population: racial and ethnic minorities (Policy Alert 2005).

The inequalities found in the U.S. education system present an unavoidable irony. Education can indeed be a great equalizer, but at present, it is not. This tool of upward mobility is most likely to be placed in the hands of those who are already located in advantaged positions. Thus, rather than creating opportunity, the current educational system more accurately sustains the status quo. Our greatest educational challenge, then, may be to devise both an ideology and policies that can reset the educational mission of our nation.

Table 22.5 Top 10 Fastest-growing U.S. Occupations and Their Educational Requirements

Rank	**Educational Requirements**
1. Biomedical engineers	Bachelor's degree
2. Network systems & data communications analysts	Bachelor's degree
3. Home health aides	On-job-training
4. Personal and home care aides	On-job-training
5. Financial examiners	Bachelor's degree
6. Medical scientists	Doctoral degree
7. Physician assistants	Master's degree
8. Skin care specialists	Post-secondary vocational degree
9. Biochemists and biophysicists	Doctoral degree
10. Athletic trainers	Bachelor's degree

Source: U.S. Department of Labor. 2010. *Occupational Outlook Handbook, 2010–11 Edition*. http://www.bls.gov/oco/oco2003.htm#education.

LEARNING MORE ABOUT IT

Samuel Bowles and Herbert Gintis offer a classic critique of the American educational system in *Schooling in Capitalist America: Educational Reform and the Contradictions of Economic Life* (New York: Basic Books, 1976). A more recent analysis can be found in Jonathan Kozol's books, *Savage Inequalities* (New York: Crown, 1991) and *Amazing Grace: The Lives of Children and the Conscience of a Nation* (New York: Crown, 1995). Kozol examines the growing resegregation of our public schools in *The Shame of the Nation: The Restoration of Apartheid Schooling in America* (New York: Crown, 2005). You can read an interview with Kozol about this recent work at Campusprogress.org (http://www.campusprogress.org/features/552/five-minutes-with-jonathan-kozol).

In *The Chosen: The Hidden History of Admission and Exclusion at Harvard, Yale and Princeton* (Boston: Houghton Mifflin, 2005), Jerome Karabel examines the history of admissions (and organizational gatekeeping) at these Ivy League institutions from 1900 through 2005.

In *Race in the Schools: Perpetuating White Dominance?*, Judith Blau offers a compelling case for the value of pluralism in American schools (Boulder, CO: Lynne Rienner Publishers, 2003).

Douglas Downey offers an assessment of the theory that racial/ethnic differences in school performance are the result of minority groups developing an oppositional view of schooling in his article "Black/White Differences in School Performance: The Oppositional Culture Explanation" (*Annual Review of Sociology* 34:107–126, 2008).

Shannon Harper and Barbara Reskin review the history and impact of affirmative action in the areas of education and employment in their 2005 article "Affirmative Action at School and on the Job" (*Annual Review of Sociology,* 31:357–380).

GOING ONLINE FOR SOME SECOND THOUGHTS

To learn the educational requirements for specific occupations, see the Department of Labor's Occupational Outlook Handbook: http://www.bls.gov/oco/oco2003.htm#education.

The following organizations and sites can help you learn more about various issues in education:

Institute for Higher Education (dedicated to improving access to and success in secondary education): http://www.ihep.org

National Center for Public Policy and Higher Education (responsible for issuing the Measuring Up reports): http://www.highereducation.org (or access the report directly at http://measuringup2008.highereducation.org/)

National Assessment Governing Board (responsible for issuing the Nation's Report Card on academic achievement across various subject areas): http://www.nagb.org

National Institute for Literacy: http://www.nifl.gov

U.S. Department of Education: http://www.ed.gov

NOW IT'S YOUR TURN

1. Access a college catalog from each of the following categories: (a) an Ivy League college, (b) a 4-year state college, and (c) a local community college. Compare the mission statement contained in each school's catalog. Also compare each school's programs of study and the types of courses it offers. Use your data to prepare a discussion regarding the equal opportunity philosophy of U.S. colleges and universities.

2. Access the most recent edition of the *World Almanac*. Obtain information on the following four items: high school graduation rate by state, student–teacher ratio per state, per-capita personal income per state, and state revenues for the public schools. Identify those states that represent the top five and the bottom five of each data category. Is there any overlap in these top five and bottom five groups? Speculate on your findings.

3. Imagine that you and your spouse and two small children are currently residing in New Jersey. You have just received two job offers. One would require a move to California, the other a move to New Mexico. Assume that education prospects for your children are a major consideration for you. Access the website for the National Center for Public Policy and Higher Education (http://www.highereducation.org/about/about.shtml).
   - Click on the link for the latest edition of *Measuring Up* (last icon to right at top of page; The National Report Card on Education) and then click on the link for Online State Report Cards.
   - Use the Compare States link to access data that will help you decide whether you should remain in New Jersey or move west to California or New Mexico. Share and defend your decision.

SOCIAL INSTITUTIONS: RELIGION

ESSAY 23

Conventional Wisdom Tells Us . . . We Are One Nation Under God

God bless America . . . it's an invocation frequently heard across the United States. Yet in light of our country's long-standing commitment to the separation of church and state, "God bless America" is also a prayer that can make some uncomfortable. Are we united or divided with regard to the place of God in our nation? This essay explores the issue.

One nation under God . . . most recognize these words as part of the U.S. Pledge of Allegiance. For well over 100 years, American schoolchildren have recited the pledge as part of their daily school ritual. The earliest version of the pledge, penned in 1892, was quite simple: "I pledge allegiance to my Flag, and to the Republic for which it stands: one Nation indivisible, with Liberty and Justice for all."

The first significant change to the pledge was made in 1923. The place: the first National Flag Conference in Washington, D.C. At this time, "my flag" was changed to "the flag of the United States." (The change was designed to avoid potential confusion among members of the growing immigrant population in the United States.) A year later, the target of the pledge was further specified as "the flag of the United States of America." Finally, the addition of the words "under God" was authorized in 1954. The campaign to add these two words was led by the Knights of Columbus and was a byproduct of our national fear of communism. President Eisenhower authorized the change, arguing that it highlighted religion's significant place in American history and culture:

> In this way we are reaffirming the transcendence of religious faith in America's heritage and future; in this way we shall constantly strengthen those spiritual weapons which forever will be our country's most powerful resource in peace and war. (UnderGodProCon, 2006)

To be sure, religion and the American way form a natural pairing. The Pilgrims' arrival in the New World was motivated by the pursuit of religious freedom. The Puritans too actively sought to establish strong ties between politics and the tenets of Protestantism. In fact, in some colonies, the Puritans successfully established official religions. And, of course, many credit the very spirit and growth of U.S. capitalism to the strong influence of Protestantism—especially its core value of the work ethic.

Many of our Founding Fathers supported the overlap between government and religion. George Washington asserted that it was impossible to govern without God and the Bible. Ben Franklin contended that no nation could rise without the aid of God. And for centuries, our secular government has peacefully coexisted with the sacred. Presidents invoke God's help as they are sworn into office, a practice started by George Washington (Davis 1990). Witnesses in our courts swear on Bibles to tell the truth. Public ceremonies and congressional sessions start with religious invocations. Since the early 1950s we have had an annual declaration of a national day of prayer. Our national currency and motto both attest to our trust in God.

Of course, there are limits to the secular–sacred relationship. Despite early efforts by the Puritans, the United States has no official religion. Indeed, the Constitution mandates the separation of church and state. While some Founding Fathers supported the infusion of religion into politics, others opposed this development. President James Madison warned that failing to honor the separation between church and state would make us vulnerable to the religious conflicts that had plagued European nations. Jefferson advocated for a "wall of separation" between church and state. Every 10 years, the U.S. government conducts a census of the entire population, yet Public Law 94–521 *prohibits* questions about religion. And consider that in the spring of 2010, a federal district court ruled that a national day of prayer was unconstitutional.[1] To be sure, we are rather diligent about keeping a separation between church and state. Only 14% of Americans report that their religious beliefs are the main influence on their political thinking (Pew Research Center 2008). We remain vigilant about any religious undermining of our civil rights.

Nonetheless, we are a nation of believers, making us, in sociological terms, high in **religiosity.** Religiosity refers to an individual's religious beliefs as opposed to our active participation in religious groups. A 2008 Pew study found that 92% of Americans report believing in God or a higher universal spirit and roughly six in 10 Americans report that religion is very important in their lives. Interestingly enough, "belief" runs even to those we might expect *not* to believe: 21% of self-identified **atheists** (those who deny the existence of God) and more than 50% of self-identified **agnostics** (those who claim the existence of God cannot be known) claim to have a belief in God or a higher spirit (Pew Research Center 2008). Still, despite this near-universal belief in a higher force, most Americans take a nondogmatic view of religion and their faith. Instead, most Americans express religious tolerance, believing there are many paths to salvation. So while we may pledge our allegiance to one nation under God, we are open to who or what that God looks like. This sentiment is echoed in the variety of religious traditions found in the United States.

Religiosity
Refers to individuals' beliefs as opposed to their actual participation in religious groups

Atheists
Individuals who deny the existence of God

Agnostics
Individuals who maintain that the existence of God cannot be known

[1]On the other hand, in the spring of 2010, the Supreme Court *overturned* a lower court ruling that ordered the removal of a cross from a National Park location.

Your Thoughts . . .

A recent poll by the Pew Forum on Religion and Public Life (2010a) found that atheists and agnostics scored the highest in correctly answering questions about various religions and major religious figures (21 correct out of 32 questions). Protestants and Catholics scored the lowest (16 and 15 correct, respectively). Other religions were somewhere in between. Do these findings surprise you?

In a culture that values personal freedom and autonomy and in a nation steeped in immigrant history, the religious diversity that is found in the United States is not surprising. This diversity is clearly seen in the array of religious affiliations found here. **Religious affiliation** refers to formal group membership in a religious community. In recent years, formal group membership has been on the wane. Today, about 16% of Americans say they are unaffiliated with any particular faith—a figure that is more than double the percentage who report having no affiliation as children. For young adults (18–29), 25% lack any current affiliation (Pew Research Center 2008, 2010d). Still, despite this trend, the overwhelming majority of Americans self-report some religious affiliation. As shown in Table 23.1, while Christian traditions still dominate the American society, religious diversity is alive in the United States.

Religious affiliation Formal group membership in a religious community

Diversity also aptly describes American's religious practices and engagement. A majority of Americans (54%) say they attend religious services fairly regularly, but only 39% do so on a weekly basis. As might be expected, attendance varies greatly by affiliation. So while only 34% of mainline Protestants attend weekly services, 42% of Catholics, 58% of Evangelicals, and 75% of Mormons report doing so. It must be noted, however, that some researchers maintain that attendance figures are unrealistically high, the effect of a

Table 23.1 Major Religious Traditions in the U.S.

Among all adults	%
Christian	**78.4**
Protestant	51.3
Evangelical churches	*26.3*
Mainline churches	*18.1*
Hist. black churches	*6.9*
Catholic	23.9
Mormon	1.7
Jehovah's Witness	0.7
Orthodox	0.6
Other Christian	0.3
Other religions	**4.7**
Jewish	1.7
Buddhist	0.7
Muslim*	0.6
Hindu	0.4
Other world religions	<0.3
Other faiths	1.2
Unaffiliated	**16.1**
Don't know/Refused	**0.8**
	100%

Source: U.S. Religious Landscape Survey, Pew Research Center's Forum on Religion & Public Life, © 2008, Pew Research Center. http://pewforum.org/.

*From "Muslim Americans: Middle Class and Mostly Mainstream," Pew Research Center, 2007.

Due to rounding, figures may not add to 100 and nested figures may not add to the subtotal indicated.

Social desirability bias
Measurement error that results when respondents answer questions in a way that makes them look good

social desirability bias in survey research (i.e., answering questions in a way that makes respondents look good). When researchers count people in attendance at a sample of churches, weekly attendance rates are found to be around 20% (Hartford Institute for Religion Research 2006).

Sixty percent of Americans envision God as a person with whom they can have a relationship; 40% see God as an impersonal force. As Figure 23.1 shows, 75% of Americans say they pray at least weekly; 58% report praying daily. And many report that their prayers are answered: 31% get answers at least once a month while 19% find their prayers answered at least once a week. About two-thirds see their faith's sacred texts as the word of God, but within this group, there is a rather even divide regarding whether the texts should be taken literally. More than a quarter of Americans say their faith's sacred texts are written by mortals (Pew Research Center 2008).

Some insight into the array of religious traditions and practices can be gleaned from understanding how religions are structured. Sociologists distinguish among several different kinds of religious groups that vary along their level of *social organization* and *social integration.* A **church** is a long-standing religious organization with a formalized social structure and routinized interactions. Churches typically enjoy strong ties to society. When some members of churches are dissatisfied with the established beliefs or practices, they might leave to form denominations and sects. **Denominations** are groups that break away from their "mother" churches and establish their own unique identities via their specific beliefs, traditions, and rituals. United Methodists, Presbyterians, and Episcopalians are all examples of Protestant denominations. **Sects** can be seen as the product of more conflictual breakaways; sects establish themselves as independent rival religious organizations. **Cults** are new and relatively small and unorganized religious groups whose beliefs set them apart from more established or mainstream groups. Cults are frequently met with suspicion by mainstream society. Yet note that many major religious groups of today started as cults (e.g., the Catholic Church and the Mormon Church).

Church
A long-standing religious organization with a formalized social structure

Denominations
Groups that break away from their "mother" churches and establish their own unique identities via their specific beliefs, traditions, and rituals

Sects
The product of conflictual breakaways from churches; groups that establish themselves as independent rival religious organizations

Cults
New and relatively small and unorganized religious groups whose beliefs set them apart from more established or mainstream groups

The variety of structures and patterns of interaction afford Americans a rather full menu of religions. Forty-four percent of the affiliated want their religion to preserve its traditional beliefs and practices. As such, these individuals may be most "at home" in established religious groups like churches with highly bureaucratic structures. **Bureaucracies** strive to achieve rationality and efficiency via a hierarchical authority structure, a clear division of labor, reliance on written rules/procedures, and authoritarian interaction. But there are also many of the affiliated (47%) who want their religions to adjust to the times. For these individuals, bureaucratic

Bureaucracies
Organizations that strive to achieve rationality and efficiency via a hierarchical authority structure, a clear division of labor, reliance on written rules/procedures, and impersonal interaction

Figure 23.1 Frequency of Prayer by Selected Social Characteristics

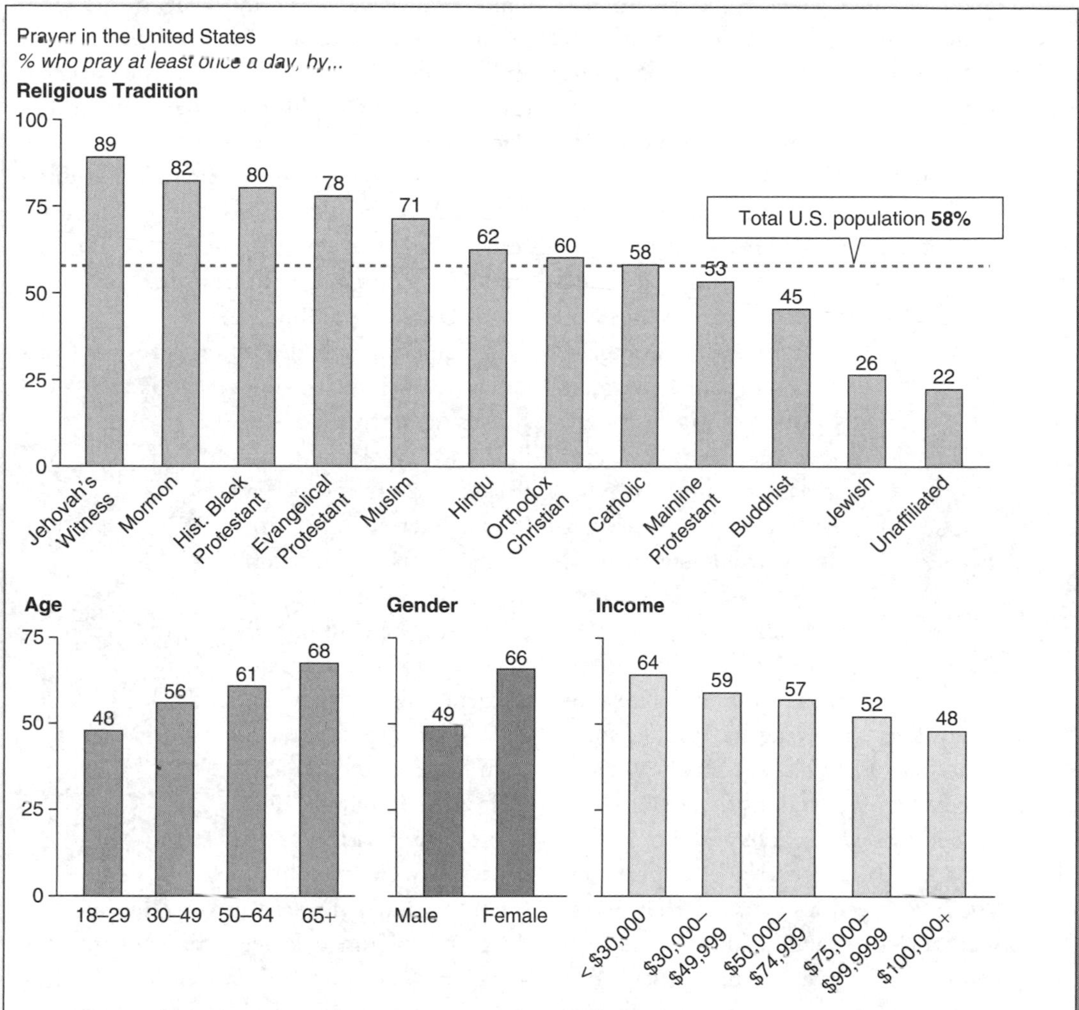

Prayer in the United States
% who pray at least once a day, by...

Religious Tradition

Jehovah's Witness 89
Mormon 82
Hist. Black Protestant 80
Evangelical Protestant 78
Muslim 71
Hindu 62
Orthodox Christian 60
Catholic 58
Mainline Protestant 53
Buddhist 45
Jewish 26
Unaffiliated 22

Total U.S. population **58%**

Age

18–29 48
30–49 56
50–64 61
65+ 68

Gender

Male 49
Female 66

Income

< $30,000 64
$30,000–$49,999 59
$50,000–$74,999 57
$75,000–$99,9999 52
$100,000+ 48

Data from the Pew Forum U.S. Religious Landscape Survey, conducted May 8 to Aug. 13, 2007, among more than 35,000 Americans age 18 and older; released in 2008.

Question wording: People practice their religion in different ways. Outside of attending religious services, do you pray several times a day, once a day, a few times a week, once a week, a few times a month, seldom, or never?

Pew Forum on Religion & Public Life. *U.S. Religious Landscape Survey, June 2008*

Source: Pew Research Center's Forum on Religion & Public Life, © 2010, Pew Research Center. Available at http://pewforum.org/.

structures can pose serious impediments to change.

The bureaucratic structure of more established churches can also contradict the emotionally rich experiences many hope to achieve via their spiritual pursuits. The desire for spiritually moving religious experiences is fueling the fast-growing Pentecostal movement. **Pentecostals** engage in "spirit-filled" worship (speaking in tongues, faith healing) and are found in most Christian traditions (Goodstein 2005). In the United States, Pentecostals

Pentecostals
Those who engage in "spirit-filled" worship (speaking in tongues, faith healing)

account for 4.4% of the adult population and 8.5% of all Protestants (Smith 2008). Former vice presidential candidate Sarah Palin is perhaps one of the most famous Pentecostals in the United States.

Other Americans desire a closer alignment between their religion and government, seeking faith-based policies and laws. These individual are surely an important force behind the political activist stance associated with evangelical and fundamentalist religious groups in the United States. Currently, 42% of Americans describe themselves as evangelical (Newport 2005). **Evangelicals** don't share a common religious group or church; rather, they share a *style or approach* to faith that transcends any one denomination. Consequently, it is possible to speak of evangelical Baptists, evangelical Lutherans, and even evangelical Catholics (19% of Catholics say they are evangelical). Still, some would argue that *practically* speaking, Evangelicals are essentially white, non-Catholic Christians who self-identify as evangelicals and that 30% of Americans meet these criteria (Newport and Carroll 2005). Evangelicals view the Holy Scriptures as authoritative and infallible—the only path to salvation. A core evangelical belief is that we must be born again—that is, we must have our hearts changed by Jesus. Evangelicals are also committed to evangelizing—spreading the word and bringing others to Christ (Frontline 2004). Evangelicals are more likely to identify themselves as politically conservative and tend to align themselves with the Republican Party (Pew Research Center 2008). They also tend to see themselves as an embattled subgroup in American society, one that must fight back against a hostile popular culture and media (Greenberg and Berktold 2004). Their "under siege" mentality has spurred evangelicals to develop and flex their political muscle; they are credited with having a significant hand in the elections of our most recent Republican presidents: Ronald Reagan, George H. W. Bush, and our "born again" president, George W. Bush (who received 79% of white evangelical votes in 2004). Some credit recent welfare reform policies, tax cut policies, and attacks on the U.S. judiciary as evidence of the muscle of right-wing religious groups (Hough 2005). Most recently however, younger evangelicals (aged 18–29), dissatisfied with the returns from their support of the GOP and concerned about the secular agenda of the Tea Party, are starting to shift away from the Republican Party (Cox 2007; Smith 2010).

Evangelicals
Individuals who view Holy Scriptures as authoritative and infallible and who are committed to proselytizing, a.k.a. born again

A significant subset of evangelicals are Christian **fundamentalists**. About one-quarter of evangelicals identify themselves this way (Greenberg and Berktold 2004). Christian fundamentalists differ from evangelicals on several fronts: They insist that the Bible must be read literally, they tend to be intolerant of alternate religious views, and they advocate separatism from those who don't share their views (Frontline 2004). These characteristics can be particularly troublesome in a nation that prides itself on religious freedom and tolerance. Indeed, fundamentalists of any faith—Christian and non-Christian alike—appear to be falling out of favor worldwide. The decline in fundamentalism is attributed to its association with extremism, violence, and political agendas (Goodstein 2005). Since 9/11, the world has become more aware of certain fundamentalist groups like the Taliban and, more importantly, more aware of such groups' reliance on various terrorism campaigns to forward their political and religious causes.

Fundamentalists
A subset of evangelicals who insist that the Holy Scriptures must be read literally and who advocate separation from those who don't share their views

Many of the Founding Fathers argued that the separation of church and state was the only way for religion to flourish in America. On this point they seem correct. Our legal and cultural tradition of separation of church and state has created a favorable environment for religious diversity. The United States has the greatest number of religious groups of any nation of the world (Adherents 2005). The Hartford Institute estimates that there are 335,000 religious congregations in the United States. Mormons are one of the fastest-growing religious groups in the country today and are currently tied with those of the Jewish faith in terms of their percentage of the adult population. Currently, it is estimated that there are 2.5 million Muslims in the United States (representing .8% of our population; Pew Research Center 2009b). (Because the U.S. Census is prohibited from asking questions about religious affiliation, data about the religious affiliations of Americans must be estimated from various sources.) **Ecclesias**, official state religions, are simply not part of the U.S. landscape.

Ecclesias
Official state religions

Still, despite the diversity of religions found here, America has long been regarded as a Christian and, more specifically, a Protestant nation. Indeed, as indicated in Table 23.1, this perception is still accurate (although the percentages of Protestants are decreasing). Furthermore, while many religious groups are found in America, the religious landscape of the United States is nonetheless dominated by a handful of traditional, mainstream Christian churches. Despite their presence in the United States, the Buddhist, Hindu, Jehovah's Witness, Muslim, and other world religious traditions constitute only 5% or less of the population in nearly every state of the nation (Pew Research Center 2008).

Your Thoughts . . .

With the nomination of Elena Kagan to the Supreme Court, some expressed concern over the fact that, for the first time in history, there would be no Protestant justices on the court. Should the religious composition of the Supreme Court be a relevant issue when selecting judges?

Many believed that the terrorist attacks on September 11 strengthened America's strong Christian ties. The event, it was said, promoted an "us against Islam" mentality. Yet since 9/11, research shows that there has been a growing interest in *interfaith* worship and cooperation (FACT 2006). This development sees Christian and non-Christian religious groups engaging each other in dialogue to promote mutual fellowship and understanding. This interfaith movement can be seen as the latest manifestation of the United States' longstanding commitment to religious freedom and tolerance. It also bodes well in light of the changing face of religion in the United States.

BOX 23.1 SECOND THOUGHTS ABOUT THE NEWS: ISLAMOPHOBIA?

As we approached the ninth anniversary of 9/11, a new controversy took center stage: the debate over building a Muslim cultural center two blocks from Ground Zero in a former Burlington Coat Factory building—a building that has stood empty for most of the post–9/11 decade. Opponents to the proposed center argue that Ground Zero is a sacred place and therefore an inappropriate location for the center. Defenders of the proposed center remind critics that the United States values both the separation of church and state and the principle of religious tolerance.

(Continued)

(Continued)

Despite the current highly charged atmosphere surrounding the proposed center, the fact is that Muslims have been utilizing the location without incident for quite some time. Perhaps the center's activities have gone unnoticed because there have been no traditional "call to pray" sounds typical of mosques. (And developers say there never will be such sounds at the center.) It was only in the last weeks of the summer that demonstrations took place at the site. The debate over the location heated up as first local and then national politicians started weighing in on the issue. Indeed, some pundits feel candidates are strategically placing themselves for the next round of elections. Former New York Mayor Rudy Giuliani and former New York Governor Paterson would like to see the center relocate to another site. (While Governor, Paterson offered alternate state lands.) Current Mayor Bloomberg defends the building of the center on the principles of free enterprise and cultural diversity. And after weeks of silence, President Obama reminded Americans in the late summer of 2010 of our nation's commitment to religious freedom, including the right of groups to construct places of worship.

The president's comments drew much attention and criticism. Surely negative reactions to his comments have fed an interesting development: In late August of 2010, a Pew poll found that nearly one in five Americans now erroneously believe President Obama to be Muslim. (He is in fact a practicing Christian.) The controversy has led some observers to wonder if opponents to the center are Islamophobic and if the tragic events of 9/11 will preclude the religious tolerance that has been extended to other religious groups residing in the United States. Muslim Americans are concerned that they are being placed in the untenable position of being asked to choose between their country and their religion. This point is underscored by recent Congressional activity. On March 10, 2011, New York Representative Peter King opened House hearings on the response of the American Muslim community to extremists. Most of the opening-day session, however, was devoted to discussions of whether the hearings themselves are un-American.

One final point bears mentioning. Despite its conventional association with tradition, the real hallmark of religion is change. The variety of religious groups both in the United States and worldwide is powerful testimony of religion's capacity for responsive adaptation. Above all else, religion is functional—it serves basic human and social needs. Two important functions of religion are providing answers to the meaning of our existence and offering comfort to those in need. The inherent diversity of people—the diversity of their life issues and needs—guarantees that one religion will not fit all. Consistent with this dynamic face of religion is the fact that more than two-fifths of Americans are **"switchers"**—those who change their religious preferences or affiliation over time. While switching tends to occur before the age of 30, it can be a repeated pattern—that is, some move away from and then back to the religions of their childhood. Those who have changed their affiliations report many reasons for doing so (Pew Research Center 2009c). This shopping for churches is consistent with American consumerism (Hough 2005). The search for a church that fits also accounts for the remarkable growth of megachurches.

Switchers
Those who change their religious preferences or affiliation over time

Megachurches are Protestant congregations with at least 2,000 members in weekly attendance. In a typical week, more than 5 million Americans attend one of the thousand-plus megachurches across

Megachurches
Protestant congregations with a least 2,000 members in weekly attendance

America. More than two-thirds of these attendees have been members of these churches for less than 5 years, many are switchers, and roughly one-quarter do not regard the megachurch as their exclusive or "home" church. Megachurch attendees also tend to be younger, better educated, and wealthier than members of traditional churches (Thumma and Bird 2009). Some argue that megachurches are megabusinesses that use entrepreneurial approaches for growing congregations (Kroll 2003). The innovative use of technology (visual presentations, podcasts, and Internet communications) and the array of spiritual options/experiences make megachurches powerful rivals to smaller, more traditional church groups. Indeed, it is the ability to customize one's religious experience that attracts so many to megachurches (Thumma and Bird 2009).

To be sure, humans are active seekers of religious truths and meaningful existences. Cults form in response to changing times and newly emerging ideas and beliefs. Sects develop when groups grow dissatisfied with the status quo or when they desire greater exclusivity. Denominations offer new variations on established themes. Megachurches grow as individuals are more interested in finding a church that works for them and less interested in maintaining loyalty to a particular faith or denomination (*NewsHour* 2006). We may be one nation under God, but diversity is the insignia of religion in America.

LEARNING MORE ABOUT IT

Two recent books look at the role of religion in the lives of young adults: Robert Wuthnow's *After the Baby Boomers: How Twenty- and Thirty-Somethings Are Shaping the Future of American Religion* (Princeton: Princeton University Press, 2007) and Christian Smith's (with Patricia Snell) *Souls in Transition* (New York: Oxford University Press, 2009).

D. Michael Lindsay examines Evangelicals' rise in political, economic, and social power in his book *Faith in the Halls of Power: How Evangelicals Joined the American Elite* (New York: Oxford University Press, 2007).

In their 2008 article "Religion and Science: Beyond the Epistemological Conflict Narrative," John Evans and Michael Evans examine the relationship between religion and science and challenge the traditional assumption that any conflict between the two fields is based on competing truth claims about the world (*Annual Review of Sociology* 34: 87–105).

Michael Emerson and David Hartman offer an informative review of religious fundamentalism in their article "The Rise of Religious Fundamentalism" (*Annual Review of Sociology* 32: 127–144, 2006).

GOING ONLINE FOR SOME SECOND THOUGHTS

A review of the faith-based presidency of George W. Bush is presented by the PBS Frontline series (2004) *The Jesus Factor*: http://www.pbs.org/wgbh/pages/frontline/shows/jesus/.

The Pew Forum on Religion and Public Life offers an array of timely religion related topics: http://pewforum.org/.

NOW IT'S YOUR TURN

1. Use a "convenience sample" (people close at hand or easily accessed) to conduct an informal survey in which you ask your respondents to (a) self-describe their religious identification and (b) identify the top 10 religions in America. Compare the results you get with those found in the latest edition of the *Statistical Abstracts* ("Self-Described Religious Identification," http://www.census.gov/compendia/statab/2011/tables/11s0075.pdf) as well as in Table 23.1 of this essay. What insights about religion or religious awareness in the United States are offered by this exercise?

2. Conduct another informal survey (see above) where you ask your respondents to reflect on the relative importance of their religious versus national identity when making important life decisions: for example, serving in the military, raising children, getting married, or endorsing candidates in local, state, or national elections.

Conclusion

Why Does Conventional Wisdom Persist?

Love knows no reason. . . . Beauty is only skin deep. . . . Honesty is the best policy. . . . Education is the great equalizer. These statements represent just a sampling of the conventional wisdom that we so often hear throughout our lives.

In the introduction to this book, we noted that many of these adages contain some elements of truth. Within certain settings or under certain conditions, conventional wisdom can prove accurate. Yet throughout *Second Thoughts,* we have also noted that social reality is generally much more involved and much more complex than conventional wisdom would have us believe. Traditional adages and popular sayings rarely provide us with a complete picture of the broader social world. Knowing this, one might ask why individuals continue to embrace conventional wisdom. Given the limited usefulness of such assertions and tenets, why do such adages persist?

The Positive Functions of Conventional Wisdom

Conventional wisdom represents a people's attempt at "knowing." Such adages promise some insight into what is actually occurring. In this way, a culture's conventional wisdom comes to serve a variety of positive social functions; such adages can induce many productive outcomes for those who invoke them. Here, we speak specifically to five positive social functions served by conventional wisdom.

First, by providing an explanation for an unexpected, unusual, or disturbing occurrence, *conventional wisdom helps social members confront the unknown and dispel the fear the unknown can generate.* When conventional wisdom proclaims that immigrants are ruining this country, it provides members of a society with a tangible explanation for their increasing inability to make ends meet. Similarly, when conventional wisdom advises us to fear strangers, it offers up the usual suspects for the perennial and always frightening problems of crime, violence, loneliness, and chaos.

By identifying the causes of looming social problems, conventional wisdom not only dispels fear but also implies a hopeful resolution. As we noted in Essay 7, naming a problem's cause can increase our sense of control and encourage us to believe that a solution cannot be far away. Furthermore, identifying the cause of frightening social conditions offers a protective shield to the broader population. For example, consider a frequently heard popular belief: "Homeless people are mentally ill." Here, conventional wisdom offers a reason for a phenomenon most Americans find foreign and frightening. In addition, identifying mental illness as the instigator of the homeless condition gives most "sane" people a secure guarantee—homelessness could never happen to them. This piece of wisdom, like so many other adages, locates the source of a problem within the individual. Thus, as long as other social members distinguish themselves from the "problemmed" individual, they can protect themselves from that problem.

Second, *conventional wisdom also can function to maintain social stability.* Consider the common belief, "Every dog has its day." This adage urges people to be patient, to keep striving, or to leave revenge to fate—all attributes necessary for peaceful coexistence. Similarly, consider the adage, "Education is the great equalizer." This belief provides an incentive for citizen commitment to an institution whose greatest social contribution may be the consignment to the population of national customs, norms, and values. In the same way, conventional wisdom that warns that "united we stand, divided we fall" can effectively squelch protest or disagreement. Such a sentiment can enhance cooperation and dedication to a particular group or goal.

In these examples and others like them, conventional wisdom steers a population toward behaviors that maintain smooth social operations. It keeps societies balanced by making constructive effort a matter of common knowledge.

Third, under certain conditions, *conventional wisdom can function to legitimate the actions of those who invoke it.* Often, speakers will create or tap popular adages with a specific goal in mind. In such cases, conventional wisdom takes on the guise of political, religious, or social rhetoric. "Wisdom" emerges as strategically selected and stylized speech delivered to influence an individual or group.

As rhetoric, conventional wisdom proves effective in instituting policy or law because it promotes a vision of sound common sense. For example, we witness politically conservative members of the U.S. Congress forwarding wisdom such as "Welfare is ruining this country" or "Welfare breeds dependency." They do so because such rhetoric projects a prudent justification for shaving federal contributions to this cause. Similarly, politicians often espouse wisdom that claims that "capital punishment deters murderers" or "affirmative action programs favor unqualified minorities." They offer these claims despite factual evidence to the contrary because such rhetoric effectively employs popular assumptions in the service of the speaker's special interests.

Fourth, at yet another level, *conventional wisdom can strengthen or solidify a social group's identity.* Conventional wisdom often underscores shared values or attributes. In so doing, such beliefs can enhance collective identity. Adages such as "Great minds think alike" or "Like father, like son" and sayings such as "The apple doesn't fall far from the tree" or "Birds of a feather flock together" bond individuals by accentuating their similarities. Such wisdom unites individuals by underscoring the common ground they share.

In some cases, however, note that conventional wisdom supports solidarity by creating a "them-versus-us" milieu. Such approaches may unite the members of one group by accentuating the group's hatred or fear of others. For example, whites who feel threatened by the influx of nonwhite immigrants to the United States may readily espouse wisdom that advises individuals to "stick with their own kind." As like-minded individuals rally around such wisdom, white group solidarity can be heightened. Similarly, males who find it difficult to accept growing numbers of females in the workplace may rally around the traditional adage, "A woman's place is in the home." Using conventional wisdom to legitimate their fears, the threatened group can comfortably join in opposition.

Fifth, it is important to note that *conventional wisdom is often created or tapped as a tool for power maintenance.* When certain religious traditions defined "the love of money as the root of all evil," such wisdom effectively maintained the divide between the "haves" and the "have nots." Dissuading the masses from engaging in the struggle for material goods allows those in power to maintain their control over limited resources. (Note that such reasoning led social philosopher Karl Marx to refer to religion as "the opiate of the people." Marx argued that religion promoted a passive acceptance among the poor of an unfair economic structure.) Similarly, adages such as "You've come a long way, baby" or "Good things come to those who wait" serve to dampen efforts toward gender, racial, or ethnic equality. If common consensus suggests that minority group goals have been satisfactorily achieved or addressed, then the continued struggle toward true equality becomes difficult to sustain.

Conventional Wisdom as Knowledge

Whatever its functions, conventional wisdom appears to offer individuals an intelligence boost—a phenomenon social psychologist David Myers (2002:16) refers to as the "I-knew-it-all-along" effect. No matter what happens, there exists a conventional wisdom to cover or explain social behaviors; there always exists a saying or belief that predicts all outcomes. Herein lies conventional wisdom's most troubling feature: A society's "common knowledge" simultaneously proclaims contradictory "facts." For example, conventional wisdom assures us that "haste makes waste" while at the same time warning us that "he who hesitates is lost." Whereas one adage suggests that "too many cooks spoil the broth," another claims that "many hands make light work."

All in all, conventional wisdoms abound for every possible behavior and outcome. Such claims form a stockpile of knowledge to which socialization affords us access. Once introduced to a culture's conventional wisdom, social actors draw on this stockpile of ancient and contemporary adages to make almost any discovery seem like common sense. Thus, when I discover that separation intensifies my romantic attraction, I confirm the phenomenon by saying, "Absence makes the heart grow fonder." If, instead, separation dampens the fires of my romance, I confirm my experiences by noting, "Out of sight is out of mind." Indeed, conventional wisdom allows me to confirm any of my impressions and experiences—whatever they might be—and thereby frames those experiences as if they constitute a general norm or the ultimate truth.

The drawbacks of conventional wisdom are heightened by the fact that once introduced, these wisdoms take on lives of their own. Such adages become a "taken for granted" part of our culture; they state what is known, implicitly suggesting that such topics need not be further considered. In this way, conventional wisdom constitutes tenacious knowledge—information that endures even if there's no empirical evidence to support it. The mere passing of time, the longevity of an idea or belief, becomes a sufficient indicator of a tenet's veracity. Facts or observations that contradict the adage lose out to the test of time.

The staying power of conventional wisdom may be tied to dimensions of wisdom per se. Indeed, equating conventional adages with wisdom may help accentuate their appeal. Wisdom is a highly valued commodity in our society. It is born of good judgment and experience. Furthermore, it can offer us a sense of inner peace—an ability to live with what we know. Wisdom cannot be taught; courses in wisdom are not part of the college curriculum. Indeed, our formal education experiences often convince us that wisdom is not to be found in books or in research. Rather, wisdom emerges from the ordinary, the common, the everyday. In the final analysis, the wise person is one who has "lived."

In turning to our experiences for wisdom, however, we return to a problem cited earlier and explored at some length in this book's introduction—namely, the limitations of experientially based knowledge. If we base our wisdom solely on *personal* experiences, we will probably build dubious knowledge. Although our speculations about various social topics typically start with personal experiences, such experiences may not offer us the best empirical evidence for verification.

Our experiences are subjective and therefore vulnerable to distortion and personal bias. They require a correction factor, one that can control for distortions emerging from personal prejudices or sloppy thinking. Social scientific inquiry offers one such correction factor. The sociological approach to knowledge follows a set of standardized rules and procedures that can maximize our chances of obtaining valid and reliable knowledge.

Although sociology may start with what we already know, good sociology does not end there. Rather, good sociology explores commonsense notions about the social world by collecting and comparing varied reports, observations, and other data in the interest of building an all-encompassing picture of reality.

In Closing

Joseph Story, an early associate justice of the U.S. Supreme Court, is quoted as saying, "Human wisdom is the aggregate of all human experience, constantly accumulating, selecting and reorganizing its own materials." Story's statement suggests that true wisdom requires a wealth of experience. It is in that spirit that we prepared *Second Thoughts.* In each essay, we proposed a bit of wisdom that requires us to consider myriad experiences and facts. We advocated an approach to knowledge that remains open ended, a stance that treats new information as an opportunity to rethink what we know. In this way, we cast all social actors as perpetual students of their environments—students who regularly question assumptions and who seek to see beyond themselves.

LEARNING MORE ABOUT IT

The flaws inherent in relying on conventional wisdom for knowledge are effectively portrayed in a 1949 classic by sociologist Paul Lazarsfeld, entitled "The American Soldier: An Expository Review" (*Public Opinion Quarterly* 13 (3): 378–380). A similarly striking demonstration comes from social psychologist Karl Halvor Teigen in his 1986 article, "Old Truths or Fresh Insights? A Study of Students' Evaluations of Proverbs" (*Journal of British Social Psychology* 25 (1): 43–50).

In a related vein, anthropologist Claude Levi-Strauss explores the origins of myth in his book, *The Raw and the Cooked: Introduction to a Science of Mythology* (New York: Harper & Row, 1964).

Howard Kahane and Nancy Cavender provide a detailed exploration of both the tools and pitfalls of everyday reasoning and problem solving in *Logic and Contemporary Rhetoric: The Use of Reason in Everyday Life,* 9th edition (Belmont, CA: Wadsworth, 2002).

Glossary

Achieved status Status earned or gained through personal effort

Age structure The distribution that results from dividing a population according to socially defined, age-based categories: childhood, adolescence, young adulthood, middle age, and old age

Agnostics Individuals who maintain that the existence of God cannot be known

Aid to Families with Dependent Children (AFDC) A New Deal program created by the Social Security Act of 1935 to provide federal assistance to needy dependent children; replaced in the late 1990s by the Temporary Assistance to Needy Families (TANF) Program

Aliens Citizens of a foreign country. The United States distinguishes four legal statuses for aliens: legal immigrants, temporary legal migrants, refugees, and unauthorized migrants

Anticipatory socialization Socialization that prepares a person to assume a future role

Appearance norms Society's generally accepted standards of appropriate body height, body weight, distribution or shape, bone structure, skin color, etc.

Ascribed status Status assigned, given, or imposed without regard to persons' efforts or desires

Assimilation The process by which immigrant groups come to adopt the dominant culture of their new homeland as their own

Atheists Individuals who deny the existence of God

Binge drinking The consumption of five or more drinks in a row

Biofuels Renewable sources of energy derived from living organisms or from metabolic byproducts; they are considered carbon dioxide neutral

Blended families Families created when those who remarry already have children from their previous marriages

Bradley Effect Discrepancies in poll results and actual outcomes,usually linked to respondents' deception with reference to sensitive topics

Boundary construction Social partitioning of life experience or centers of interaction

Bureaucracies Organizations that strive to achieve rationality and efficiency via a hierarchical authority structure, a clear division of labor, reliance on written rules/procedures, and impersonal interaction

Cap-and-trade system A strategy for combating global warming that uses the power of the market to create financial incentives for reducing emissions

Carbon footprint Term used by environmentalists to convey an individual's personal contribution to carbon dioxide emissions

Charter schools Schools that receive public money but are free of most public regulation *if* they produce performance outcomes stipulated in their charter contracts

Chronic stress The relatively enduring problems, conflicts, and threats that individuals face on a daily basis

Church A long-standing religious organization with a formalized social structure

Civil lawsuits Private legal proceedings for the enforcement of a right or the redressing of a wrong

Class action suits Litigation in which one or more persons brings a civil lawsuit on behalf of other similarly situated individuals

Climate The average weather conditions of a region over time

Climate change Any significant change in climate (e.g., temperature, precipitation, wind, etc.) that lasts for decades or longer. *See also* **Global warming**

Conflict theorists Sociologists who analyze social organization and social interactions by attending to the differential resources controlled by different sectors of a society

Conventional wisdom Body of assertions and beliefs that is generally recognized as part of a culture's common knowledge

Cults New and relatively small and unorganized religious groups whose beliefs set them apart from more established or mainstream groups

Cultural capital Attributes, knowledge, or ways of thinking that can be converted or used for economic advantage

Cultural inconsistency A situation in which actual behaviors contradict cultural goals; an imbalance between ideal and real culture

Cultural value A general sentiment regarding what is good or bad, right or wrong

Culture The values, beliefs, symbols, objects, customs, and conventions that characterize a group or society's way of life

Culture against people When beliefs, values, or norms of a society lead to destructive or harmful patterns of behavior

Cyberbullying The repeated harassment, mistreatment, or humiliation executed by one or more persons toward another with the specific use of cell phones or cameras and the Internet

Davis-Moore thesis The view that social inequality is beneficial to the overall functioning of society

***De facto* segregation** Segregation due to actual behavioral practices

***De jure* segregation** Segregation that is imposed by the law

Denominations Groups that break away from their "mother" churches and establish their own unique identities via their specific beliefs, traditions, and rituals

Deviance Any act that violates a social norm

Deviant lies Falsehoods always judged to be wrong by a society; they represent a socially unacceptable practice, one that destroys social trust

Dialectic A process by which contradictions and their solutions lead participants to more advanced thought

Digital divide The formation of an information underclass—a portion of the population that cannot afford to access and capitalize on the things that new technologies have to offer

Direct communication Face-to-face or physically copresent exchange

Discrimination Unfavorable treatment of individuals on the basis of their group membership

Dyad A group of two

e-philanthropy Online or "cyber giving"

Ecclesias Official state religions

Ecology The study of the interaction between organisms and the natural environment

Emotional regulation The ability to maintain a positive affect

Endogamy Marriage that is restricted by shared group membership ("in-group" phenomenon)

Ethnocentrism Tendency to view one's own cultural experience as a universal standard

Evangelicals Individuals who view Holy Scriptures as authoritative and infallible and who are committed to proselytizing, a.k.a. born again

Expert advocate An attorney who has made at least five oral arguments or a firm that has made at least 10 oral arguments before the Supreme Court

Extended family A unit containing parent(s), children, and other blood relatives such as grandparents or aunts and uncles

Family Two or more people residing together and related by birth, marriage, or adoption

Family household Household where one or more persons are related to the householder by marriage, birth, or adoption

Families of orientation The families into which we are born

Families of procreation The families we establish via marriage and children

Fear of strangers A dread or suspicion of those who look, behave, or speak differently from oneself; such fears can ultimately make the world seem unfamiliar and dangerous

Formal social control Social control that is state centered

Functional analysis Analysis that focuses on the interrelationships among the various parts of a society; it is ultimately concerned with the ways in which such interrelationships contribute to social order

Fundamentalists A subset of evangelicals who insist that the Holy Scriptures must be read literally and who advocate separation from those who don't share their views

Gemeinschaft An environment in which social relationships are based on ties of friendship and kinship

Gender scripts The articulation of gender norms and biases

Gender socialization Process by which individuals learn the culturally approved expectations and behaviors for males and females

Gender typing Gender-based expectations and behaviors

Gesellschaft An environment in which social relationships are formal, impersonal, and often initiated for specialized or instrumental purposes

Global warming A rise in the earth's temperature due to the increase in greenhouse or heat-trapping gases in the atmosphere

Green card holders *See* **Legal immigrants**

Greenhouse effect Radiated heat trapped in the earth's atmosphere by an insulating blanket of certain gas molecules (e.g., carbon dioxide, methane, and nitrous oxide)

Happiness gap The discrepancy in reported happiness attributed to marital status: married individuals report higher levels of general happiness than the never married

High-poverty schools Schools where 75% or more of students are eligible for free or reduced-price lunches

Household All the people who occupy a housing unit

Ideal culture Values, beliefs, and norms each society claims as central to its modus operandi; the aspirations, ends, or goals of our behaviors

Identity Essential characteristics that both link us to and distinguish us from other social players and thus establish who we are

Illegal aliens Foreigners in the United States without valid visas

Immigrant groups Groups containing individuals who have left their homelands in pursuit of a new life in a new country

Impression management Process by which individuals manipulate or maneuver their public images to elicit desired reactions

In-group A group whose members possess a strong sense of identity and group loyalty and hold their group in high esteem

Income The amount of money earned via an occupation or investments during a specific period of time

Infant mortality rates The number of deaths per 1,000 live births for children under 1 year of age

Informal social control Social control initiated by any party; no special status or training is required

Intergenerational upward mobility Social status gains by children vis-à-vis their parents

Labeling theory Maintains that the names or labels we apply to people, places, or circumstances influence and direct our interactions and the emerging reality of the situation

Latent dysfunctions Less-than-obvious negative outcomes or consequences of a social practice or pattern

Latent functions Unintended benefits or consequences of a social practice or pattern

"Legacy" admission The practice by some colleges and universities that gives preference to applicants who are relatives of alumni

Legal immigrants Foreigners granted visas that allow them to live and work permanently in the United States, also known as legal permanent residents or green card holders

Life chances Odds of obtaining desirable resources, positive experiences, and opportunities for a long and successful life

Life expectancy The average number of years that a specified population can expect to live

Linguistic profiling Identifying a person's race from the sound of his or her voice

Literacy The ability to use printed and written information to function in society

Longitudinal data Data collected at multiple points in time

Looking-glass self Process by which individuals use the reactions of other social members as mirrors by which to view themselves and develop an image of who they are

Low-income families Families with income above the federal poverty level but below the level deemed necessary for meeting basic needs

LPRs (Legal Permanent Residents) *See* **Legal immigrants**

Macro-level analysis Social analysis that focuses on broad, large-scale social patterns as they exist across contexts or through time

Manifest dysfunctions Obvious negative outcomes or consequences of a social practice or pattern

Manifest functions Obvious consequences or benefits of some social process, practice, or pattern

Marriage A socially approved economic and sexual union

Master status A single social status that overpowers all other social positions occupied by an individual; it directs the way in which others see, define, and relate to an individual

Mediated communication An indirect connection funneled through a mechanical medium

Megachurches Protestant congregations with a least 2,000 members in weekly attendance

Meta-analysis A statistical technique used to assemble and summarize a large collection of quantitative studies addressing a single problem or issue

Minorities *See* **Social minority**

Monogamy An exclusive union (e.g., one man married to one woman at one time)

Moral entrepreneurs Individuals who seek to legally regulate behaviors that they consider morally reprehensible

Mortality rates Document the number of deaths per each 1,000 or (10,000 or 100,000) members of the population

Multiculturalism A perspective that accentuates rather than dilutes ethnic and racial differences

Naturalized citizens Legal immigrants (at least 18 years of age) who have lived in the United States for at least 5 years, paid application fees, undergone a background check, and passed English and civic tests

Negative life events Major and undesirable changes in one's day-to-day existence, such as the loss of a spouse, divorce, or unemployment

New communication technologies Developments such as fiber optics, the Internet, rapid satellite transmissions, and virtual reality imaging

Nonfamily household A householder living alone or living with nonrelative others

Nontraditional female occupation One in which women comprise 25% or less of total employment

Normal lies Falsehoods that are socially acceptable practice linked to productive social outcomes; individuals rationalize and legitimate normal lies as the means to a noble end—that is, the good of one's family, colleague, or country

Norms Social rules or guidelines that direct behavior, they are the "shoulds" and "should nots" of social action, feelings, and thought

Norms of homogamy Social rules that encourage interaction between individuals occupying similar social statuses

Nuclear family A self-contained, self-satisfying unit composed of father, mother, and children

Obesity A condition of having a body mass index of 30 or higher or being about 30 pounds overweight

Occupational prestige The respect or recognition that one's occupational position commands

Occupational prestige scale Relative ratings of select occupations as collected from a representative national sample of Americans

Operationalization The way in which a researcher defines and measures the concept or variable of interest

Out-group Group that is considered undesirable and is thought to hold values and beliefs foreign to one's own

Paired testing Hiring experiments where two equally qualified candidates of different races each apply for the same job

Pay gap The discrepancy between women's and men's earnings; a ratio calculated when women's earnings are divided by men's earnings

Pentecostals Those who engage in "spirit-filled" worship (speaking in tongues, faith healing); found in most Christian traditions

Poor Those with incomes below the minimal level needed to meet basic needs

Population The totality of a collection of individuals, institutions, events, or objects about which one wishes to generalize

Postmodern theory An approach that destabilizes or deconstructs fixed social assumptions and meanings

Poverty An economic state or condition in which one's annual income is below that judged necessary to support a predetermined minimal standard of living

Poverty threshold The federal government's designation of the annual income a family requires to meet its basic needs

Power The ability of groups and/or individuals to get what they want even in the face of resistance

Prejudice A prejudgment directed toward members of certain social groups

Primary deviance Isolated violations of norms

Primary socialization The earliest phase of social training wherein we learn basic social skills and form the core of our identities

Public assistance programs Programs directed exclusively at the eligible poor

Race A group of individuals who share a common genetic heritage or obvious physical characteristics that members of a society deem socially significant; *see also* **Social minority**

Racial steering Practice in which real estate agents direct prospective buyers toward neighborhoods that largely match the buyer's race or national origin

Racism Prejudice and discrimination based on the belief that one race is superior to another

Real culture Values, beliefs, and norms actually executed or practiced; behaviors or the means actually used to pursue a society's ends

Refugees and asylees Foreigners allowed to live in the United States because of fear of persecution in their home countries; refugees may become legal permanent residents after living in the United States for a year

Reification Process by which the subjective or abstract erroneously comes to be treated as objective fact or reality

Reliable measure A measure that yields consistent or stable results

Religiosity Refers to individuals' beliefs as opposed to their actual participation in religious groups

Religious affiliation Formal group membership in a religious community

Replacement-level fertility The fertility level needed for a population to continually renew itself without growing

Representative sample Refers to a group or subset of elements that mirrors the characteristics of the larger population of interest

Rituals Set of actions that take on symbolic significance

Role conflict Conflict that occurs when individuals occupy two or more social positions that carry opposing demands

Sample A portion or subset of the population of interest

Secondary deviance Occurs when labeled individuals come to view themselves according to what they are called. The labeled individual incorporates the impression of others into her or his self-identity.

Sects The product of conflictual breakaways from churches; groups that establish themselves as independent rival religious organizations

Self-esteem Personal judgments individuals make regarding their own self-worth

Self-fulfilling prophecy A phenomenon whereby that which we believe to be true, in some sense, becomes true for us

Serial monogamy The practice of successive, multiple marriages—that is, over the course of a lifetime, a person enters into consecutive monogamous unions

Sex segregation (in the work sphere) The separation of male and female workers by job tasks or occupational categories

Social context The broad social and historical circumstances surrounding an act or an event

Social control The process of enforcing social norms

Social construction of reality A process by which individuals create images, ideas, and beliefs about society based on their social interactions

Social desirability bias Measurement error that results when respondents answer questions in a way that makes them look good

Social engineering Using the law to satisfy social wants and to bring about desired social change

Social functions Intended and unintended social consequences of various behaviors and practices

Social indicators Quantitative measures or indices of social phenomena

Social institution Set of behavior patterns, social roles, and norms combining to form a system that fulfills an important need or function for a society

Social insurance programs Programs that require payroll contributions from the future beneficiaries; neither program eligibility nor benefits are linked to financial need

Social location One's social position as indicated by one's total collection of social statuses—that is, age, education, gender, income, race, and so forth

Social minority A group regarded as subordinate or inferior to a majority or dominant group

Social patterns General trends that can be seen when we look beyond any one or two cases

Social policies Officially adopted plans of action

Social relations The types of connections and patterns of interaction that structure the broader society

Social reproduction theory The view that existing social, cultural, and economic arrangements work to reproduce in future generations the social class divisions of the present generation

Social scripts The shared expectations that govern those interacting within a particular setting or context

Social status The position or location of an individual with reference to characteristics such as age, education, gender, income, race, and religion

Social strain A social event or trend that disrupts the equilibrium of a society's social structure

Social structure The organization of a society; the ways in which social statuses, resources, power, and mechanisms of control combine to form a framework of operations

Social support network A network that consists of family, friends, agencies, and resources that actively assist individuals in coping with adverse or unexpected events

Socialization The process by which we learn the norms, values, beliefs, and symbols of a social group and our own place in that social group

Socioeconomic status A particular social location defined with reference to education, occupation, and financial resources; *see also* **Poverty and wealth**

Sociological imagination The ability to see and evaluate the personal realm in light of the broader social/cultural and historical arenas

Status set The total collection of statuses occupied by a social actor

Stereotypes Generalizations applied to *all* members of a group; *see also* **Gender typing**

Strategies of action Means and methods social actors use to achieve goals and fulfill needs

Stratification system The hierarchical ranking of individuals with regard to control of a society's resources, privileges, and rewards

Structural functionalism A theoretical approach that stresses social order and that views society as a collection of interdependent parts that function together to produce consensus and stability

Survey research Data collection technique using a carefully designed set of questions

Switchers Those who change their religious preferences or affiliation over time

Symbols Arbitrary signs that come to be endowed with special meaning and gain the ability to influence behaviors, attitudes, and emotions

Task-oriented stress Short-term stress that accompanies particular assignments or settings

Techniques of neutralization Methods of rationalizing deviant behavior

Temporary Assistance to Needy Families (TANF) A federal program created in 1996 by the Personal Responsibility and Work Opportunity Reconciliation Act to provide temporary assistance and work opportunities to needy families with children; also referred to as a "welfare-to-work" program

Temporary legal migrants Foreigners in the United States for specific purposes (e.g., visiting, studying, or working)

Tolerance Inaction in the face of some offense; an informal social control mechanism

Tracking A practice by which students are divided into groups or classes based on their perceived ability

Unauthorized migrants Also known as illegal aliens; foreigners in the United States without valid visas

Uniform Crime Reports Index for Serious Crime FBI-issued statistics on crime in the United States

Unitary schools Those deemed to have eliminated a dual system of education based on racial inequality

Valid measure One that accurately captures or measures the concept or property of interest to the researcher

Value-free researcher One who keeps personal values and beliefs out of the collection and interpretation of data

Victimization studies Produce statistics based on victims' self-reports and not the reports of police

Wealth Totality of money and resources controlled by an individual or a family

Weather The atmospheric conditions of a given moment and place

Working poor Individuals who have spent at least 27 weeks in the labor force but have incomes below the poverty level

References

Aaron, K. 2009. "Predatory Lending: A Decade of Warnings." The Center for Public Integrity. http://www.publicintegrity.org/investigations/economic_meltdown/articles/entry/1309/.

AARP. 2003. *Beyond 50.03: A Report to the Nation on Independent Living and Disability.* http://www.aarp.org/health/doctors-hospitals/info-11–2003/aresearch-import-753.html.

AARP. 2009. *Internet Use Among Midlife and Older Adults: An AARP Bulletin Poll.* http://assets.aarp.org/rgcenter/general/bulletin_internet_09.pdf.

AARP. 2010. "Connecting and Giving: A Report on How Mid-life and Older Americans Spend Their Time, Make Connections and Build Communities." http://assets.aarp.org/rgcenter/general/connecting_giving.pdf.

AARP Services, Inc., and Focalyst. 2007. "The Sky's the Limit: Travel Trends among the Baby Boom Generation & Beyond." http://www.aarp.org/travel/trips/info-2007/travel_trends.html.

Abel, R. 2002. "Judges Write the Darnedest Things: Judicial Mystification of Limitations on Tort Liability." *Texas Law Review* 80: 1547–48.

Abramson, A., and M. Silverstein. 2006. *Images of Aging in America, 2004.* http://assets.aarp.org/rgcenter/general/images_aging.pdf.

Abwender, D. A., and K. Hough. 2001. "Interactive Effects of Characteristics of Defendant and Mock Juror on U.S. Participants' Judgment and Sentencing Recommendations." *Journal of Social Psychology* 141 (5): 603–15.

Achenreiner, G. B., and D. R. John. 2003. "The Meaning of Brand Names to Children: A Developmental Investigation." *Journal of Consumer Psychology* 13 (3): 205–19.

Adams, R. G., and G. Allan, eds. 1998. *Placing Friendship in Context.* Cambridge, UK: Cambridge University Press.

Adelson, R. 2004. "Detecting Deception." *Monitor on Psychology* 35 (7): 70.

Adherents.com. 2005. "Largest Religious Groups in the United States of America." http://www.adherents.com/rel_USA.html#religions.

Administration on Aging. 2010. "Profile of Older Americans: 2009." http://www.aoa.gov/AoARoot/Aging_Statistics/Profile/index.aspx.

Advisory Committee on Student Financial Assistance. 2008. "Shifts in College Enrollment Increase Projected Losses in Bachelor's Degrees." *Policy Bulletin* (May 1). http://www2.ed.gov/about/bdscomm/list/acsfa/mofpolicybulletin.pdf.

Affleck, G., H. Tennen, and A. Apter. 2000. "Optimism, Pessimism and Daily Life with Chronic Illness." In *Optimism and Pessimism,* edited by E. C. Chang, 580–602. Washington, DC: APA Books.

AFL-CIO. 2010. "Trends in CEO Pay." http://www.aflcio.org/corporatewatch/paywatch/pay/index.cfm.

Agency for Healthcare Research and Quality. 2005. "Women's Health Care in the United States: Selected Findings from the 2004 National Healthcare Quality and Disparities Reports." Fact Sheet. AHRQ Publication No. 05-P021 (May). http://www.ahrq.gov/qual/nhqrwomen/nhqrwomen.htm.

———. 2008. "Pharmaceutical Research: Women Are Dispensed More Drugs Than Men During Their Reproductive Years." *Research Activities* (November). http://www.ahrq.gov/research/nov08/1108RA24.htm.

———. 2009. "Women's Health: More Women Than Men Are Hospitalized for Chest Pain With No Known Cause." *Research Activities* (January). http://www.ahrq.gov/research/jan09/0109RA11.htm.

Aginsky Consulting Group. 2007. "Cosmetics Market Research Summary." (October). http://www.aginskyconsulting.com/downloads/ACG%20Industry%20Summary%20Reports%20-%20Q4–07/ACG%20Russian%20Cosmetics%20Market%20Overview.pdf.

Ahlburg, D., and C. DeVita. 1992. "New Realities of the American Family." *Population Bulletin* 47 (2). Washington, DC: Population Reference Bureau.

Alba, R. 2009. *Blurring the Color Line: The New Chance for a More Integrated America.* (Cambridge, MA: Harvard University Press.

Alba, R., and V. Nee. 2005. *Remaking the American Mainstream: Assimilation and Contemporary Immigration.* Cambridge, MA: Harvard University Press.

Alcoff, L. M. 2005. "A Response to Garcia." *Philosophy and Social Criticism* 31 (4): 419–22.

Alper, S., D. Tjosvold, and K. S. Law. 2000. "Conflict Management, Efficacy, and Performance in Organizational Teams." *Personnel Psychology* 53 (3): 625–42.

Altheide, D. 2002. *Creating Fear: News and the Construction of Crisis.* New York: Aldine de Gruyter.

———. 2009. *Terror, Post 9/11 and the Media.* New York: Peter Lang Publishing.

Alzheimer's Association. 2010. "What Is Alzheimer's." http://www.alz.org/alzheimers_disease_what_is_alzheimers.asp?type=more_information#early.

Amabile, T. M., and L. G. Kabat. 1982. "When Self-Descriptions Contradict Behavior." *Social Cognition* 1: 311–35.

Amato, P. 2001. "What Children Learn From Divorce." *Population Today* (January). http://www.prb.org/Articles/2001/WhatChildrenLearnFromDivorce.aspx.

Amato, P., A. Booth, D. Johnson, and S. Rogers. 2007. *Alone Together: How Marriage in America Is Changing.* Cambridge, MA: Harvard University Press.

Amato, P. R., and F. Fowler. 2002. "Parenting Practices, Child Adjustment and Family Diversity." *Journal of Marriage and Family* 64 (3): 703–16.

American Association of Retired Persons (AARP). 2009. "Internet Use Among Midlife and Older Adults: An AARP Bulletin Poll." http://assets.aarp.org/rgcenter/general/bulletin_internet_09.pdf.

———. 2010. "Connecting and Giving: A Report on How Mid-life and Older Americans Spend Their Time, Make Connections and Build Communities." http://assets.aarp.org/rgcenter/general/connecting_giving.pdf.

American Bar Association. 2009. "Enrollment and Degrees Awarded, 1963–2009 Academic Years." http://www.abanet.org/legaled/statistics/charts/stats%20-%201.pdf.

American Cancer Society. 2009. "Cancer Facts & Figures 2009." http://www.cancer.org/docroot/STT/content/STT_1x_Cancer_Facts__Figures_2009.asp?from=fast.

American Civil Liberties Union. 2006. "Federal Court Strikes Down NSA Warrantless Surveillance Program." http://www.aclu.org/safefree/nsaspying/26489prs20060817.html.

American Medical Association. 2006. "Table 4: Women Residents by Specialty 2005." http://www.ama-assn.org/ama/pub/about-ama/our-people/member-groups-sections/women-physicians-congress/statistics-history/table-4-women-residents-specialty-2005.shtml.

"American Patriots." 2006. *The Nation* (July 2). http://www.thenation.com/article/american-patriots.

American Psychological Association. 2005. "Emotional Fitness in Aging: Older Is Happier." *Psychology Matters.* APA Online. http://www.apa.org/research/action/emotional.aspx.

———. 2010. "Socioeconomic Status and Health Fact Sheet." http://www.apa.org/about/gr/issues/socioeconomic/ses-health.aspx.

American Society of Plastic Surgeons. 2010. "2009 Quick Facts." http://www.plasticsurgery.org/Documents/Media/statistics/2009quickfacts-cosmetic-surgery-minimally-invasive-statistics.pdf.

American Tort Reform Association. 2007. "ATRA at a Glance." http://www.atra.org/about/.

Ameriprise Financial. 2006. "The Five Stages of Retirement." *Corporate Training and Development Advisor* 11 (3): 5.

Amodeo, N. P. 2001. "Be More Cooperative to Be More Competitive." *Journal of Rural Cooperation* 29 (2): 115–24.

Anderson, C.A., A. Shibuya, N. Ihori, E. L. Swing, B. J. Bushman, A. Sakamoto, H. R. Rothstein, and M. Saleem. 2010. "Violent Video Game Effects on Aggression, Empathy, and Pro-Social Behavior in Eastern and Western Countries: A Meta-Analytic Review." *Psychological Bulletin* 136 (2): 151–73.

Anderson-Clark, T. N., R. J. Green, and T. B. Henley. 2008. "The Relationship Between First Names and Teacher Expectations for Achievement Motivation." *Journal of Language and Social Psychology* 27 (1): 94–99.

Andreoletti, C., L. A. Zebrowitz, A. Leslie, and M. E. Lachman. 2001. "Physical Appearance and Control Beliefs in Young, Middle-Aged, and Older Adults." *Personality and Social Psychology Bulletin* 27 (8): 969–81.

Andresen, K. 2007. "Who Gives Money Online?" Network for Good Learning Center. http://www.fundraising123.org/article/who-gives-money-online.

Andrews, K. 2002. "Movement-Countermovement Dynamics and the Emergence of New Institutions: The Case of 'White Flight' Schools in Mississippi." *Social Forces* 80 (3): 911–36.

Aneshensel, C. 1992. "Social Stress: Theory and Research." *Annual Review of Sociology* 18: 15–38.

Angelini, J., D. Goh, S. Eastman, J. Rosow, T. Dodge, W. Deng, and N. Zhou. 2009. "Prominence of Characters on Television Program Websites." *The Howard Journal of Communications* 20 (3): 276–94.

Angier, N. 2000. "Do Races Differ? Not Really, Genes Show." *New York Times* (August 22). http://www.nytimes.com/library/national/science/082200sci-genetics-race.html.

Ansari, M. K. 2010. "Madoffs Seek Name Change: The Rules of Name Changes," FindLaw.com (February 26). http://blogs.findlaw.com/law_and_life/2010/02/madoffs-grandkids-daughter-in-law-seek-name-change.html.

Arai, M., and P. S. Thoursie. 2009. "Renouncing Personal Names: An Empirical Examination of Surname Change and Earnings." *Journal of Labor Economic* 27 (1): 127–135.

Aranda-Suh, E. 2009. "Bilingualism in the Job Market." (March 25). http://www.extranews.net/news.php?nid=4735&pag=1&st=0.

Archdeacon, T. 1992. "Reflections on Immigration to Europe in Light of U.S. Immigration History." *International Migration Review* 26 (Summer): 524–48.

Arichi, M. 1999. "Is It Radical? Women's Right to Keep Their Own Surnames After Marriage." *Women's Studies International Forum* 22 (4): 411–15.

Aries, P. 1962. *Centuries of Childhood: A Social History of Family Life.* Translated by R. Baldick. New York: Knopf.

Aronson, E. 1980. *The Social Animal.* San Francisco: Freeman.

Aronson, E., and V. Cope. 1968. "My Enemy's Enemy Is My Friend." *Journal of Personality and Social Psychology* 8: 8–12.

Aronson, E., and R. Thibodeau. 1992. "The Jigsaw Classroom: A Cooperative Strategy for Reducing Prejudice." In *Cultural Diversity in the Schools,* edited by J. Lynch, C. Modgil, and S. Modgil (p. 179). London: Falmer.

Arum, R., J. Roksa, and M. J. Budig. 2008. "The Romance of College Attendance: Higher Education Stratification and Mate Selection." *Research in Social Stratification and Mobility* 26 (2): 107–21.

Ashenfelter, D. 2002. "Judge Criticized for Coin-Toss Custody Ruling; The Grandparents Who Lost Are Calling It 'A Mockery of the Judicial Process.'" *Philadelphia Inquirer* (February 10).

Associated Content. 2010. "Still Missing 10-month-old Lauryn Dickens' Mother Arrested, Believed Connected to Disappearance." http://www.associatedcontent.com/article/5835438/still_missing_10monthold_lauryn_dickens.html?cat=9.

Associated Press. 2006. "If Surveys Tell the Truth, White Lies Are Necessary Evil." (July 12). http://www.boston.com/news/nation/articles/2006/07/12/if_surveys_tell_the_truth_white_lies_are_necessary_evil/.

Association of American Medical Colleges. 2006. "An Overview of Women in U.S. Academic Medicine, 2005–06." *Analysis in Brief* 6 (7). https://www.aamc.org/download/100876/data/overview_of_women_in_the_us.pdf.

Aud, S., W. Hussar, M. Planty, T. Snyder, K. Bianco, M. A. Fox, L. Frohlich, et al. 2010. *The Condition of Education 2010* (NCES 2010–028). Washington, DC: National Center for Education Statistics, Institute of Education Sciences, U.S. Department of Education. http://nces.ed.gov/pubs2010/2010028_1.pdf.

Auerbach, J. A. 2003."Passing Father's Surname on to Child Questioned." *Family Law* 228 (71): 10.

Auwarter, A., and M. Aruguete. 2008. "Effects of Student Gender and Socioeconomic Status on Teacher Perceptions." *Journal of Educational Research* 101 (4): 242–46.

Axelrod, R. 2006. *The Evolution of Cooperation.* New York: Penguin.

Ayers, D. 2010. "Putting the Community Back into the College." *Academe* 96 (3): 9–11.

Ayres, M. E. 2010. "Book Review: *Working After Welfare: How Women Balance Jobs and Family in the Wake of Welfare Reform.* By Kristin S. Seefeldt, Kalamazoo, MI, W.E. Upjohn Institute for Employment Research, 2008, 171 pp." *Monthly Labor Review Online* 133 (5). http://www.bls.gov/opub/mlr/2010/05/bookrevs.htm.

Azaryahu, M., and R. Kook. 2002. "Mapping the Nation: Street Names and Arab-Palestinian Identity: Three Case Studies." *Nations and Nationalism* 8 (2): 195–213.

Babbie, E. 1994. *What Is Society? Reflections on Freedom, Order, and Change.* Thousand Oaks, CA: Pine Forge.

———. 1998. *Observing Ourselves: Essays in Social Research.* Prospect Heights, IL: Waveland.

———. 2009. *The Practice of Social Research,* 12th ed. Belmont, CA: Wadsworth.

Babwin, D. 2010. "Cameras Make Chicago Most Closely Watched US City." Associated Press. http://abcnews.go .com/US/wireStory?id=10295920.

Back, M. D., S. S. Schmukle, and B. Egloff. 2008. "Becoming Friends by Chance." *Psychological Science* 19 (5): 439–40.

Badr, L. K., and B. Abdallah. 2001. "Physical Attractiveness of Premature Infants Affects Outcome at Discharge from NICU." *Infant Behavior and Development* 24 (1): 129–33.

Baker, D. 2009. "The Problem: Lack of Accountability." *Politico* (June 18). http://dyn.politico.com/printstory .cfm?uuid=EFB8697C-18FE-70B2-A8DE834F523A77A2.

Baker, K., and A. Raney. 2007. "Equally Super? Gender-Role Stereotyping of Superheroes in Children's Animated Programs." *Mass Communication & Society* 10 (1): 25–41.

Baldas, T. 2003. "Saying Goodbye to Desegregation Plans." *National Law Journal* 25: 85.

Balliet, D. 2010. "Communication and Cooperation in Social Dilemmas: A Meta-Analytic Review." *Journal of Conflict Resolution* 54 (1): 39–57.

Baltzell, E. D. 1987. *The Protestant Establishment: Aristocracy and Caste in America.* New Haven, CT: Yale University Press.

Bamburg, J. 1994. *Raising Expectations to Improve Student Learning.* Oak Brook, IL: North Central Regional Educational Laboratory.

Banse, R. 1999. "Autonomic Evaluation of Self and Significant Others: Affective Priming in Close Relationships." *Journal of Social and Personal Relationships* 16 (6): 803–21.

Barak, A. 2008. "Down the Rabbit Hole: The Role of Place in the Initiation and Development of Online Relationships." In *Psychological Aspects of Cyberspace: Theory, Research, Applications,* edited by A. Barak, 163–84. New York: Cambridge University Press.

Barnes, J. A. 1994. *A Pack of Lies: Toward a Sociology of Lying.* New York: Cambridge University Press.

Barrett, L. L. 2008. *Healthy @ Home.* AARP Foundation. http://assets.aarp.org/rgcenter/il/healthy_home.pdf.

Bartfeld, J. 2000. "Child Support and the Postdivorce Economic Well-Being of Mothers, Fathers, and Children." *Demography* 37 (2): 203–13.

Bartlett, C. P., C. L. Vowels, and D. A. Saucier. 2008. "Meta-Analyses of the Effects of Media Images on Men's Body-Image Concerns." *Journal of Clinical and Social Psychology* 27 (3): 279–310.

Basow, S. 1992. *Gender: Stereotypes and Roles,* 3rd ed. Monterey, CA: Brooks/Cole.

Baum, S., and J. Ma. 2007. "Education Pays: The Benefits of Higher Education for Individuals and Society." College Board, Trends in Higher Education Series. http://www.collegeboard.com/prod_downloads/about/news_info/cbsenior/yr2007/ed-pays-2007.pdf.

Baum, S., and P. Steele. 2010. "Who Borrows Most? Bachelor's Degree Recipients with High Levels of Student Debt." College Board, Trends in Higher Education Series. http://advocacy.collegeboard.org/sites/default/files/Trends-Who-Borrows-Most-Brief.pdf.

BBC h2g2. 2010. "Why People Lie." http://www.bbc.co.uk/dna/h2g2/alabaster/A996942.

BBC News. 2008a. "'Ridiculous' Name Banned in Italy." http://news.bbc.co.uk/2/hi/7686564.stm.

———. 2008b. "S Africa Chinese 'Become Black.'" http://news.bbc.co.uk/2/hi/africa/7461099.stm.

Beaky, L. 2010. "A College Education? Or Diminished Expectations?" *Academe* 96 (3): 19–21.
Beaman, R., K. Wheldall, and C. Kemp. 2006. "Differential Teacher Attention to Boys and Girls in the Classroom." *Educational Review* 58 (3): 339–66.
Becker, H. 1963. *The Outsiders.* Glencoe, IL: Free Press.
Becker, J. A. H., A. J. Johnson, E. A. Craig, E. S. Gilchrist, M. M. Haigh, and L. T. Lane. 2009. "Friendships Are Flexible, not Fragile: Turning Points in Geographically-Close and Long-Distance Friendships." *Journal of Social and Personal Relationships* 26 (4): 347–69.
Bell, M. M. 2007. *An Invitation to Environmental Sociology.* Thousand Oaks, CA: Sage.
Bellah, R., R. Madssen, W. Sullivan, A. Swidler, and S. Tipton. 1985. *Habits of the Heart: Individualism and Commitment in American Life.* Berkeley: University of California Press.
Bem, S. L. 1993. *The Lenses of Gender: Transforming the Debate on Sexual Inequality.* New Haven, CT: Yale University Press.
Bendick, M., and A. Nunes. 2011. "Developing the Research Basis for Controlling Bias in Hiring." *Journal of Social Issues* 67(forthcoming).
Bendor, J., and P. Swistak. 1997. "The Evolutionary Stability of Cooperation." *American Political Science Review* 91 (June): 290–307.
Bennett, N. G., D. E. Bloom, and P. H. Craig. 1992. "American Marriage Patterns in Transition." In *The Changing American Family: Sociological and Demographic Perspectives,* edited by S. J. South and S. E. Tolnay, 89–108. Boulder, CO: Westview.
Benokraitis, N. 2010. *Marriages and Families: Changes, Choices, and Constraints,* 7th ed. Upper Saddle River, NJ: Prentice Hall.
Beresin, E. V. 2009. "The Impact of Media Violence on Children and Adolescents: Opportunities for Clinical Intervention." American Academy of Child Adolescent Psychiatry, *The DevelopMentor.* http://www.aacap.org/cs/root/developmentor/the_impact_of_media_violence_on_children_and_adolescents_opportunities_for_clinical_interventions.
Berger, P. 1963. *Invitation to Sociology.* New York: Anchor Books.
Berscheid, E. 1981. "An Overview of the Psychological Effects of Physical Attractiveness and Some Comments upon the Psychological Effects of Knowledge on the Effects of Physical Attractiveness." In *Logical Aspects of Facial Form,* edited by W. Lucker, K. Ribbens, and J. A. McNamera. Ann Arbor: University of Michigan Press.
———. 1982. "America's Obsession With Beautiful People." *U.S. News and World Report* (January 11): 59–61.
Berscheid, E., and E. Hatfield. 1983. *Interpersonal Attraction,* 2nd ed. Reading, MA: Addison-Wesley.
Besnard, P., and G. Desplanques. 1993. *La Cote des Prénoms en (1994).* Paris: Balland.
Best, J. 2001. *Damned Lies and Statistics: Untangling Numbers From the Media, Politicians, and Activists.* Berkeley: University of California Press.
———. 2004. *More Damned Lies and Statistics.* Berkeley: University of California Press.
———. 2008. *Stat-Spotting: A Field Guide to Identifying Dubious Data.* Berkeley: University of California Press.
Bevans, R. 2009. "Who Knows Baby Best? Investigating Connotative Gender Information, Gender Processing, and Gender Identification by Adults." *Dissertation Abstracts International, B: Sciences and Engineering* 69 (12): 7829.
Bhagat, V. 2007. "The State of e-Philanthropy." http://www.civicus.org/new/media/State-E-Philanthropy-in-2007.doc.
Bhanot, R., and J. Jovanovic. 2005. "Do Parents' Academic Gender Stereotypes Influence Whether They Intrude on Their Children's Homework?" *Sex Roles* 52 (9–10): 597–607.
Biancie, S., M. Milkie, L. Sayer, and J. Robinson. 2000. "Is Anyone Doing the Housework? Trends in the Gender Division of Household Labor." *Social Forces* 79 (1): 191–229.
Bidwell, C., and N. Friedkin. 1988. "The Sociology of Education." In *Handbook of Sociology,* edited by N. Smelser, 449–71. Newbury Park, CA: Sage.
Bienstock, J. L., and D. Laube. 2005. "The Recruitment Phoenix: Strategies for Attracting Medical Students Into Obstetrics and Gynecology." *Obstetrics & Gynecology,* Part 1 of 2, 105 (5): 1125–27.
Bishop, J. E. 1986. "'All for One . . . One for All?' Don't Bet on It." *Wall Street Journal* (December 4): 31.

Black, D. 1976. *The Behavior of Law.* New York: Academic Press.

———. 1998. *The Social Structure of Right and Wrong.* San Diego, CA: Academic Press.

Black, D., and J. Smith. 2006. "Estimating the Returns to College Quality With Multiple Proxies for Quality." *Journal of Labor Economics* 24 (3): 701–728.

Black, K., J. Marola, A. Littman, J. Chrisler, and W. Neace. 2009. "Gender and Form of Cereal Box Characters: Different Medium, Same Disparity." *Sex Roles: A Journal of Research* 60 (11–12): 882–89.

Blackstone, W. (1765–1769) 1979. *Commentaries on the Laws of England.* Chicago: University of Chicago Press.

Blackwell, D. L., and D. T. Lichter. 2004. "Homogamy Among Dating, Cohabitating and Married Couples." *Sociological Quarterly* 45 (4): 719–37.

Blake, R. R., and J. S. Moulton. 1979. "Intergroup Problem Solving in Organizations: From Theory to Practice." In *The Social Psychology of Intergroup Relations,* edited by W. G. Austin and S. Worschel, 151–163.

Blau, J. 2003. *Race in the Schools: Perpetuating White Dominance.* Boulder, CO: Lynne Rienner Publishers.

Bloemraad, I., A. Korteweg, and G. Yurdakul. 2008. "Citizenship and Immigration: Multiculturalism, Assimilation and Challenges to the Nation State." *Annual Review of Sociology* 34: 153–179.

Bloom, K., K. Moore-Schoenmakers, and N. Masataka. 1999. "Nasality of Infant Vocalizations Determines Gender Bias in Adult Favorability Ratings." *Journal of Nonverbal Behavior* 23 (3): 219–36.

Blossfeld, H. P. 2009. "Educational Assortative Marriage in Comparative Perspective." *Annual Review of Sociology* 35: 513–30.

Blum, S. D. 2005. "Five Approaches to Explaining 'Truth' and 'Deception' in Human Communication." *Journal of Anthropological Research* 61 (3): 289–315.

Boardman, J. 2004. "Stress and Physical Health: The Role of Neighborhoods as Mediating and Moderating Mechanisms." *Social Science and Medicine* 58: 2473–83.

Boase, J., J. B. Horrigan, B. Wellman, and L. Rainie. 2006. *The Strength of Internet Ties.* http://www.pewinternet.org/Reports/2006/The-Strength-of-Internet-Ties.aspx.

Bodnar, J. 1976. "Socialization and Adaptation: Immigrant Families in Scranton." *Pennsylvania History,* 43 (Apr): 147–63.

Bonilla-Silva, E., C. Goar, and D. Embrick. 2006. "When Whites Flock Together: The Social Psychology of White Habitus." *Critical Sociology* 32 (2/3): 229–53.

Bonner, J. 1984. *Research Presented in "The Two Brains."* Public Broadcasting System Telecast.

Books, S., and A. McAninch. 2006. "Jonathan Kozol's *Savage Inequalities:* A 15-Year Reconsideration." *Educational Studies* 40 (1).

Boshera, R. 2003. "The $6,000 Solution." *Atlantic Monthly* (January/February): 91–95.

Bourdieu, P. 1977a. "Cultural Reproduction and Social Reproduction." In *Power and Ideology in Education,* edited by J. Karabel and A. H. Halsey, 487–511. New York: Oxford University Press.

———. 1977b. *Outline of a Theory of Practice.* Cambridge: Cambridge University Press.

———. 1984. *Distinction: A Social Critique of the Judgment of Taste.* Cambridge, MA: Harvard University Press.

Boushey, H. 2002. "'This Country Is Not Woman-Friendly or Child-Friendly': Talking About the Challenge of Moving from Welfare-to-Work." *Journal of Poverty* 6 (2): 81–115.

Bowles, S., and H. Gintis. 1976. *Schooling in Capitalist America: Educational Reform and the Contradictions of Economic Life.* New York: Basic Books.

———. 1982. "The Crisis of Liberal Democratic Capitalism: The Case of the U.S." *Politics and Society* 11: 51–59.

———. 2002. "Schooling in Capitalist America Revisited." *Sociology of Education* 75 (1): 1–18.

Boyle, D. E., N. Marshall, and W. Robeson. 2003. "Gender at Play: Fourth-Grade Girls and Boys on the Playground." *American Behavioral Scientist* 46 (10): 1326–45.

Brady, D., and C. Palmeri. 2007. "The Pet Economy." *Bloomberg BusinessWeek* (August 6). http://www.businessweek.com/magazine/content/07_32/b4045001.htm.

Bramlett, M., and W. Mosher. 2002. "Cohabitation, Marriage, Divorce, and Remarriage in the United States." National Center for Health Statistics. *Vital Health Statistics* 23 (22).

Brand, J., and C. Halaby. 2006. "Regression and Matching Estimates of the Effects of Elite College Attendance on Educational and Career Achievement." *Social Science Research* 35 (3): 749–770.

Brandenburger, A. M., and B. J. Nalebuff. 1996. *Co-Opetition.* New York: Doubleday.

Branigan, T. 2001. "Women: In the Eye of the Beholder: Black and Asian Women Are Spending Thousands on Plastic Surgery—To Look More Caucasian." *Guardian* 2: 10.

Brewer, G., and J. Archer. 2007. "What Do People Infer From Facial Attractiveness?" *Journal of Evolutionary Psychology* 5 (1–4): 39–49.

Briggle, A. 2008. "Real Friends: How the Internet Can Foster Friendship." *Journal of Ethics and Information Technology* 10 (1): 71–79.

Brinkman, I. 2004. "Language, Names and War: The Case of Angola." *African Studies Review* 47 (3): 143–63.

Brophy, J. 1983. "Research on the Self-Fulfilling Prophecy and Teacher Expectations." *Journal of Educational Psychology* 75: 631–61.

Brown, J. F. 1994. "35 Black Men Quizzed in Union, S.C." *Afro-American* (November 12): A1.

Brown, J., and C. Pardun. 2004. "Little in Common: Racial and Gender Differences in Adolescents' Television Diets." *Journal of Broadcasting & Electronic Media* 48 (2): 266–78.

Browne, M. W. 1997. "Naming of 6 Elements to End Long Disputes." *New York Times* (March 4): C5.

BRS Labs. 2009. "Harris Poll Shows 96 Percent of Americans Support Use of Video Surveillance to Counteract Terrorism." www.brslabs.com/files/Harris_Poll_survey_release_FINAL.doc.

Brunet, G., and A. Bideau. 2000. "Surnames: History of the Family and History of Populations." *History of the Family* 5 (2): 153–61.

Bruning, J. L., N. K. Polinko, J. I. Zerbst, and J. T. Buckingham. 2000. "The Effect on Expected Job Success of the Connotative Meanings of Names and Nicknames." *Journal of Social Psychology* 140 (2): 197–201.

Bryan, J. H., and N. H. Walbek. 1970. "Preaching and Practicing Generosity." *Child Development* 41: 329–53.

Bryner, J. 2007. "Gay Unions Sanctioned in Medieval Europe." *Livescience.com* (August 27).

Brynin, M., A. M. Oerez, and S. Longhi. 2009. "The Social Significance of Homogamy." In *Changing Relationships,* edited by M. Brynin and J. Ermisch, 73–90. New York Routledge.

Buchmann, C., T. Prete, and A. McDaniel. 2008. "Gender Inequalities in Education." *Annual Review of Sociology* 34: 319–37.

Bukowski, M., M. Moya, S. de Lemus, S., and A. Szmajke. 2009. "Selective Stereotype Activation: The Joint Impact of Interpersonal Goals and Task Context." *European Journal of Social Psychology* 39 (2): 317–24.

Bumpass, L., and H. Lu. 2000. "Trends in Cohabitation and Implications for Children's Family Contexts in the United States." *Population Studies* 54 (1): 29–41.

Bureau of Justice Statistics. 2010a. "Key Facts at a Glance: Trends in Victimization Rates by Age." http://bjs.ojp.usdoj.gov/content/glance/tables/vagetab.cfm.

———. 2010b. "Criminal Victimization in the United States, 2007 Statistical Tables." http://bjs.ojp.usdoj.gov/content/pub/pdf/cvus0701.pdf.

———. 2010c. "Key Facts at a Glance: Federal Tort Trials." http://bjs.ojp.usdoj.gov/content/glance/tables/torttrialtab.cfm.

Burnard, T. 2001. "Slave Naming Patterns: Onomastics and the Taxonomy of Race in Eighteenth-Century Jamaica." *Journal of Interdisciplinary History* 31 (3): 325–46.

Burton-Chellew, M. N., A. Ross-Gillespie, and S. A. West. 2010. "Cooperation in Humans: Competition between Groups and Proximate Emotions." *Evolution and Human Behavior* 31 (2): 104–08.

Business Wire. 2002. "71% of Online Donors Give Weekdays from 9 a.m. to 6 p.m., Kintera Study Reports; Workplace Giving Important Factor in Rise of e-Philanthropy." *Business Wire* (November 14).

Buss, D. M., Shackelford, T. K., Kirkpatrick, L. A., and Larsen, R. J. 2001. "A Half Century of Mate Preferences: The Cultural Evolution of Values." *Journal of Marriage and the Family* 63 (2): 491–503.

Bussey, K. 1999. "Children's Categorization and Evaluation of Different Types of Lies and Truths." *Child Development* 70 (6): 1338–47.

Cadell, D. 2010. *The Friendship Factor and Church Commitment: The Impact of Social Integration and Consensus.* Saarbrücken, Germany: VDM Publishers.

Cadinu, M., and C. Reggiori. 2002. "Discrimination of a Low-Status Outgroup: The Role of Ingroup Threat." *European Journal of Social Psychology* 32 (4): 501–15.

Calhoun, C. 1998. “Community Without Propinquity Revisited: Communications Technology and the Transformation of the Urban Public Sphere.” *Sociological Inquiry* 68 (3): 373–97.

Calhoun, L. G., and R. G. Tedeschi. 2001. “Posttraumatic Growth: The Positive Lessons of Loss.” In *Meaning, Construction, and the Experience of Loss,* edited by R. A. Neimeyer, 157–72. Washington, DC: American Psychological Association.

Callan, P. 2007. “College Affordability: Colleges, States Increase Financial Burden on Students and Families.” *Measuring Up. The National Report Card on Higher Education.* http://measuringup.highereducation.org/commentary/collegeaffordability.cfm.

Callan, P. M. 2008. “The 2008 National Report Card: Modest Improvements, Persistent Disparities, Eroding Global Competitiveness.” The National Center for Public Policy and Higher Education. http://measuringup2008.highereducation.org/commentary/callan.php.

Camarota, S. A. 2007. “Immigrants in the United States, 2007: A Profile of America’s Foreign-Born Population.” Center for Immigration Studies. *Backgrounder* (November). http://www.cis.org/articles/2007/back1007.pdf.

Campos, P. 1998. *Jurismania: The Madness of American Law.* New York: Oxford University Press.

Caplan, L. J., and C. Schooler. 2007. “Socioeconomic Status and Financial Coping Strategies: The Mediating Role of Perceived Control.” *Social Psychology Quarterly* 70 (1): 43–58.

Carey, K., and M. Roza. 2008. “School Funding’s Tragic Flaw.” Education Sector. Center on Reinventing Public Education. http://www.educationsector.org/usr_doc/Tragic_Flaw_may14_combo.pdf.

Carley, K. M., and D. Krackhardt. 1996. “Cognitive Inconsistencies and Non-Symmetric Friendship.” *Social Networks* 18 (1): 1–27.

Carlson, D. K. 2005a. “Americans and Guns: Danger or Defense?” http://www.gallup.com/poll/14509/americans-guns-danger-defense.aspx.

Carmalt, J. H., J. Cawley, K. Joyner, and J. Sobal. 2008. “Body Weight and Matching with a Physically Attractive Romantic Partner.” *Journal of Marriage and Family* 70 (5): 1287–96.

Carpusor, A., and W. Loges. 2006. “Rental Discrimination and Ethnicity in Names.” *Journal of Applied Social Psychology* 36 (4): 934–52.

Carr, D., and M. A. Friedman. 2005. “Is Obesity Stigmatizing? Body Weight, Perceived Discrimination, and Psychological Well-Being in the United States.” *Journal of Health and Social Behavior* 3 (2): 244–59.

Carr, N. 2007. “Courting Middle-Class Parents to Use Public Schools.” *Education Digest* 72 (6): 35–41.

Carroll, J. 2007. “Americans: 2.5 Children Is ‘Ideal’ Family Size.” http://www.gallup.com/poll/27973/americans-25-children-ideal-family-size.aspx.

Cash, T. F., and L. H. Janda. 1984. “The Eye of the Beholder.” *Psychology Today* (December): 46–52.

Cash, T. F., and T. Pruzinsky. 1990. “The Psychology of Physical Appearance: Aesthetics, Attributes, and Images.” In *Body Images: Development, Deviance, and Change,* edited by T. F. Cash and T. Pruzinsky, 51–79. New York: Guilford.

———, eds. 2004. *Body Image: A Handbook of Theory, Research, and Clinical Practice.* New York/London: Guilford.

Cataldi, E. F., J. Laird, and A. Kewal-Ramani. 2009. “Dropout and Completion Rates in the United States: 2007.” National Center for Education Statistics. http://nces.ed.gov/pubs2009/dropout07/index.asp.

Catanzarite, L. 2003. “Race-Gender Composition and Occupational Pay Degradation.” *Social Problems* 50 (1): 14–37.

Cavanagh, G. F. 2005. *American Business Values: A Global Perspective.* Englewood, NJ: Prentice Hall.

CBS News. 2003. “Sex Crimes Cover-Up by Vatican.” CBS Evening News (August 6). http://www.cbsnews.com/stories/2003/08/06/eveningnews/main566978.shtml?tag=mncol;lst;1.

———. 2010. “No Vacation Nation.” http://www.cbsnews.com/video/watch/?id=6532684n.

CBS News Poll. 2006. “Religion in America.” (April 6–9). www.cbsnews.com/htdocs/CBSNews_polls/religion_041306.pdf.

Center for American Women and Politics. 2009a. “Fact Sheet, Voter Turnout.” http://www.cawp.rutgers.edu/fast_facts/voters/turnout.php.

———. 2009b. “CAWP Analysis of the 2008 Women’s Vote.” Women’s Vote Watch. http://www.cawp.rutgers.edu/fast_facts/voters/wvwatch/wvwatch_analysis.php.

———. 2010. "Fact Sheet: Levels of Office." http://www.cawp.rutgers.edu/fast_facts/levels_of_office/index.php.

Center for Justice and Democracy. 2007. "Mythbuster: Debunking Myths about Tort System Costs." http://www.centerjd.org/archives/issues-facts/MB_2007costs.pdf.

Center for Law and Social Policy. 2007. "Child Support Substantially Increases Economic Well-Being of Low- and Moderate-Income Families," Research Fact Sheet. http://www.clasp.org/publications/CS_Economic_FS.pdf.

Center for Responsive Politics. 2010. "Lobbying Database" OpenSecrets.org. http://www.opensecrets.org/lobby/index.php.

———. 2010a. "The Food Stamp Dependent Care Deduction: Help for Families with Child Care Costs" (March 23). http://www.cbpp.org/cms/index.cfm?fa=view&id=3130.

———. 2010b. "Policy Basics: Where Do Our Federal Tax Dollars Go?" (Updated April 14). http://www.cbpp.org/cms/index.cfm?fa=view&id=1258.

Centers for Disease Control and Prevention. 2008a. "Fact Sheets: Underage Drinking." http://www.cdc.gov/alcohol/quickstats/underage_drinking.htm.

———. 2008b. Youth Risk Behavior Surveillance—United States 2007. http://www.cdc.gov/mmwr/preview/mmwrhtml/ss5704a1.htm.

———. 2009a. "NCHS Health E-Stat: Prevalence of Overweight, Obesity and Extreme Obesity among Adults: United States, Trends 1960–62 through 2005–2006." http://www.cdc.gov/nchs/data/hestat/overweight/overweight_adult.htm.

———. 2009b. "U.S. Obesity Trends." http://www.cdc.gov/obesity/data/trends.html#Race.

———. 2009c. "Youth Violence: National Statistics." http://www.cdc.gov/ViolencePrevention/youthviolence/stats_at-a_glance/leading_cause.html.

———. 2009d. "Injury Prevention & Control: Motor Vehicle Safety." http://www.cdc.gov/MotorVehicleSafety/Child_Passenger_Safety/CPS-Factsheet.html.

———. 2009e. "Youth Violence: Facts at a Glance." http://www.cdc.gov/violenceprevention/pdf/YV_DataSheet_Summer2009-a.pdf.

———. 2009f. "Suicide: Facts at a Glance." http://www.cdc.gov/violenceprevention/pdf/Suicide-DataSheet-a.pdf.

———. 2009g. "Coping with Stress." http://www.cdc.gov/Features/HandlingStress/.

———. 2009h. "Health-Related Quality of Life: Nationwide Trend." http://apps.nccd.cdc.gov/HRQOL/TrendV.asp?State=1&Measure=7&Category=3&submit1=Go.

———. 2009i. "Fact Sheet. Adult Cigarette Smoking in the United States: Current Estimate." http://www.cdc.gov/tobacco/data_statistics/fact_sheets/adult_data/cig_smoking/index.htm.

———. 2009j. "FastStats: Suicide and Self-Inflicted Injury." http://www.cdc.gov/nchs/fastats/suicide.htm.

———. 2010a. "FastStats: Obesity and Overweight." http://www.cdc.gov/nchs/fastats/overwt.htm.

———. 2010b. "Childhood Overweight and Obesity." http://www.cdc.gov/obesity/childhood/index.html.

———. 2010c. "Health, United States, 2009: In Brief." http://www.cdc.gov/nchs/data/hus/hus09_InBrief.pdf.

———. 2010d. "Women and Heart Disease Fact Sheet." http://www.cdc.gov/dhdsp/data_statistics/fact_sheets/docs/fs_women_heart.pdf.

———. 2010e. "Traumatic Occupational Injuries." http://www.cdc.gov/niosh/injury/.

Central Florida News 13. 2008. "Florida Man Offers Rights to Name His Unborn Baby to Fight Gas Prices." http://whtq.com/morning_show/davidpartin.html.

Cerulo, K. A. 1997. "Re-Framing Sociological Concepts for a Brave New (Virtual?) World." *Sociological Inquiry* 67 (1): 48–58.

———. 1998. *Deciphering Violence: The Cognitive Structure of Right and Wrong.* New York: Routledge.

———. 2002. "Individualism . . . Pro Tem: Reconsidering U.S. Social Relations." In *Culture in Mind: Toward a Sociology of Culture and Cognition,* edited by K. A. Cerulo, 135–71). New York: Routledge.

———. 2006. *Never Saw It Coming: Cultural Challenges to Envisioning the Worst.* Chicago: University of Chicago Press.

———. 2008. "Social Relations, Core Values and the Polyphony of the American Experience: Robin Williams Got It Right." *Sociological Forum* 23 (2): 852–860.

———. 2009. "Non-Humans in Social Interaction." *Annual Review of Sociology* 35: 531–52.

Cerulo, K. A., and J. M. Ruane. 1997. "Death Comes Alive: Technology and the Re-Conception of Death." *Science as Culture* 6 (28): 444–66.

———. 1998. "Coming Together: New Taxonomies for the Analysis of Social Relations." *Sociological Inquiry* 68 (3): 398–425.

Cerulo, K. A., J. M. Ruane, and M. Chayko. 1992."Technological Ties That Bind: Media-Centered Primary Groups." *Communication Research* 19 (1): 109–29.

Chan, K. B. 2002. "Coping with Work Stress, Work Satisfaction, and Social Support: An Interpretive Study of Life Insurance Agents." *Southeast Asian Journal of Social Science* 30 (3): 657–85.

Chaney, B., and L. Lewis. 2007. "Public School Principals Report on Their School Facilities: Fall 2005." National Center for Education Statistics, NCES 2007–007. Washington, DC: U.S. Department of Education.

Chayko, M. 2002. *Connecting: How We Form Social Bonds and Community in the Internet Age.* Albany: SUNY Press.

Chia, R. C., and L. J. Alfred. 1998. "Effects of Attractiveness and Gender on the Perception of Achievement Related Variables." *Journal of Social Psychology* 138 (4): 471–78.

Child Trends Data Bank. 2010. "Family Structure." http://www.childtrendsdatabank.org/alphalist?q=node/231.

Children Now. 2004. *Fall Colors: Prime Time Diversity Report 2003–2004.* http://www.childrennow.org/uploads/documents/fall_colors_2003.pdf

Children's Defense Fund. 2007. "Each Day in America." http://www.childrensdefense.org/site/PageServer?pagename=research_national_data_each_day.

———. 2008. "The State of America's Children 2008." http://www.childrensdefense.org/child-research-data-publications/data/state-of-americas-children-2008-report.pdf.

———. 2009. "Moments in America for Children." December 2009. http://www.childrensdefense.org/child-research-data-publications/moments-in-america-for-children.html.

———. 2010. "The State of America's Children 2010: Child Health." http://www.childrensdefense.org/child-research-data-publications/data/the-state-of-americas-children-2010-report-health.pdf.

Chira, S. 1992. "Bias Against Girls Is Found Rife in Schools, with Lasting Damage." *New York Times* (February 12): A1.

Chomsky, A. 2007. *"They Take Our Jobs!" and 20 Other Myths about Immigration.* Boston: Beacon Press.

Chow, E., D. Coulombe, V. Garcia, D. Vuu, and J. Wade. 2009. "Age and Gender in Online Dating Websites: An Analysis of User Profiles on Mingles.com." *Culture, Power, and Cyberspace* (May 23). http://anthrocyber.blogspot.com/2009/05/age-and-gender-in-online-dating.html.

Christopher, A. N. 1998. "The Psychology of Names: An Empirical Examination." *Journal of Applied Social Psychology* 28 (13): 1173–95.

Chronicle of Higher Education. 2006. "Number of Students Transferring from Community Colleges to Elite Institutions Plunges." *Chronicle of Higher Education* 52 (44): 38.

Clark, R. 2007. "From Margin to Margin? Females and Minorities in Newbery and Caldecott Medal-Winning and Honor Books for Children." *International Journal of Sociology of the Family* 33 (2): 263–83.

Clark, W. 2009. "Changing Residential Preferences across Income, Education, and Age: Findings from the Multi-City Study of Urban Inequality" *Urban Affairs Review* 44 (3): 334–55.

Clarke, E. H. 1873. *Sex in Education: Or a Fair Chance for Girls.* Boston: J.R. Osgood.

Clifford, M. M., and E. H. Walster. 1973. "The Effects of Physical Attractiveness on Teacher Expectation." *Sociology of Education* 46: 245–58.

Clinton Foundation. 2005. "Press Release: Bush-Clinton Katrina Fund Passes $1 Million in Online Donations." http://www.clintonfoundation.org/news/news-media/0906[illegible]-hur-usa-pr-bush-clinton-katrina-fund-passes-one-million-dollars-in-online-donations.

Cochran, J. 2005. "A Simple Case of Complexity." *CQ Weekly* (January 31).

Cockerham, W. 2009. *Medical Sociology,* 11th ed. Upper Saddle River, NJ: Prentice Hall.

———. 2010. *Sociology of Mental Disorder*, 7th ed. Upper Saddle River, NJ: Prentice Hall.

Coffey, B., and P. A. McLaughlin 2009. "Do Masculine Names Help Female Lawyers Become Judges? Evidence from South Carolina." *American Law and Economics Review* 11 (1): 112–33.

Cohen, L., and C. B. Gray. 2003. "The Blaine Game: School Vouchers and State Constitutions." *The Taubman Center Report—Our 15th Anniversary.* John F. Kennedy School of Government, Harvard University, Cambridge, MA.

Cohen, S. 2004. "Social Relationships and Health." *American Psychologist* 59: 676–84.

Colantonio, S. E., V. Fuster, M. Ferreyras, and J. G. Lascano. 2006. "Isonymic Relationships in Ethno-Social Categories (Argentinian Colonial Period) Including Illegitimate Reproduction." *Journal of Biosocial Science* 38 (3): 381–89.

College Board. 2009a. "Trends in College Pricing 2009." http://www.trends-collegeboard.com/college_pricing/.

———. 2009b. "Trends in Student Aid 2010." http://trends.collegeboard.org/downloads/student_aid/highlights.pdf.

———. 2010. "Community Colleges." Trends in Higher Education Series. http://professionals.collegeboard.com/profdownload/trends-2010-community-colleges-one-page.pdf.

Colley, A., and J. Maltby. 2008. "Impact of the Internet on our Lives: Male and Female Personal Perspectives." *Computers in Human Behavior* 24 (5): 2005–13.

Collier, J. L. 1991. *The Rise of Selfishness in America.* New York: Oxford University Press.

Collier, R. 2005. "State Bypasses Kyoto, Fights Global Warming: California Tries to Cut Emissions on Its Own." *San Francisco Chronicle* (February 17). http://www.sfgate.com/cgi-bin/article.cgi?f=/c/a/2005/02/17/MNG1IBCSUS1.DTL.

Collins, J., and B. Wellman. 2010. "Small Town in the Internet Society: Chapleau is No Longer an Island." *American Behavioral Scientist* 53 (9): 1344–66.

Collins, P. H. 1990. *Black Feminist Thought.* New York: Routledge.

———. 2001. "Conflict Theory of Educational Stratification." In *Schools and Society: A Sociological Approach,* edited by J. Ballantine and J. Spade, 34–40. Belmont, CA: Wadsworth/Thomson.

Colwell, M., and E. Lindsey. 2005. "Preschool Children's Pretend and Physical Play and Sex of Play Partner: Connections to Peer Competence." *Sex Roles: A Journal of Research* 52 (7–8): 497–509.

Combs, A., ed. 1992. *Cooperation: Beyond the Age of Competition.* Philadelphia: Gordon and Breach.

Condron, D. 2008. "An Early Start: Skill Grouping and Unequal Reading Gains in the Elementary Years." *The Sociological Quarterly* 4 (2): 363–94.

Connecticut Law Tribune. 2010. "*Profits per Partner Still Alive as Law Firm Metric.*" (June 4). http://www.law.com/jsp/law/careercenter/lawArticleCareerCenter.jsp?id=1202459196241&Profits_Per_Partner_Still_Alive_as_Law_Firm_Metric.

Cooley, C. H. 1902. *Human Nature and Social Order.* New York: Scribner.

———. 1909. *Social Organization.* New York: Charles Scribner.

Coontz, S. 2000. *The Way We Never Were: American Families and the Nostalgia Trap.* New York: Basic Books.

———. 2006. *Marriage, a History: From Obedience to Intimacy, or How Love Conquered Marriage.* New York: Penguin.

———. 2007. "The Origins of Modern Divorce." *Family Process* 46 (1): 7–16.

Coopersmith, S. 1967. *Antecedents of Self-Esteem.* San Francisco: Freeman.

Cosaro, W. 2005. *The Sociology of Childhood.* Thousand Oaks, CA: Sage.

Coser, L. 1956. *The Functions of Social Conflict.* Glencoe, IL: Free Press.

———. 1963. *Sociology through Literature: An Introductory Reader.* Englewood Cliffs, NJ: Prentice Hall.

Cowe, R. 1990. "New Ice Cream Plans to Lick Rivals." *Guardian II* (April 2): 3.

Cox, D. 2007. "Young White Evangelicals: Less Republican, Still Conservative." The Pew Forum on Religion & Public Life. http://pewforum.org/Politics-and-Elections/Young-White-Evangelicals-Less-Republican-Still-Conservative.aspx.

Crane, D. R., and T. B. Heaton. 2007. *Handbook of Families and Poverty.* Thousand Oaks, CA: Sage.

Critser, G. 2003. *Fat Land: How Americans Became the Fattest People in the World.* Boston: Houghton Mifflin.

Crocker, P. 1985. "The Meaning of Equality for Battered Women Who Kill Men in Self-Defense." *Harvard Women's Law Journal* 8: 121–53.

Cross, S., and B. Bagilhole. 2002. "Girls' Jobs for the Boys? Men, Masculinity and Non-Traditional Occupations." *Gender, Work and Organization* 9 (2): 204–26.

Cross, S. E., and Vick, N. V. 2001. "The Interdependent Self-Construal and Social Support: The Case of Persistence in Engineering." *Personality and Social Psychology Bulletin* 27 (7): 820–832.

Cuklanz, L. M., and S. Moorti. 2006. "Television's 'New' Feminism: Prime-Time Representations of Women and Victimization." *Critical Studies in Media Communication* 23 (4): 302–21.

Cummings, J. 2003. "When Poverty Cripples." *NEA Today* 21 (4): 22.

Cummins, J. 2007. "Pedagogies for the Poor? Realigning Reading Instruction for Low-Income Students With Scientifically Based Reading Research." *Educational Researcher* 36 (9): 564–72.

Currie, D. 2007. "Clearing the Air: Did the EPA really cover up New York's 9/11 air pollution?" *Weekly Standard* 12 (July 16): 41. http://www.theweeklystandard.com/Content/Public/Articles/000/000/013/852oexrf.asp.

Cyberbullying Research Center. 2010. "Research." http://www.cyberbullying.us/research.php.

Dai, D. 2002. "Incorporating Parent Perceptions: A Replication and Extension Study of the Internal-External Frame of Reference Model of Self-Concept Development." *Journal of Adolescent Research* 17 (6): 617–45.

Dalphonse, S. 1997. "Choosing to Be Childfree: Broadening the Definition of Family." *The Population Connection Reporter* (May/June).

Daniels, R. 1990. *A History of Immigration and Ethnicity in American Life.* New York: Harper Perennial.

Davis, K. 1990. *Don't Know Much about History.* New York: Harper Collins.

Davis, K., and W. Moore. 1945. "Some Principles of Stratification." *American Sociological Review* 27 (1): 5–19.

Davis-Black, A., and J. Broschak. 2009. "Outsourcing and the Changing Nature of Work." *Annual Review of Sociology* 35: 321–340.

Dawkins, C. 2005. "Evidence on the Intergenerational Persistence of Residential Segregation by Race." *Urban Studies* 42 (3): 545–55.

Day, J., and B. Downs. 2007. "Examining the Gender Earnings Gap: Occupational Differences and the Life Course." Population Association of America 2007 Annual Meeting, New York. http://paa2007.princeton.edu/.

De Dreu, C. K. W., L. R. Weingart, and S. Kwon. 2000. "Influence of Social Motives on Integrative Negotiation: A Meta-Analytic Review and Test of Two Theories." *Journal of Personality and Social Psychology* 78 (5): 889–905.

Degher, D., and G. Hughes. 1992. "The Identity Change Process: A Field Study of Obesity." *Deviant Behavior* 2: 385–401.

DeNavas-Walt, C., B. Proctor, and J. Smith. 2009. "Income, Poverty, and Health Insurance Coverage in the United States: 2008." *Current Population Reports*, P60–236. Washington, DC: U.S. Government Printing Office. http://www.census.gov/newsroom/releases/archives/income_wealth/cb09–141.html.

Denton, N., and D. Massey. 1989. "Racial Identity Among Caribbean Hispanics: The Effect of Double Minority Status on Residential Segregation." *American Sociological Review* 54: 790–808.

DePaulo, B. M., D. A. Kashy, and S. E. Kirkendol. 1996. "Lying in Everyday Life." *Journal of Personality and Social Psychology* 70 (May): 979–95.

DeSantis, A., and W. Kayson. 1997. "Defendants' Characteristics of Attractiveness, Race, Sex, and Sentencing Decisions." *Psychological Reports* 81 (October): 679–83.

De Schipper, S., C. Hirschberg, and G. Sinha. 2002. "Blame the Name." *Popular Science* 261 (1): 36.

Deutsch, M. 2000. "Cooperation and Competition." In *The Handbook of Conflict Resolution: Theory and Practice,* edited by M. Deutsch and P. T. Coleman, 21–40. San Francisco: Jossey-Bass.

Deutsch, M., and R. M. Krauss. 1960. "The Effect of Threat on Interpersonal Bargaining." *Journal of Abnormal and Social Psychology* 1: 629–36.

Deutscher, E., and D. Messner. 2005. "Europe's Response to World Politics." *Society* 42 (6): 59–63.

Dewar, D. 2000. "Gender Impacts on Health Insurance Coverage: Findings for Unmarried Full-Time Employees." *Women's Health Issues* 10 (5): 268–77.

Diamond, J. 2005. "Evolutionary Biology, Geography and Skin Colour." *Nature* 435 (7040): 283–84.

Dill, K., and K. Thill. 2007. "Video Game Characters and the Socialization of Gender Roles: Young People's Perceptions Mirror Sexist Media Depictions." *Sex Roles: A Journal of Research* 57 (11–12): 851–64.

Dion, K. K. 1979. "Physical Attractiveness and Interpersonal Attraction." In *Love and Attraction,* edited by M. Cook and G. Wilson, 3–9. New York: Pergamon.

———. 2001. "Cultural Perspectives on Facial Attractiveness." In *Facial Attractiveness: Evolutionary, Cognitive, and Social Perspectives,* edited by G. Rhodes and L. A. Zebrowitz, 239–59. Greenwich, CT: Ablex.

Dion, K. K., and E. Berscheid. 1974. "Physical Attractiveness and Peer Perception Among Children." *Sociometry* 37: 1–12.

Dion, K. K., E. Berscheid, and E. Walster. 1972. "What Is Beautiful Is Good." *Journal of Personality and Social Psychology* 24: 285–90.

Dion, K. K., and K. Dion. 1996. "Cultural Perspectives on Romantic Love." *Personal Relationships* 3: 5–17.

Dion, K. L. 1979. "Intergroup Conflict and Intragroup Cohesiveness." In *The Social Psychology of Intergroup Relations,* edited by W. G. Austin and S. Worschel, 211–224. Monterey, CA: Brooks/Cole.

Dittmar, H. 2005. "Vulnerability Factors and Processes Linking Sociocultural Pressures and Body Dissatisfaction." *Journal of Social and Clinical Psychology* 24 (8): 1081–87.

Dizard, J. E., and H. Gadlin. 1990. *The Minimal Family.* Amherst: University of Massachusetts Press.

Dohnt, H. K., and M. Tiggemann. 2006. "Body Image Concerns in Young Girls: The Role of Peers and Media Prior to Adolescence." *Journal of Youth and Adolescence* 35 (2): 141–51.

Donohue, B. 2002. "Their Own Names Are What They Fear." *Star Ledger* (March 3): A1.

Downey, D. 2008. "Black/White Differences in School Performance: The Oppositional Culture Explanation." *Annual Review of Sociology* 34: 107–126.

Downey, D., and A. Vogt-Yuan. 2005. "Sex Differences in School Performance during High School: Puzzling Patterns and Possible Explanations." *The Sociological Quarterly* 46 (2): 299–321.

Doyle, R. 2002. "Calculus of Happiness." *Scientific American* 287 (5): 32–33.

Driskell, J. E., J. H. Johnston, and E. Salas. 2001. "Does Stress Training Generate to Novel Settings?" *Human Factors* 43 (1): 99–110.

Drucker, P. 1993. "The Rise of the Knowledge Society." *Wilson Quarterly* (Spring): 52–71.

Dubner, S. J. 2009. "The Cost of Fearing Strangers." The Opinion Pages. http://freakonomics.blogs.nytimes.com/2009/01/06/the-cost-of-fearing-strangers/.

DuBois, D. L., R. D. Feiner, S. Brand, A. M. Adan, and E. G. Evans. 2008. "A Prospective Study of Life Stress, Social Support, and Adaptation in Early Adolescence." *Child Development* 63 (3): 542–57.

Du Bois, W. E. B. (1903) 1982. *The Souls of Black Folks.* New York: Penguin.

Dudley, D. 2002. "Forever Cool." *AARP Modern Maturity* (September 5). http://www.globalaging.org/health/us/forevercool.htm.

Duina, F. 2011. *Winning: Reflections on an American Obsession.* Princeton, NJ: Princeton University Press.

Durkheim, É. (1938) 1966. *The Rules of Sociological Method,* 8th ed., trans. S. A. Solovay and J. H. Mueller. New York: Free Press.

Dye, J. 2005. "Fertility of American Women: June 2004." *Current Population Reports,* P20–555. Washington, DC: U.S. Census Bureau.

Eaton, D. K., L. Kann, S. Kinchen, S. Shanklin, J. Ross, J. Hawkins, W. A. Harris, et al. 2008. "Youth Risk Behavior Surveillance—United States, 2007." http://www.cdc.gov/mmwr/preview/mmwrhtml/ss5704a1.htm.

Economic Policy Institute. 2007. *The State of Working America 2006/2007.* Ithaca, NY: Cornell University Press.

———. 2010. *The State of Working America 2008/2009.* http://www.stateofworkingamerica.org/pages/about.

Edin, K., and L. Lein. 1997a. "Work, Welfare, and Single Mothers' Economic Strategies." *American Sociological Review* 62 (2): 253–66.

———. 1997b. *Making Ends Meet: How Single Mothers Survive Welfare and Low-Wage Work.* New York: Russell Sage.

Education Trust. 2008. "Core Problems: Out-of-Field Teaching Persists in Key Academic Courses and High-Poverty Schools." http://www.edtrust.org/sites/edtrust.org/files/publications/files/SASSreportCoreProblem.pdf.

Ehrenberg, R. G. 2007. "How Governments Can Improve Access to College." *Chronicle of Higher Education,* 53 (31, April 6): 47.

Ehrenreich, B. 2001. *Nickel and Dimed: On (Not) Getting by in America.* New York: Owl Books.

Eichenwald, K., and R. Atlas. 2003. “2 Banks Settle Accusations They Aided in Enron Fraud.” *New York Times* (July 29): A1.

Ekman, P. 2009. *Telling Lies: Clues to Deceit in the Marketplace, Politics, and Marriage*, 3rd ed. New York: W. W. Norton and Company.

Ekman, P., and M. G. Frank. 1993. “Lies That Fail.” In *Lying and Deception in Everyday Life*, edited by M. Lewis and C. Saarni, 184–200. New York: Guilford.

Ekman, P., M. O’Sullivan, W. V. Friesen, and K. R. Scherer. 1991. “Face, Voice and Body in Detecting Deceit.” *Journal of Nonverbal Behavior* 15 (2): 125–35.

Elles, L. 1993. *Social Stratification and Socioeconomic Inequality*. Westport, CT: Praeger.

Ellis, H. C. 1972. “Motor Skills in Learning.” In *Fundamentals of Human Learning and Cognition*, edited by H.C. Ellis. Dubuque, IA: Wm. C. Brown.

Elvish, R., I. James, and D. Milne. 2010. “Lying in Dementia Care: An Example of a Culture That Deceives in People’s Best Interests.” *Aging & Mental Health* 14 (3): 255–62.

Emerson, M., and D. Hartman. 2006. “The Rise of Religious Fundamentalism.” *Annual Review of Sociology* 32: 127–44.

Emerson, M., and D. Sikkink. 2006. “School Choice and Residential Segregation in U.S. Schools: The Role of Parents’ Education.” *Ethnic and Racial Studies* 31 (2): 267–293.

Employee Benefits Research Institute. 2010. “Retirement Confidence Survey.” http://www.ebri.org/surveys/rcs/2010/.

Environmental Protection Agency. 2010a. “2010 U.S. Greenhouse Gas Inventory Report.” http://www.epa.gov/climatechange/emissions/usinventoryreport.html.

———. 2010b. “Climate Change Indicators in the United States.” http://www.epa.gov/climatechange/indicators.html.

———. 2010c. “Coastal Zones and Sea Level Rise.” http://epa.gov/climatechange/effects/coastal/index.html.

———. 2010d. “Emissions.” Climate Change. http://www.epa.gov/climatechange/fq/emissions.html#q3.

———. 2010e. “Methane.” http://www.epa.gov/methane/.

Epstein, R. 2007. “The Truth about Online Dating.” *Scientific American* (February/March): 8.

Equal Employment Opportunity Commission. 2007. “Race-Based Charges FY 1997–FY 2006.” http://www.eeoc.gov/stats/race.html. http://www.eeoc.gov/eeoc/statistics/enforcement/race.cfm.

Erian, M., C. Lin, N. Patel, A. Neal, and R. E. Geiselman. 1998. “Juror Verdicts as a Function of Victim and Defendant Attractiveness in Sexual Assault Cases.” *American Journal of Forensic Psychology* 16 (3): 25–40.

Ericson, R. V., and A. Doyle. 2006. “The Institutionalization of Deceptive Sales in Life Insurance: Five Sources of Moral Risk.” *British Journal of Criminology* 46 (6): 993–1010.

Espenshade, T. J. 2009. *No Longer Separate, Not Yet Equal: Race and Class in Elite College Admission and Campus Life*. Princeton, NJ: Princeton University Press.

Espino, R., and M. M. Franz. 2002. “Latino Phenotypic Discrimination Revisited: The Impact of Skin Color on Occupational Status.” *Social Science Quarterly* 83 (2): 612–23.

ESPN.com. 2010. “Major League Rosters.” http://espn.go.com/mlb/stats/rosters/_/sort/null/order/false.

Evelyn, J. 2004. “Community Colleges at a Crossroads.” *Chronicle of Higher Education* (April 30). http://chronicle.com/article/Community-Colleges-at-a/16147.

Etzioni, A. 1994. *The Spirit of Community: The Reinvention of American Society*. New York: Touchstone Books.

———. 1996. *The New Golden Rule*. New York: Basic Books.

Ezzati, M., A. B. Friedman, S. C. Kulkarni, and C. J. L. Murray. 2008. “The Reversal of Fortunes: Trends in County Mortality and Cross-County Mortality Disparities in the United States.” *PLoS Med* 5 (4): e66. doi:10.1371/journal.pmed.0050066 http://www.plosmedicine.org/article/info%3Adoi%2F10.1371%2Fjournal.pmed.0050066.

Fallows, J. 1983. “Immigration: How It Is Affecting Us.” *Atlantic Monthly* 252 (5): 45–48.

Faw, B. 2005. “Katrina Exposes New Orleans’ Deep Poverty.” MSNBC News. http://www.msnbc.msn.com/id/9163091/41412000.

Feagin, J. R. 1975. *Subordinating the Poor*. Englewood Cliffs, NJ: Prentice Hall.

Federal Election Commission. 2009. "Presidential Campaign Finance." http://www.fec.gov/DisclosureSearch/mapApp.do.

Federal Interagency Forum on Aging-Related Statistics. 2006. "Older Americans Update 2006: Key Indicators on Wellness." Washington, DC: U.S. Government Printing Office. http://www.aoa.gov/agingstatsdotnet/Main_Site/Data/2006_Documents/OA_2006.pdf.

———. 2008. "Older Americans 2008: Key Indicators of Well-Being." http://www.agingstats.gov/agingstatsdotnet/main_site/default.aspx.

Federal Interagency Forum on Child and Family Statistics. 2009. "America's Children: Key National Indicators of Well-Being, 2009." http://www.childstats.gov/americaschildren/famsoc1.asp.

Federal Trade Commission. 2007a. "Annual Report to Congress for FY 2006: The National Do Not Call Registry." http://www.ftc.gov/os/2007/04/P034305FY 2006RptOnDNC.pdf.

———. 2007b. "Spam." http://www.ftc.gov/bcp/edu/microsites/spam/index.html.

———. 2010. "FTC Approves Two Reports to Congress on the National Do Not Call Registry." http://www.ftc.gov/opa/2010/01/donotcall.shtm.

Feeley, T. H. 2002. "Evidence of Halo Effects in Student Evaluations of Communication Instruction." *Communication Education* 51 (3): 225–236.

Feinson, R. 2004. *The Secret Universe of Names.* New York: Overlook.

Feldman, B. 2003. "'Savage Inequalities' Revisited." *Dollars and Sense* (January/ February). http://www.thirdworldtraveler.com/Education/SavageInequal_Revisited.html.

Felson, R. B., and M. Reed. 1986. "Reference Groups and Self-Appraisals of Academic Ability and Performance." *Social Psychology Quarterly* 49: 103–09.

———. 1987. "The Effect of Parents on the Self-Appraisals of Children." *Social Psychology Quarterly* 49: 302–08.

Fernback, J. 2007. "Beyond the Diluted Community Concept: A Symbolic Interactionist Perspective on Online Social Relations." *New Media and Society* 9 (1): 49–69.

Festinger, L., S. Schachter, and K. Back. 1950. *Social Pressures in Informal Groups: A Study of Human Factors in Housing.* New York: Harper and Brothers.

Fiese, B., and G. Skillman. 2000. "Gender Differences in Family Stories: Moderating Influence of Parent Gender Role and Child Gender." *Sex Roles* 43 (5–6): 267–83.

Figueroa, C. 2003. "Self-Esteem and Cosmetic Surgery: Is There a Relationship Between the Two?" *Plastic Surgical Nursing* 23 (1): 21–25.

Finch, J. 2008. "Naming Names: Kinship, Individuality and Personal Names." *Sociology* 42 (4): 709–25.

Finkelstein, E. A., and L. Zuckerman. 2008. *The Fattening of America: How the Economy Makes Us Fat, If It Matters and What to Do about It* . New York: John Wiley & Sons.

Finlinson, A. R., A. M. B. Austin, and R. Pfister. 2000. "Cooperative Games and Children's Positive Behaviors." *Early Child Development and Care* 164: 29–40.

Fischer, C. 1982. *To Dwell Among Friends: Personal Networks in Town and City.* Chicago: University of Chicago Press.

Fischer, D. H. 1977. *Growing Old in America.* New York: Oxford University Press.

Foucault, M. 1971. *The Order of Things: An Archeology of Human Sciences.* New York: Pantheon.

Fowler, J. H., and N. A. Christakis. 2010. "Cooperative Behavior Cascades in Social Networks." *Proceedings of the National Academy of Sciences of the United States of America* 107 (12): 5334–38.

Fox, V., and M. Quit. 1980. *Loving, Parenting, and Dying: The Family Cycle in England and America, Past and Present.* New York: Psychohistory Press.

Frank, J. B. 2002. *The Paradox of Aging in Place in Assisted Living.* Westport, CT: Bergin and Garvey.

Frankel, T. 2005. *Trust and Honesty: America's Business Culture at a Crossroad.* New York: Oxford University Press.

Frankenberg, E., C. Lee, and G. Orfield. 2003a. "A Multiracial Society With Segregated Schools: Are We Losing the Dream?" The Civil Rights Project, Harvard University, Cambridge, MA. http://pages.pomona.edu/~vis04747/h21/readings/AreWeLosingtheDream.pdf.

Frankenberg, E., G. Siegel-Hawley, and J. Wang. 2010. "Choice Without Equity: Charter School Segregation and the Need for Civil Rights Standards." The Civil Rights Project/Proyecto Derechos Civiles at UCLA, Los Angeles, CA. www.civilrightsproject.ucla.edu.

Frazier, P. A., A. P. Tix, C. D. Klein, and N. J. Arikian. 2000. "Testing Theoretical Models of the Relations Between Social Support, Coping, and Adjustments to Stressful Life Events." *Journal of Social and Clinical Psychology* 19: 314–35.

Freeman, C. 2004. "Trends in Educational Equity of Girls and Women: 2004." *Education Statistics Quarterly* 6 (4). http://nces.ed.gov/programs/quarterly/vol_6/6_4/8_1.asp#2.

Freeman, N. 2007. "Preschoolers' Perceptions of Gender Appropriate Toys and Their Parents' Beliefs About Genderized Behaviors: Miscommunication, Mixed Messages, or Hidden Truths?" *Early Childhood Education Journal* 34 (5): 357–66.

Freng, S., and D. Webber. 2009. "Turning Up the Heat on Online Teaching Evaluations: Does 'Hotness' Matter?" *Teaching of Psychology* 36 (3): 189–93.

Freudenheim, B. 1988. "Who Lives Here, Go-Getter or Grouch?" *New York Times* (March 31): 15–16.

Friedman, L. 1975. *The Legal System: A Social Science Perspective.* New York: Russell Sage.

Frontline. 2004. "'Evangelical'—What It Means." http://www.pbs.org/wgbh/pages/frontline/shows/jesus/evangelicals/whatitmeans.html.

Fuchs, C. 2010. "Theoretical Foundations of Defining the Participatory, Co-Operative, Sustainable Information Society." *Information, Communication and Society* 13 (1): 23–47.

Fukuoka, Y. 1998. "Japanese Alias vs. Real Ethnic Name: On Naming Practices Among Young Koreans in Japan." Paper Presented at the Annual Meeting of the International Sociological Association.

Fuligni, A., and S. Pedersen. 2002."Family Obligation and the Transition to Young Adulthood." *Developmental Psychology* 38 (5): 856–68.

Funiciello, T. 1990. "The Poverty of Industry." *Ms.* (November/December): 32–40.

Furstenburg, F., Jr. 1976. "Pre-Marital Pregnancy and Marital Instability." *Journal of Social Issues* 32 (1): 67–86.

Furstenburg, F. F., Jr., and K. G. Talvitie. 1980. "Children's Names and Parental Claims: Bonds Between Unmarried Fathers and Their Children." *Journal of Family Issues* 1: 31–57.

Gallup, G., Jr. 1976–2003. *The Gallup Poll: Public Opinion.* Wilmington, DE: Scholarly Resources.

———. 2004–2010. *The Gallup Poll: Public Opinion.* Wilmington, DE: Scholarly Resources.

Gallup Poll. 2009. "Race Relations." http://www.gallup.com/poll/1687/Race-Relations.aspx.

Ganahl, D., T. Prinsen, and S. B. Netzley. 2003. "A Content Analysis of Prime Time Commercials: A Contextual Framework of Gender Representation." *Sex Roles* 49 (9/10): 545–51.

Gannon, S. 2007. "Laptops and Lipsticks: Feminising Technology." *Learning, Media, & Technology* 32 (1): 53–67.

Gans, H. 1971. "The Uses of Poverty: The Poor Pay for All." *Social Policy* (Summer): 20–24.

Garcia, S. D. 1998. "Appearance Anxiety, Health Practices, Metaperspectives and Self-Perception of Physical Attractiveness." *Journal of Social Behavior and Personality* 13 (2): 12–15.

Gardyn, R. 2002. "The Mating Game." *American Demographics* (July/August): 33–37.

Gates, H. L., Jr. 2007. *Finding Oprah's Roots.* New York: Random House.

———. 2009a. *In Search of Our Roots: How 19 Extraordinary African Americans Reclaimed Their Past,* New York: Crown Publishing.

———. 2009b. "Michelle's Great-Great-Great-Granddaddy—and Yours." The Root. http://www.theroot.com/views/michelle-s-great-great-great-granddaddy-and-yours?GT1=38002.

Gelles, R. J., D. R. Loseke, and M. M. Cavanaugh, eds. 2004. *Current Controversies on Family Violence.* Newbury Park, CA: Sage.

General Social Survey. 2004. *Cumulative Datafile 1972–2004.* http://sda.berkeley.edu/cgi-bin/hsda?harcsda+gss04.

———. 2007. *General Social Surveys, 1972–2006.* Cumulative File. http://www.norc.org/GSS+Website/Data+Analysis/.

———. 2010. *Cumulative Codebook. 1972–2010.* http://www.icpsr.umich.edu: 8080/GSS. http://www.norc.uchicago.edu/GSS+Website/.

George, J. 2006. "Access to Justice, Costs, and Legal Aid." *American Journal of Comparative Law Supplement* 54 (Fall): 293.

Gerbner, G., L. Gross, M. Morgan, N. Signorielli, and J. Shanahan, J. 2002. "Growing Up With Television: Cultivation Processes." In *Media Effects: Advances in Theory and Research,* 2nd ed., edited by J. Bryant and D. Zillmann, 43–67. Mahwah, NJ: Lawrence Erlbaum.

Gergen, K. 1971. *The Concept of Self.* New York: Holt, Rinehart and Winston.

Gergen, M. M., K. J. Gergen, K. March, and C. Ellis. 2003. "The Body and the Physical Self." In *Inner Lives and Social Worlds: Readings in Social Psychology,* edited by J. A. Holstein and J. F. Gubrium, 301–40. London: Oxford University Press.

Gerhards, J., and S. Hans. 2009. "From Hasan to Herbert: Name-Giving Patterns of Immigrant Parents Between Acculturation and Ethnic Maintenance." *American Journal of Sociology* 114 (4): 1102–28.

Gerull, F., and R. Rapee. 2002. "Mother Knows Best: The Effects of Maternal Modeling on the Acquisition of Fear and Avoidance Behavior in Toddlers." *Behavior Research and Therapy* 40 (3): 279–87.

Gettleman, T. E., and J. K. Thompson. 1993. "Actual Differences and Stereotypical Perceptions in Body Image and Eating Disturbance: A Comparison of Male and Female Heterosexual and Homosexual Sample." *Sex Roles* 29 (7/8): 545–62.

Gibson, A. L., and J. C. Quick. 2008. "Best Practices for Work Stress and Well-Being: Solutions for Human Dilemmas in Organizations." In *Building More Effective Organizations: HR Management and Performance in Practice,* edited by R. J. Burke and C. L. Copper, 84–109. New York: Cambridge University Press.

Giddens, A. 1992. *The Transformation of Intimacy, Sexuality, Love and Eroticism in Modern Societies.* Stanford, CA: Stanford University Press.

Gilbert, D. 2003. *The American Class Structure in an Age of Growing Inequality.* Belmont, CA: Wadsworth/Thompson.

Gilman, S. L. 2008. *Fat: A Cultural History of Obesity.* Cambridge, UK: Polity Press.

Gimlin, D. 2002. *Body Work: Beauty and Self-Image in American Culture.* Berkeley: University of California Press.

Giordano, P. 2003. "Relationships in Adolescence." *Annual Review of Sociology* 29: 257–81.

Glaser, C. L. 2010. *Rational Theory of International Politics: The Logic of Competition and Cooperation.* Princeton, NJ: Princeton University Press.

Glassman, M. 2000. "Mutual Aid Theory and Human Development: Sociability as Primary." *Journal for the Theory of Social Behavior* 30 (4): 391–412.

Glassner, B. 2000. *The Culture of Fear: Why Americans Are Afraid of the Wrong Things.* New York: Basic Books.

Glenn, D. 2009. "Why Needy Students Miss Out on Elite Colleges." *Chronicle of Higher Education* 55 (19): A4.

Glenn, N. D., and M. Supancic. 1984. "The Social and Demographic Correlates of Divorce and Separation in the United States." *Journal of Marriage and Family* 46: 563–75.

Glick, P., and L. Sung-Ling. 1986. "Recent Changes in Divorce and Remarriage." *Journal of Marriage and Family* 48: 737–47.

Goffman, E. 1959. *The Presentation of Self in Everyday Life.* New York: Anchor.

———. 1963. *Stigma: Notes on the Management of Spoiled Identity.* Englewood Cliffs, NJ: Prentice Hall.

———. 1974. *Frame Analysis.* Cambridge, MA: Harvard University Press.

Goldberg, S., and M. Lewis. 1969. "Play Behavior in the Year-Old Infant: Early Sex Differences." *Child Development* 40: 21–31.

Goldiner, D. 2001. "What's in a Name? At Least 500G Couple Hopes." *New York Daily News* (July 27).

Goleman, D. 1985. *Vital Lies, Simple Truths.* New York: Simon and Schuster.

Gonzalez Faraco, J. C., and M. D. Murphy. 1997. "Street Names and Political Regimes in an Andalusian Town (Almonte, Spain)." *Ethnology* 36 (Spring): 123–48.

Good, J., J. Woodzicka, and L. Wingfield. 2010. "The Effects of Gender Stereotypic and Counter-Stereotypic Textbook Images on Science Performance." *The Journal of Social Psychology* 150 (2): 132–47.

Gooding, G. E., and R. M. Kreider. 2010. "Women's Marital Naming Choices in a Nationally Representative Sample." *Journal of Family Issues* 31 (5): 681–701.

Goodstein, L. 2005. "More Religion, but Not the Old-Time Kind." *New York Times* (January 9) 4:1, 4.

Goodwin, P. Y., W. D. Mosher, and A. Chandra. 2010. "Marriage and Cohabitation in the United States: A Statistical Portrait Based on Cycle 6 (2002) of the National Survey of Family Growth." National Center for Health Statistics. *Vital Health Statistics* 23(28). http://www.cdc.gov/nchs/data/series/sr_23/sr23_028.pdf.

Gordon, T., J. Holland, E. Lahelma, and T. Tolonen. 2005. "Gazing With Intent: Ethnographic Practice in Classrooms." *Qualitative Research* 5 (1): 113–131.

Gouldner, H., and M. S. Strong. 1987. *Speaking of Friendship: Middle-Class Women and Their Friends.* New York: Greenwood.

Graham, R. 2011. "The Science of Gray Hair." InteliHealth. http://www.intelihealth.com/IH/ihtIH/c/9023/24253/348513.html?d=dmtContent.

Grall, T. 2009. "Custodial Mothers and Fathers and Their Child Support: 2007." *Current Population Reports.* http://www.census.gov/prod/2009pubs/p60–237.pdf.

Grant, M. J., C. M. Button, T. E. Hannah, and A. S. Ross. 2002. "Uncovering the Multidimensional Nature of Stereotype Inferences: A Within-Participants Study of Gender, Age, and Physical Attractiveness." *Current Research in Social Psychology* 8 (2): 19–38.

Grassl, W. 1999. "The Reality of Brands: Towards an Ontology of Marketing." *American Journal of Economics and Sociology* 58 (2): 313–59.

Grave, S., L. M. Ward, and J. S. Hyde. 2008. "The Role of Media in Body Image Concerns Among Women: A Meta-Analysis of Experimental and Correlational Studies." *Psychological Bulletin* 134 (3): 460–76.

Graves, J. 2006. "What We Know and What We Don't Know: Human Genetic Variation and the Social Construction of Race." Is Race "Real"? Social Science Research Council. http://raceandgenomics.ssrc.org/Graves/.

Greenfield, L. 2006. *Thin.* San Francisco: Chronicle Books.

Greenhouse, L. 2006. "Dwindling Docket Mystified Supreme Court." *New York Times* (December 7) http://www.nytimes.com/2006/12/07/washington/07scotus.html?_r=1&pagewanted=2.

Greimel, H. 2007. "Japan Changes Name of Iwo Jima." Associated Press (June 20). http://www.usatoday.com/news/world/2007–06–20-iwo-jima_N.htm.

Greven, P. J. 1970. *Four Generations: Population, Land, and Family in Colonial Andover.* Ithaca, NY: Cornell University Press.

Groesz, L. M., M. P. Levine, and S. K. Murnen. 2002. "The Effects of Experimental Presentation of Thin Media Images on Body Satisfaction: A Meta-Analytic Review." *International Journal of Eating Disorders* 31 (1): 1–16.

Grover, K. J., C. S. Russell, and W. Schumm. 1985. "Mate Selection Processes and Marital Satisfaction." *Family Relations* 34: 383–86.

Gueguen, N., M. Dufourcq-Brana, and A. Pascual. 2005. "The First Name: An Attribute of Self-Identity That Affects One's Evaluation of Oneself and That of Others." *Cahiers Internationoux de Psychologie Sociale* 65 (1): 33–44.

Gueguen, N., M. Lourel, C. Charron, J. Fischer-Lokou, and L. Larny. 2009. "A Web Replication of Snyder, Decker, and Berschied's (1977) Experiment on the Self-Fulfilling Nature of Social Stereotypes." *The Journal of Social Psychology* 149 (5): 600–02.

Guillen, M. 2001. "Is Globalization Civilizing, Destructive or Feeble? A Critique of Five Key Debates in the Social Science Literature." *Annual Review of Sociology* 27: 235–60.

Guimond, S., S. Dif, and A. Aupy. 2002. "Social Identity, Relative Group Status and Intergroup Attitudes: When Favorable Outcomes Change Intergroup Relations . . . for the Worse." *European Journal of Social Psychology* 32 (6): 739–60.

Gumbel, A. 2002. "Fast Food Nation: An Appetite for Litigation." *Independent/UK* (June 4). http://www.commondreams.org/headlines02/0604–01.htm.

Gunnthorsdottir, A. 2001. "Physical Attractiveness of an Animal Species as a Decision Factor for Its Preservation." *Anthrozoos* 14 (4): 204–15.

Gusfield, J. 1963. *Symbolic Crusade: Status Politics and the American Temperance Movement.* Urbana: University of Illinois Press.

Haas, A., and S. W. Gregory Jr. 2005. "The Impact of Physical Attractiveness on Women's Social Status and Interactional Power." *Sociological Forum* 20 (3): 449–71.

Hacker, A. 2003a. *Mismatch: The Growing Gulf Between Women and Men.* New York: Scribner.

———. 2003b. *Two Nations: Black and White, Separate, Hostile, Unequal.* New York: Scribner.

Hacking, I. 1995. "The Looping Effect of Human Kinds." In *Causal Cognition: A Multidisciplinary Debate,* edited by D. Sperber, D. Premack, A. J. Premack, and J. Premack, 351–94. New York: Oxford University Press.

———. 1999. *The Social Construction of What?* Cambridge, MA: Harvard University Press.

Haden, S. C., and A. Scarpa. 2008. "Community Violence Victimization and Depressed Mood: Moderating Effects of Coping and Social Support." *Journal of Interpersonal Violence* 23 (9): 1213–34.

Hagan, J. 2008. *Migration Miracle: Faith, Hope and Meaning on the Undocumented Journey.* Cambridge, MA: Harvard University Press.

Hagan, J., R. MacMillan, and B. Wheaton. 1996. "New Kid in Town: Social Capital and the Life Course Effects of Family Migration on Children." *American Sociological Review* 61 (3): 368–85.

Halbert, D. J. 2002. "Citizenship, Pluralism, and Modern Public Sphere." *Innovation* 15 (1): 33–42.

Hall, W. 1986. "Social Class and Survival on the S.S. *Titanic.*" *Social Science and Medicine* 22: 687–90.

Hampton, K. N. 2010. "Internet Use and the Concentration of Disadvantage: Glocalization and the Urban Underclass." *American Behavioral Scientist* 53 (8): 1111–32.

Hansen, K. 2005. *Not-So-Nuclear Families: Class, Gender, and Networks of Care.* New Brunswick, NJ: Rutgers University Press.

Hareven, T. 1978. "The Dynamics of Kin in an Industrial Community." In *Turning Points: Historical and Sociological Essays on the Family,* edited by J. Demos and S. Boocock, 151–82. Chicago: University of Chicago Press.

Hargreaves-Heap, S. P., and Y. Varoufakis. 2005. *Game Theory: A Critical Text.* London: Routledge.

Harper, B., and M. Tiggemann. 2008. "The Effect of Thin Ideal Media Images on Women's Self-Objectification, Mood, and Body Image." *Sex Roles* 58 (9–10): 649–57.

Harper, S., and B. Reskin. 2005. "Affirmative Action at School and on the Job." *Annual Review of Sociology* 31: 357–380.

Harrington, B. 2009. *Deception: From Ancient Empires to Internet Dating.* Stanford, CA: Stanford University Press.

Harrington, M., and H. Hsi. 2007. "Women Lawyers and Obstacles to Leadership." MIT Workplace Center. (Spring). http://web.mit.edu/workplacecenter/docs/law-report_4–07.pdf.

Harris, A., K. Jamison, and M. Trujillo. 2008. "Disparities in the Educational Success of Immigrants: An Assessment of the Immigrant Effect for Asians and Latinos." *The Annals of the American Academy of Political and Social Science* 620: 90–114.

Harris, T. G. 1978. *Introduction to E. H. Walster and G. W. Walster, a New Look at Love.* Reading, MA: Addison-Wesley.

Harris Interactive. 2006. "Over Thirty Million Adults Claim to Be Victims of Domestic Violence." http://www.ncdsv.org/images/Over33MillionAdultsClaimtobeVictimDV.pdf.

———. 2008. "Leisure Time Plummets 20% in 2008—Hits New Low." http://www.harrisinteractive.com/vault/Harris-Interactive-Poll-Research-Time-and-Leisure-2008–12.pdf.

———. 2009. "Man of Many Sweaters, Cliff Huxtable Is America's Favorite TV Dad." http://www.harrisinteractive.com/vault/Harris-Interactive-Poll-Research-Fav-TV-Dad-2009–06.pdf.

———. 2010a. "Substantial Numbers Still Willing to Donate Time and Money." http://www.harrisinteractive.com/vault/HI-Harris-Poll-Charitable-Giving-2010–11–04.pdf.

———. 2010b. "Annual Happiness Index Finds One-Third of Americans Are Very Happy." http://www.harrisinteractive.com/vault/HI-Harris-Poll-Happiness-Index-2010–05–20.pdf.

Harrison, K. 2000. "Television Viewing, Fat Stereotyping, Body Shape Standards, and Eating Disorder Symptomatology in Grade School Children." *Communication Research* 27 (5): 617–40.

———. 2006. "Fast and Sweet: Nutritional Attributes of Television Food Advertisements with and without Black Characters." *The Howard Journal of Communications* 17 (4): 249–64. .

Harrison, K., and J. Cantor. 1997. "The Relationship between Media Consumption and Eating Disorders." *Journal of Communication* 47 (1): 40–67.

Hartford Institute for Religion Research. 2006. "Fast Facts: How Many People Go to Church Each Sunday?" http://hirr.hartsem.edu/research/fastfacts/fast_facts.html.

Hartman, H. 1976. "Capitalism, Patriarchy, and Job Segregation by Sex." *Signs* 1 (3): 137–70.

Harvey, K. 2007. "Hooray for Gray!" http://www.eldr.com/article/culture/hooray-gray.

Hattie, J., and H. Timperley. 2007. "The Power of Feedback." *Review of Educational Research* 77 (1): 81–112.

Haub, C. 2003. "The U.S. Birth Rate Falls Further." Population Reference Bureau. http://www.prb.org/Articles/2003/TheUSBirthRateFallsFurther.aspx.

———. 2007. "Is Fertility Rising in Countries With Low Birth Rates?" Population Reference Bureau. http://www.prb.org/Articles/2007/IsFertilityRisinginLowBirthRateCountries.aspx.

———. 2008. "Tracking Trends in Low Fertility Countries: An Uptick in Europe?" Population Reference Bureau. http://www.prb.org/Articles/2008/tfrtrendsept08.aspx/.

Hauser, C., and A. O'Connor. 2007. "Virginia Tech Shooting Leaves 33 Dead." *New York Times* (April 16) http://www.nytimes.com/2007/04/16/us/16cnd-shooting.html?pagewanted=2.

Hausman, C. 2000. *Lies We Live By: Defeating Double-Talk and Deception in Advertising, Politics, and the Media.* New York: Routledge.

Hawkins, B., R. A. Tuff, and G. Dudley. 2006. "African American Women, Body Composition, and Physical Activity." *Journal of African American Studies* 10 (1): 44–56.

Hawkins, D., and A. Booth. 2005. "Unhappily Ever After: Effects of Long-Term, Low-Quality Marriages on Well-Being." *Social Forces* 84 (1): 445–65.

Hebl, M. R., and T. F. Heatherton. 1998. "The Stigma of Obesity in Women: The Difference Is Black and White." *Personality and Social Psychology* 24 (4): 417–26.

Heeren, J. W. 1999. "Emotional Simultaneity and the Construction of Victim Unity." *Symbolic Interaction* 22 (2): 163–79.

Heimann, L. B. 2002. "Social Proximity in Swedish Mother-Daughter and Mother-Son Interactions in Infancy." *Journal of Reproductive and Infant Psychology* 20 (1): 37–43.

Henry J. Kaiser Family Foundation and Health Research & Educational Trust. 2009. *Employer Health Benefits 2009 Annual Survey.* http://ehbs.kff.org/?CFID=29339258&CFTOKEN=30893308&jsessionid=6030f8357dc76211bea37c3c2354225b1415.

Hensler, D. 2001. "Revisiting the Monster: New Myths and Realities of Class Action and Other Large Scale Litigation." *Duke Journal of Comparative and International Law* 11: 179–213.

Hetherington, E. M., and J. Kelly. 2002. *For Better or for Worse: Divorce Reconsidered.* New York: W. W. Norton.

Hewstone, M., M. Rubin, and H. Willis. 2002. "Intergroup Bias." *Annual Review of Psychology* 53: 575–604.

Hill, C., and G. Winston. 2010. "Low-Income Students and Highly Selective Private Colleges: Geography, Searching, and Recruiting." *Economics of Education Review* 29 (4): 495–503.

Hinduja, S., and J. W. Patchin. 2010a. "Cyberbullying: Identification, Prevention, and Response." Cyberbullying Research Center. http://www.cyberbullying.us/Cyberbullying_Identification_Prevention_Response_Fact_Sheet.pdf.

———. 2010b. "State Cyberbullying Laws: A Brief Review of State Cyberbullying Laws and Policies." Cyberbullying Research Center. http://cyberbullying.us/Bullying_and_Cyberbullying_Laws.pdf.

Hitchens, C. 2010. The Great Catholic Cover-Up. *Washington Post* (March 16), http://newsweek.washingtonpost.com/onfaith/guestvoices/2010/03/the_great_catholic_cover-up.html.

Hoag, C. 2010. "LA Unveils $578 Million School, Costliest in the Nation." http://www.huffingtonpost.com/2010/08/22/robert-f-kennedy-communit_n_690497.html#s129588.

Hochschild, A. R. 2001 [1997]. *The Time Bind: When Work Becomes Home and Home Becomes Work.* New York: Owl Books.

Hochschild, A. R. 2003a. *The Commercialization of Intimate Life: Notes From Home and Work.* Berkeley: University of California Press.

———. 2003b. *The Second Shift: Working Parents and the Revolution at Home.* New York: Penguin.

Holman, T. B., J. H. Larson, and J. A. Olsen. 2001. *Premarital Predictors of Marital Quality or Breakup: Research, Theory and Practice.* New York/London: Kluwer Academic/Plenum.

Holtz, R., and N. Miller. 2001. "Intergroup Competition, Attitudinal Projection, and Opinion Certainty: Capitalizing on Conflict." *Group Processes and Intergroup Relations* 4 (1): 61–73.

Holtzman, D. H. 2006. *Privacy Lost: How Technology Is Endangering Your Privacy.* San Francisco: Jossey-Bass.

Holzer, H. J. 2007, September. "The Labor Market and Young Black Men: Updating Moynihan's Perspective." The Urban Institute. http://www.urban.org/UploadedPDF/1001111_moynihan_perspective.pdf.

Hoover, J., S. Hastings, and G. Musambira. 2009. "'Opening a Gap' in Culture: Women's Uses of the Compassionate Friends Website." *Women and Language* 32 (1): 82–90.

Hope, S., B. Rodgers, and C. Power. 1999. "Marital Status Transitions and Psychological Distress: Longitudinal Evidence From a National Population Sample." *Psychological Medicine* 29 (2): 381–89.

Hopkins, C. 2010. "Regional Cap-and-Trade System Sees Early Successes." NationalJournal.com. http://www.nationaljournal.com/njonline/no_20100302_2443.php.

Hopkins, N., and C. Rae. 2001. "Intergroup Differentiation: Stereotyping as a Function of Status Hierarchy." *Journal of Social Psychology* 141 (3): 323–33.

Hormone Foundation. 2008. "Stress and Your Health." http://www.hormone.org/Resources/Other/loader.cfm?csModule=security/getfile&pageid=1119.

Horrigan, J., and L. Rainie. 2006. "The Internet's Growing Role in Life's Major Moments." Pew Internet & American Life Project. http://www.pewinternet.org/PPF/r/181/report_display.asp.

Hou, F., and J. Myles. 2008. "The Changing Role of Education in the Marriage Market: Assortative Marriage in Canada and the United States Since the 1970s." *Canadian Journal of Sociology* 33 (2): 337–66.

House, J. S., V. Strecher, H. L. Metzner, and C. A. Robbins. 1986. "Occupational Stress and Health Among Men and Women in the Tecumseh Community Health Study." *Journal of Health and Social Behavior* 31 (2): 123–40.

Houser, A. 2007. "Women & Long Term Care: Research Report." AARP Policy and Research. http://www-static-w2-md.aarp.org/research/longtermcare/trends/fs77r_ltc.html.

Houser, A., and M. J. Gibson. 2008. "Valuing the Invaluable: The Economic Value of Family Caregiving, 2008 Update." AARP Public Policy Institute. http://assets.aarp.org/rgcenter/il/i13_caregiving.pdf.

Houston, J. M., J. Kinnie, B. Lupo, C. Terry, and S. S. Ho. 2000."Competitiveness and Conflict Behavior in Simulation of a Social Dilemma." *Psychological Reports* 86 (3, Pt. 2): 1219–25.

Houweling, T. A., A. E. Caspar, W. N. Looman, and J. P. Mackenbach. 2005. "Determinants of Under-5 Mortality among the Poor and the Rich: A Cross-National Analysis of 43 Developing Countries." *International Journal of Epidemiology* 34 (6): 1257–65.

Howell, M. 1986. "Women, the Family Economy, and Market Production." In *Women and Work in Pre-Industrial Europe,* edited by B. Hanawalt, 198–222. Bloomington: Indiana University Press.

Hoxby, C. 2001. "The Return to Attending a Highly Selective College: 1960 to the Present." In *Forum Future: Exploring the Future of Higher Education. 2000 Papers. Forum Strategy Series (Vol 3).*, edited by M. Devlin and J. Meyerson, 13–42.

Huang, M. 2009. "Classroom Homogeneity and the Distribution of Student Math Performance: A Country-Level Fixed-Effects Analysis." *Social Science Research* 38 (4): 781–791.

Huffington Post. 2010. "LeBron James' Jersey Number May Change for Michael Jordan." (March 2). http://www.huffingtonpost.com/2010/03/02/lebron-james-jersey-numbe_n_482042.html.

Hughes, E. C. 1945. "Dilemmas and Contradictions of Status." *American Journal of Sociology* 50 (5): 353–59.

Hull, K. 2006. *Same-Sex Marriage.* New York: Cambridge University Press.

Hunt, S. 2000. "'Winning Ways': Globalization and the Impact of the Health and Wealth Gospel." *Journal of Contemporary Religion* 15 (3): 331–47.

Hunter, M. L. 2002. "If You're Light You're Alright: Light Skin Color as Social Capital for Women of Color." *Gender and Society* 16 (2): 175–93.

Hurd, L. C. 2000. "Older Women's Body Image and Embodied Experience: An Exploration." *Journal of Women and Aging* 12 (3–4): 77–97.

Hutchinson, S. L., C. M. Yarnal, J. Staffordson, and D. L. Kerstetter. 2008. "Beyond Fun and Friendship: The Red Hat Society as a Coping Resource for Older Women." *Ageing and Society* 28 (7): 979–99.

Indian Express. 2008. "Immigrants in UK Opting for Name Change." (January 21). http://www.indianexpress.com/news/immigrants-in-uk-opting-for-name-change/263723/0.

Information Please Almanac. 1997. Boston: Houghton Mifflin.

Insaf, S. 2002. "Not the Same by Any Other Name." *Journal of the American Academy of Psychoanalysis* 31 (3): 463–73.

Institute for Research on Poverty. 2009. "Who Is Poor?" http://www.irp.wisc.edu/faqs/faq3.htm#groups.

Institute for Women's Policy Research. 2010a. "Fact Sheet. The Gender Wage Gap By Occupation." http://www.iwpr.org/publications/pubs/the-gender-wage-gap-by-occupation.

———. 2010b. "Women and Men's Employment and Unemployment in the Great Recession." (IWPR Publication C373). http://www.iwpr.org/publications/pubs/women-and-men2019s-employment-and-unemployment-in-the-great-recession.

Institute of Medicine. 2001. *Exploring the Biological Contributions to Human Health: Does Sex Matter?* Edited by T. M. Wizemann and M. L. Pardue. Board on Health Sciences Policy. Washington, DC: National Academy Press.

Insurance Journal. 2008. "Report: Lobbyists Spend $17 Million a Day on Congress; Insurance Ranks Second." http://www.insurancejournal.com/news/national/2008/04/14/89108.htm.

International Labour Organization. 2010. "Wages and Income." http://www.ilo.org/travaildatabase/servlet/minimumwages?pageClass=org.ilo.legislation.work.web.CategorySearchPage.

Isaacs, H. 1975. *Idols of the Tribe.* Cambridge, MA: Harvard University Press.

Ismail, M. A. 2007. "Spending on Lobbying Thrives." Center for Public Integrity. http://www.publicintegrity.org/rx/report.aspx?aid=823.

Jablonski, N. 2005. "The Evolution of Human Skin and Skin Color." *Annual Review of Anthropology* 33: 585–623.

Jackson, L. A., and O. D. McGill. 1996. "Body Type Preferences and Body Type Characteristics Associated with Attractive and Unattractive Bodies by African Americans and Anglo Americans." *Sex Roles* 35 (September): 295–307.

Jacobs, N., and D. Harvey. 2010. "The Extent to Which Teacher Attitudes and Expectations Predict Academic Achievement of Final Year Students." *Educational Studies* 36 (2): 195–206.

Jacobs, P. 1985. "Keeping the Poor, Poor." In *Crisis in American Institutions,* edited by J. H. Skolnick and E. Currie, 113–123. Glenview, IL: Scott, Foresman.

Jacobs Henderson, J., and G. Baldasty. 2003. "Race, Advertising and Prime-Time Television." *Howard Journal of Communications* 14 (2): 97–113.

Jacobsen, L., and M. Mather. 2010. "U.S. Economic and Social Trends Since 2000." *Population Bulletin* 65 (1). http://www.prb.org/pdf10/65.1unitedstates.pdf.

Jacobson, D. 1989. "Context and the Sociological Study of Stress." *Journal of Health and Social Behavior* 30 (3): 257–60.

Jankowiak, W., and E. Fischer. 1992. "A Cross-Cultural Perspective on Romantic Love." *Journal of Ethnology* 31: 149–56.

Jaret, C., D. C. Reitzes, and N. Shapkina. 2005. "Reflected Appraisals and Self-Esteem." *Sociological Perspectives* 48 (3): 403–19.

Jayson, S. 2010. "Daily Stress and Worry Plummets After Age 50." *USA Today.* http://www.usatoday.com/news/health/2010–05–18-stress18online_ST_N.htm.

Jefferson, D. L., and J. E. Stake. 2009. "Appearance Self-Attitudes of African Americans and European American Women: Media Comparisons and Internalization of Beauty Ideals." *Psychology of Women Quarterly* 33: 396–409.

Jeffries, V., and E. Ransford. 1980. *Social Stratification: A Multiple Hierarchy Approach.* Boston: Allyn & Bacon.

Jehlen, A. 2009. "Study: Test Prep Grows, Curriculum Shrinks in Low-Income Schools." National Education Association. http://www.nea.org/home/37025.htm.

Jehn, K. A., and P. P. Shah. 1997. "Interpersonal Relationships and Task Performance: An Examination of Mediating Processes in Friendship and Acquaintance Groups." *Journal of Personality and Social Psychology* 72 (April): 775–90.

Jensen, M. J., J. N. Danziger, and A. Venkatesh. 2007. "Civil Society and Cyber Society: The Role of the Internet in Community Associations and Democratic Politics." *Information Society* 23 (1): 39–50.

Jensen, M., D. W. Johnson, and R. T. Johnson. 2002. "Impact of Positive Interdependence During Electronic Quizzes on Discourse and Achievement." *Journal of Educational Research* 95 (3): 161–66.

Johnson, C. 2000. "Perspective on American Kinship in the Later 1990s." *Journal of Marriage and Family* 62 (November): 623–39.

Johnson, D., and J. Wu. 2002. "An Empirical Test of Crisis, Social Selection, and Role Explanations of the Relationship Between Marital Disruption and Psychological Distress: A Pooled Time-Series Analysis of Four-Wave Panel Data." *Journal of Marriage and Family* 64 (1): 211–24.

Johnson, D. R., and L. K. Scheuble. 2002. "What Should We Call Our Kids? Choosing Children's Surnames When Parents' Last Names Differ." *Social Science Journal* 39 (3): 419–29.

Johnson, D. W., and R. T. Johnson. 1989. *Cooperation and Competition: Theory and Research.* Edina, MN: Interaction Books.

———. 2000. "The Three Cs of Reducing Prejudice and Discrimination." In *Reducing Prejudice and Discrimination: The Claremont Symposium on Applied Social Psychology,* edited by S. Oskamp, 239–68. Mahwah, NJ: Lawrence Erlbaum.

———. 2002. "Social Interdependence Theory and University Instruction: Theory Into Practice." *Swiss Journal of Psychology* 61 (3): 119–29.

———. 2005. *Joining Together: Group Theory and Group Skills.* Boston: Allyn & Bacon.

Johnson, J. 2005. "Who's Minding the Kids? Child Care Arrangements: Winter 2002." *Current Population Reports, Household Economic Studies,* P70–101. http://www.census.gov/prod/2005pubs/p70–101.pdf.

Johnson, J. L., F. T. McAndrew, and P. B. Harris. 1991. "Sociobiology and the Naming of Adopted and Natural Children." *Etiology and Sociobiology* 12: 365–75.

Johnson, R. W., D. Toohey, and J. M. Wiener. 2007. "Meeting the Long-Term Care Needs of the Baby Boomers." Discussion Paper 07–04 in the Retirement Project Series, Urban Institute. http://www.urban.org/url.cfm?ID=311451.

Jones, J. M. 2010a. "More Americans Favor Than Oppose Arizona Immigration Law." Gallup. (April 29). http://www.gallup.com/poll/127598/Americans-Favor-Oppose-Arizona-Immigration-Law.aspx.

———. 2010b. "Americans Now Divided on Who Should Lead Oil Spill Efforts." Gallup. (June 15). http://www.gallup.com/poll/140753/Americans-Divided-Lead-Oil-Spill-Efforts.aspx.

———. 2010c. "Job Creation, Securing Borders Top List of What Gov't Should Do." Gallup. (October 12). http://www.gallup.com/poll/143621/Job-Creation-Securing-Borders-Top-List-Gov.aspx.

Jones, L. 1980. *Great Expectations.* New York: Ballantine Books.

Jones, W., R. O. Hansson, and A. L. Phillips. 1978. "Physical Attractiveness and Judgments of Psychopathology." *Journal of Social Psychology* 105: 79–84.

Jones-DeWeever, A., and B. Gault. 2008. "Resilient & Reaching for More: Challenges and Benefits of Higher Education for Welfare Participants and Their Children." Institute for Women's Policy Research. http://www.iwpr.org/publications/pubs/resilient-and-reaching-for-more-challenges-and-benefits-of-higher-education-for-welfare-participants-and-their-children.

Jones-DeWeever, A., J. Peterson, and X. Song. 2003. "Before and After Welfare Reform: The Work and Well-Being of Low-Income Single Parent Families." Institute for Women's Policy Research. http://www.iwpr.org/publications/pubs/before-and-after-welfare-reform-the-work-and-well-being-of-low-income-single-parent-families.

Josephson Institute. 2008. "The Ethics of American Youth: 2008." http://charactercounts.org/programs/reportcard/2008/index.html.

Judge, T. A., C. Hurst, and L. S. Simon. 2009. "Does It Pay to Be Smart, Attractive, or Confident (or All Three)? Relationships Among General Mental Ability, Physical Attractiveness, Core Self-Evaluations, and Income." *Journal of Applied Psychology* 94 (3): 742–55.

Jurd, E. 2005. "Helping George Feel Valued." *Times Educational Supplement* (4/22) 4631: 18–19.

Kacmar, K. M., B. J. Collins, K. J. Harris, and T. A. Judge. 2009. "Core Self Evaluations and Job Performance: The Role of Perceived Work Environment." *Journal of Applied Psychology* 94 (6): 1572–80.

Kahane, H., and N. Cavender. 2002. *Logic and Contemporary Rhetoric: The Use of Reason in Everyday Life,* 9th ed. Belmont, CA: Wadsworth.

Kalick, S. M., and T. E. Hamilton. 1986. "The Matching Hypothesis Re-Examined." *Journal of Personality and Social Psychology* 51 (4): 673–82.

Kalmijn, M. 1988. "Intermarriage and Homogamy." *Annual Review of Sociology* 24: 395–421.

Kalmijn, M., and H. Flap. 2001. "Assortive Meeting and Mating: Unintended Consequences of Organized Settings for Partner Choices." *Social Forces* 79 (4): 1289–1312.

Kao, G., and J. Thompson. 2003. "Racial and Ethnic Stratification in Educational Achievement and Attainment." *Annual Review of Sociology* 29: 417–442.

Kaplan, R., and R. G. Kronick. 2006. "Marital Status and Longevity in the United States Population." *Journal of Epidemiology & Community Health* 60 (9): 760–65.

Karabel, J. 2005. *The Chosen: The Hidden History of Admission and Exclusion at Harvard, Yale and Princeton.* Boston: Houghton Mifflin.

Karl, T. R., J. M. Melillo, and T. C. Peterson, eds. 2009. *Global Climate Change Impacts in the United States.* U.S. Global Change Research Program. New York: Cambridge University Press. http://downloads.globalchange.gov/usimpacts/pdfs/climate-impacts-report.pdf.

Karraker, K., D. Vogel, and M. Lake. 1995. "Parents' Gender-Stereotyped Perceptions of Newborns: The Eye of the Beholder Revisited." *Sex Roles* 33 (9–10): 687–701.

Karylowski, J. J., M. A. Motes, H. M. Wallace, H. A. Harckom, E. M. Hewlett, S. L. Maclean, J. L. Paretta, and C. L. Vaswani. 2001. "Spontaneous Gender-Stereotypical Categorization of Trait Labels and Job Labels." *Current Research in Social Psychology* 6 (6): 77–90.

Katz, J. E., and R. E. Rice. 2002. *Social Consequences of Internet Use: Access, Involvement and Interaction.* Cambridge, MA: MIT Press.

———. 2009. "Technical Opinion: Falling into the Net: Main Street America Playing Games and Making Friends Online." *Communications of the ACM* 52 (9): 149–50.

Katz, M., ed. 1971. *School Reform Past and Present.* Boston: Little, Brown.

———. 1986. *In the Shadow of the Poorhouse: A Social History of Welfare in America.* New York: Basic Books.

Katz, M., M. Cox, L. Fleury, and T. Hacker. 2008. *Skin Deep: The Business of Beauty.* [Video.] New York: A&E Home Video.

Kay, R. 2009. "Examining Gender Differences in Attitudes Toward Interactive Classroom Communications Systems (ICCS)." *Computers & Education* 52 (4): 730–40.

Keller, M. C., and R. K. Young. 1996. "Mate Assortment in Dating and Married Couples." *Personality and Individual Differences* 21 (2): 217–21.

Keller, U., and K. Tillman. 2008. "Post-Secondary Educational Attainment of Immigrant and Native Youth." *Social Forces* 87 (1): 121–52.

Kelley, H. H., and A. J. Stahelski. 1970. "Errors in Perception of Intentions in a Mixed-Motive Game." *Journal of Experimental Social Psychology* 6: 379–400.

Kelly, A. M., M. Wall, M. E. Eisenberg, M. Story, and D. Neumark-Sztainer. 2007. "Adolescent Girls with High Body Satisfaction: Who Are They and What Can They Teach Us?" *Journal of Adolescent Health* 37 (5): 391–96.

Kenealy, P., N. Frude, and W. Shaw. 1988. "Influences of Children's Physical Attractiveness on Teacher Expectations." *Journal of Social Psychology* 128 (3): 373–83.

Kennedy, Q., M. Mather, and L. L. Carstensen. 2004. "The Role of Motivation in the Age-Related Positivity Effect in Autobiographical Memory." *Psychological Science* 15 (3): 208–214.

Kessler, R. C., R. H. Price, and C. B. Wortman. 1985. "Social Factors in Psychopathology: Stress, Social Support, and Coping Processes." *Annual Review of Psychology* 36: 531–72.

Kessler-Harris, A. 2003. *Out to Work: A History of Wage-Earning Women in the United States.* New York: Oxford University Press.

Keteyian, A. 2010. "Reform Won't Change Insurers' Profits Much." CBSNews.com Special Report. http://www.cbsnews.com/stories/2010/03/26/cbsnews_investigates/main6336685.shtml?tag=featuredPostArea.

Kifner, J. 1994. "Pollster Finds Error on Holocaust Doubts." *New York Times* (May 20): A12.

Killbourne, J. 2000. *Can't Buy Me Love: How Advertising Changes the Way We Think and Feel.* New York: Touchstone.

Kim, K. H. 2007. "Religion, Weight Perception, and weight Control Behavior." *Eating Behaviors* 8 (1): 121–31.

Kindleberger Hagan, L., and J. Kuebli. 2007. "Mothers' and Fathers' Socialization of Preschoolers' Physical Risk Taking." *Journal of Applied Developmental Psychology* 28 (1): 2–14.

Kitson, G. C., K. B. Babri, and M. J. Roach. 1985. "Who Divorces and Why?" *Journal of Family Issues* 6: 285–93.

Klein, H., and K. Shiffman. 2009. "Underrepresentation and Symbolic Annihilation of Socially Disenfranchised Groups ('Out Groups') in Animated Cartoons." *The Howard Journal of Communications* 20 (1): 55–72.

Klein, L. G., and J. Knitzer. 2007. "Promoting Effective Early Learning: What Every Policymaker and Educator Should Know." National Center for Children in Poverty. http://www.nccp.org/publications/pub_695.html.

Knauer, K., ed. 1998. *Time 75th Anniversary Celebration.* New York: Time, Inc.

Knowlton, A. R., and C. A. Latkin. 2007. "Network Financial Support and Conflict as Predictors of Depressive Symptoms Among a Highly Disadvantaged Population." *Journal of Community Psychology* 35 (1): 13–28.

Kochera, A., and T. Guterbock. 2005. *Beyond 50.05: A Report to the Nation on Liveable Communities Creating Environments for Successful Aging.* Washington, DC: AARP Public Policy Institute.

Kohn, A. 1986. *No Contest: The Case against Competition.* Boston: Houghton Mifflin.

Kollock, P. 1998. "Social Dilemmas: The Anatomy of Cooperation." *Annual Review of Sociology* 24: 183–214.

Komarovsky, M. 1962. *Blue-Collar Marriage.* New Haven, CT: Vintage.

Kovaleski, S. F. 1999. "In Jamaica, Shades of an Identity Crisis: Ignoring Health Risks, Blacks Increase Use of Skin Lighteners." *Washington Post* (August 5): A15.

Kozol, J. 1995. *Amazing Grace: The Lives of Children and the Conscience of a Nation.* New York: Crown.

———. 2001. *Ordinary Resurrections: Children in the Years of Hope.* New York: HarperCollins.

———. 2005. *The Shame of the Nation: The Restoration of Apartheid Schooling in America.* New York: Crown.

———. 2006. *Rachel and Her Children: Homeless Families in America.* New York: Crown.

Kriner, D. L., and F. X. Shen. 2010. "America's 'Casualty Gap.'" *Los Angeles Times* (May 28). http://www.uslaboragainstwar.org/article.php?id=22265.

Kristenson, M., H. R. Eriksenm, J. K. Sluiter, D. Starke, and H. Ursin. 2004. "Psychobiological Mechanisms of Socioeconomic Differences in Health." *Social Science and Medicine* 58 (8): 1511–22.

Kristof, N. 2003. "Is Race Real?" *New York Times* (July 11): A17.

Kroll, L. 2003. "Megachurches, Megabusinesses." Forbes.com. http://www.forbes.com/2003/09/17/cz_lk_0917megachurch.html.

Krosnick, J. A., and S. Presser. 2010. "Question and Questionnaire Design" In *Handbook of Survey Research,* edited by P. Marsden and J. Wright, 185–204, San Diego, CA: Elsevier.

Krugman, P. 2002. "For Richer." *New York Times Magazine* (October 20): 62–67.

Krysan, M. 2008. "Does Race Matter in the Search for Housing? An Exploratory Study of Search Strategies, Experiences, and Locations." *Social Science Research* 37 (2): 581–603.

Krysan, M., and M. Bader. 2007. "Perceiving the Metropolis: Seeing the City Through a Prism of Race." *Social Forces* 86 (2): 699–727.

Krysan, M., M. Couper, R. Farley, and T. Forman. 2009. "Does Race Matter in Neighborhood Preferences? Results from a Video Experiment." *American Journal of Sociology* 115 (2): 527–59.

Kuchler, F., and J. N. Variyam. 2003. "Mistakes Were Made: Misperception as a Barrier to Reducing Overweight." *International Journal of Obesity* 27 (7): 856–61.

Kulik, J. A., and H. I. Mahler. 1990. "Stress and Affiliation Research: On Taking the Laboratory to Health Field Settings." *Annals of Behavioral Medicine* 12 (3): 106–11.

Kulik, J. A., H. I. Mahler, and P. J. Moore. 1996. "Social Comparison and Affiliation Under Threat: Effects on Recovery from Major Surgery." *Journal of Personality and Social Psychology* 71: 967–79.

Kurtz-Costes, B., S. Rowley, A. Harris-Britt, and T. Woods. 2008. "Gender Stereotypes About Mathematics and Science and Self-Perceptions of Ability in Late Childhood and Early Adolescence." *Merrill-Palmer Quarterly* 54 (3): 386–409.

Kutner, M., E. Greenberg, Y. Jin, B. Boyle, Y. Hsu, and E. Dunleavy. 2007. *Literacy in Everyday Life: Results From the 2003 National Assessment of Adult Literacy* (NCES 2007–480). Washington, DC: U.S. Department of Education.

Kwan, A., and M. N. Trautner. 2009. "Beauty Work: Individual and Institutional Rewards: The Reproduction of Gender and Questions of Agency." *Sociology Compass* 3 (1): 49–71.

LaFountain, R., R. Schauffler, S. Strickland, C. Bromage, S. Gibson, A. Mason, and W. Raftery. 2009. *Examining the Work of State Courts: An Analysis of 2007 State Court Caseloads.* National Center for State Courts. http://www.ncsconline.org/d_research/csp/2007B_files/EWSC-2007-v21-online.pdf.

LaFraniere, S. 2009. "Name Not On Our List? Change It, China Says." *The New York Times* (April 20). http://www.nytimes.com/2009/04/21/world/asia/21china.html?_r=2&ref=todayspaper&pagewanted=all.

Lake, A. 1975. "Are We Born Into Our Sex Roles or Programmed Into Them?" *Woman's Day* (January): 25–35.

Lalwani, A. K., S. Shavitt, and T. Johnson. 2006. "What Is the Relationship Between Cultural Orientation and Socially Desirable Responding?" *Journal of Personality and Social Psychology* 90 (1): 165–78.

Lambton Health. 2002. "Ten Steps to a Healthy Body Image." http://www.lambtonhealth.on.ca/youth/bodyimage.asp.

Landers, S. J. 2009. "Forget the Gimmicks When It Comes to Dieting." *Amednews.com* (March 30). http://www.ama-assn.org/amednews/2009/03/30/hlsb0330.htm.

Langhout, R., and C. Mitchell. 2008. "Engaging Contexts: Drawing the Link between Student and Teacher Experiences of the Hidden Curriculum." *Journal of Community & Applied Social Psychology* 18 (6): 593–614.

Langlois, J. H., L. Kalakanis, A. J. Rubenstein, A. Larson, M. Hallam, and M. Smoot. 2000. "Maxims or Myths of Beauty? A Meta-Analytic and Theoretical Review." *Psychological Bulletin* 126 (3): 390–423.

Langlois, J. H., and C. W. Stephan. 1981. "Beauty and the Beast: The Role of Physical Attractiveness in the Development of Peer Relations and Social Behavior." In *Developmental Social Psychology,* edited by S. S. Brehm, S. M. Kassin, and F. X. Gibbons, 152–168. New York: Oxford University Press.

Lantz, P., J. S. House, R. P. Mero, and D. R. Williams. 2005. "Stress, Life Events, and Socioeconomic Disparities in Health: Results From the Americans' Changing Lives Study." *Journal of Health and Social Behavior* 3: 274–88.

Lanzetta, J. T. 1955. "Group Behavior Under Stress." *Human Relations* 8: 29–53.

Larose, M., and N. Wallace. 2003. "Online Giving Rose at Many Big Charities." *Chronicle of Philanthropy* (June 12). http://philanthropy.com/article/Online-Giving-Rose-at-Many-/49659/.

Larson, M. S. 2003."Gender, Race, and Aggression in Television Commercials That Feature Children." *Sex Roles* 48 (1–2): 67–75.

Laslett, B., and C. B. Warren. 1975. "Losing Weight: The Organizational Promotion of Behavior Change." *Social Problems* 23 (1): 69–80.

Latané, B., and D. C. Glass. 1968. "Social and Nonsocial Attraction in Rats." *Journal of Personality and Social Psychology* 9: 142–46.

Lauderdale, D. 2006. "Birth Outcomes for Arabic-Named Women in California Before and After September 11." *Demography* 43 (1): 185–201.

Lauzen, M., D. Dozier, and E. Cleveland. 2006. "Genre Matters: An Examination of Women Working Behind the Scenes and On-Screen Portrayals in Reality and Scripted Prime-Time Programming." *Sex Roles* 55 (7/8): 445–55.

Lauzen, M. M., D. M. Dozier, and N. Horan. 2008. "Constructing Gender Stereotypes Through Social Roles in Prime-Time Television." *Journal of Broadcasting & Electronic Media* (June). http://www.entrepreneur.com/tradejournals/article/182033209_4.html.

Lawson, T. 2000. "Are Kind Lies Better Than Unkind Truths? Effects of Perspective and Closeness of Relationship." *Representative Research in Social Psychology* 24: 11–19.

Layman, R., ed. 1995. *American Decades 1950–59.* Detroit, MI: Gale.

Layzer, J., and B. Goodson. 2006. "National Study of Child Care for Low-Income Families: Care in the Home: A Description of Family Child Care and the Experience of the Families and Children Who Use It. Wave 1 Report." U.S. Department of Health and Human Services, Administration for Children and Families. http://www.acf.hhs.gov/programs/opre/cc/nsc_low_income/reports/care_in_home/care_in_home.pdf.

Lazarsfeld, P. 1949. "The American Soldier: An Expository Review." *Public Opinion Quarterly* 13 (3): 376–404.

Lazarus, R. J. 2010 "Docket Capture at the High Court." 119 *Yale Law Journal Online* 89. Available at http://yalelawjournal.org/2010/01/24/lazarus.html.

Leahey, E., and G. Guo. 2001. "Gender Differences in Mathematical Trajectories." *Social Forces* 80 (2): 713–32.

Leary, M. R. 2010. "Affiliation, Acceptance, and Belonging: The Pursuit of Interpersonal Connection." In *Handbook of Social Psychology,* 5th ed., Vol. 2, edited by S. T. Fiske, D. T. Gilbert, and G. Lindzey, 864–96. Hoboken, NJ: John Wiley and Sons, Inc.

LeBlanc, V. R., and G. W. Bandiera. 2007. "The Effects of Examination Stress on the Performance of Emergency Medicine Residents." *Medical Education* 41 (6): 556–64.

Lee, D. A. 2006. "Dear Criminal: Smile for My Camera." *St. Petersburg Times* (September 12). http://www.sptimes.com/2006/09/16/Tampabay/Dear_criminal__Smile_.shtml.

Lee, G., and J. Bulanda. 2005. "Change and Consistency in the Relation of Marital Status to Personal Happiness." *Marriage & Family Review* 38 (1): 69–84.

Lee, J. 2010. "Naturalizations in the United States: 2009." Annual Flow Report. Office of Immigration Statistics. http://www.dhs.gov/xlibrary/assets/statistics/publications/natz_fr_2009.pdf.

Lee, J. 8. 2002. "Some States Track Parolees by Satellite." *New York Times* (January 31). www.gyre.org/news/author/Jennifer+8.+Lee/. http://www.nytimes.com/2002/01/31/technology/putting-parolees-on-a-tighter-leash.html

Lee, J., and F. Bean. 2004. "America's Changing Color Lines: Immigration, Race/Ethnicity and Multiracial Identification." *Annual Review of Sociology* 30: 221–42.

Lee, S. 2003. "The Genetics of Differences: What Genetic Discovery and the Modern Biology of 'Race' Mean for Communities of Color Fighting Health Inequities." *ColorLines* 6 (2): 25–27.

Lee, S. T. 2004. "Lying to Tell the Truth: Journalists and the Social Context of Deception." *Mass Communication and Society* 7 (1): 97–120.

Legal Services Corporation. 2010. "A Crisis in Legal Representation." http://www.lsc.gov/pdfs/a_crisis_in_legal_representation.pdf.

Lehrman, S. 2003. "The Reality of Race." *Scientific American* 288 (2): 32–34.

Leigh, P. R., ed. 2010. *International Exploration of Technology Equity and the Digital Divide: Critical, Historical and Social Perspectives.* Hershey, PA: Information Science Publishing.

Leinbach, M. D., and B. I. Fagot. 1991. "Attractiveness in Young Children: Sex-Differentiated Reactions of Adults." *Sex Roles* 25 (5–6): 269–84.

Lemay, E. P., Jr., M. S. Clark, and A. Greenberg. 2010. "What Is Beautiful Is Good because What Is Beautiful Is Desired: Physical Attractiveness Stereotyping as Projection of Interpersonal Goals " *Personality and Social Psychology Bulletin* 36 (3): 339–53.

Lemert, E. 1951. *Social Pathology: A Systematic Approach to the Theory of Sociopathic Behavior.* New York: McGraw-Hill.

Lennon, M. C. 1989. "The Structural Contexts of Stress." *Journal of Health and Social Behavior* 30 (3): 261–68.

Leonard, S. 1996. "Feeling Appealing." *Psychology Today* 29 (1): 18.

Lerer, L. 2007. "Do You Lie on Your Resume?" http://www.forbes.com/2007/02/07/leadership-resume-jobs-lead-careers-cx_ll_0207resume.html?partner=careersaol.

Levin, S., C. VanLaar, and J. Sidanius. 2003. "The Effects of Ingroup and Outgroup Friendships on Ethnic Attitudes in College: A Longitudinal Study." *Group Processes & Intergroup Relations* 6 (1): 76–92.

Levine, R. V. 1993. "Is Love a Luxury?" *American Demographics* 15 (2): 27–28.

Levinson. S. 2010. "Assessing the Supreme Court's Current Caseload: A Question of Law or Politics?" 119 *Yale Law Journal Online* 99. Available at http://yalelawjournal.org/2010/02/01/levinson.html.

Levi-Strauss, C. 1964. *The Raw and the Cooked: Introduction to a Science of Mythology.* New York: Harper & Row.

Levy, B. R., M. D. Slade, S. R. Kunkel, and S. V. Kasl. 2002. "Longevity Increased by Positive Self-Perceptions of Aging." *Journal of Personality and Social Psychology* 832 (2): 261–70.

Levy, G., A. Sadovsky, and G. Troseth. 2000. "Aspects of Young Children's Perceptions of Gender-Typed Occupations." *Sex Roles* 42 (11–12): 993–1006.

Lewis, M. 1978. *The Culture of Inequality.* New York: New American Library.

Lewis, T., and S. Cheng. 2006. "Tracking, Expectations, and the Transformation of Vocational Education." *American Journal of Education* 113 (1): 67–99.

Liddell, C., and J. Lycett. 1998. "Simon or Sipho: South African Children's Names and Their Academic Achievement in Grade One." *Applied Psychology* 47 (3): 421–37.

Lieberson, S. 2000. *A Matter of Taste.* New Haven, CT: Yale University Press.

Liebler, C. A., and G. D. Sandefur. 2002. "Gender Differences in the Exchange of Social Support With Friends, Neighbors, and Co-Workers at Midlife." *Social Science Research* 31 (3): 364–91.

Lillard, L., and J. Waite. 1995. "'Til Death Do Us Part: Marital Disruption and Mortality." *American Journal of Sociology* 100 (March): 1131–56.

Lin, N., X. Ye, and W. W. Ensel. 1999. "Social Support and Depressed Mood: A Structural Analysis." *Journal of Health and Social Behavior* 40: 344–59.

Lindner, K. 2004. "Images of Women in General Interest and Fashion Magazine Advertisements From 1955 to 2002." *Sex Roles* 51 (7/8): 409–21.

Lindsey, E., and J. Mize. 2001. "Contextual Differences in Parent–Child Play: Implications for Children's Gender Role Development." *Sex Roles* 44 (3–4): 155–176.

Lindstrom, C., and M. Lindstrom. 2006. "Social Capital, GNP per Capita, Relative Income, and Health: An Ecological Study of 23 Countries." *International Journal of Health Services* 36 (4): 679–96.

Lino, M., and A. Carlson. 2009. "Expenditures on Children by Families, 2008." U.S. Department of Agriculture, Center for Nutrition Policy and Promotion. Miscellaneous Publication No. 1528–2008. http://www.cnpp.usda.gov/Publications/CRC/crc2008.pdf.

Lips, H. 1993. *Sex and Gender: An Introduction.* Mountain View, CA: Mayfield.

Liptak. 2002."In the Name of Security, Privacy for Me, Not Thee." *New York Times* (November 24): Week in Review.

Litke, M. 2006. "Forget Child's Play, Japanese Toy Makers Target Adults." http://abcnews.go.com/Technology/story?id=2623174&page=1.

Livingston, G., and D. Cohn. 2010. "Childlessness Up Among All Women; Down Among Women With Advanced Degrees." Social and Demographic Trends, Pew Research Center. http://pewsocialtrends.org/pubs/758/rising-share-women-have-no-children-childlessness.

Lleras, C., and C. Rangel. 2009. "Ability Grouping Practices in Elementary School and African American/Hispanic Achievement." *American Journal of Education* 115 (279).

Lloyd, C. 2000. "Why Should a Baby Get the Father's Last Name?" *Salon.com* http://archive.salon.com/mwt/feature/2000/01/20/surnames/index.html.

London, A., E. Scott, K. Edin, and V. Hunter. 2004. "Welfare Reform, Work-Family Tradeoffs, and Child Well-Being." *Family Relations* 53 (2): 148–58.

Long, M., and D. Evans. 2008. "College Quality and Early Adult Outcomes." *Economics of Education Review* 27 (5): 588–602.

Lord, W. 1981. *A Night to Remember.* New York: Penguin.

Lovejoy, M. 2001. "Disturbances in the Social Body: Differences in Body Image and Eating Problems Among African American and White Women." *Gender & Society* 15 (2): 239–61.

Lu, J., D. Tjosvold, and K. Shi. 2010. "Team Training in China: Testing and Applying the Theory of Cooperation and Competition." *Journal of Applied Social Psychology* 40 (1): 101–34.

Lucas, R. 2007. "In Case You Haven't Heard." *Mental Health Weekly* 17 (15): 8.

Ludwig, J. 2005. "Perception or Reality? The Effect of Stature on Life Outcomes." http://www.shortsupport.org/News/0515.html.

Lumeng. J., P. Forrest, D. Appugliese, N. Kaciroti, R. Corwyn, and R. Bradley. 2010. "Weight Status as a Predictor of Being Bullied in Third Through Sixth Grades." *Pediatrics* 125: e1301–e1307.

Lundberg, S., S. McLanahan, and E. Rose. 2007. "Child Gender and Father Involvement in Fragile Families." *Demography* 44 (1): 79–92.

Lupart, J., and E. Cannon. 2002."Computers and Career Choices: Gender Differences in Grades 7 and 10 Students." *Gender, Technology & Development* 6 (2): 233–48.

Lutz, A., and S. Crist. 2009. "Why Do Bilingual Boys Get Better Grades in English-Only America? The Impacts of Gender, Language and Family Interaction on Academic Achievement of Latino Children of Immigrants." *Ethnic and Racial Studies* 32 (2): 346–68.

Lutz, G. 2010. "The Electoral Success of Beauties and Beast." *Swiss Political Science Review* 16 (3): 457–480.

Lynch, Z. 2009. *The Neuro Revolution: How Brain Science Is Changing Our World.* New York: St. Martin's Press.

Lynn, M., and T. Simons. 2000. "Predictors of Male and Female Servers' Average Tip Earnings." *Journal of Applied Social Psychology* 30 (2): 241–52.

Lytton, H., and D. Romny. 1991."Parents' Differential Socialization of Boys and Girls: A Meta-Analysis." *Psychological Bulletin* 109: 267–96.

MacArthur Foundation. 2009. "Facts and Fictions about An Aging America." *Contexts* 8 (4): 16–21.

Macmillan, R. 2001. "Violence and the Life Course: The Consequences of Victimization for Personal and Social Development." *Annual Review of Sociology* 27: 1–22.

Madera, J. M., K. E. Podratz, E. B. King, and M. R. Hebl. 2007. "Schematic Responses to Sexual Harassment Complainants: The Influence of Gender and Physical Attractiveness." *Sex Roles* 56 (304): 223–30.

Madrick, J. 2002. "A Rise in Child Poverty Rates Is at Risk in U.S." *New York Times* (June 13): C2.

Madrid, L. D., M. Canas, and M. Ortega-Medina. 2007. "Effects of Team Competition Versus Team Cooperation in Classwide Peer Tutoring." *Journal of Educational Research* 100 (3): 155–61.

Major League Baseball Players Association. 2009. "FAQ." http://mlb.mlb.com/pa/info/faq.jsp#average.

Manchester, J., and J. Topoleski. 2008. *Growing Disparities in Life Expectancy.* Washington, DC: Congressional Budget Office.

Manning, P. 1974. "Police Lying." *Urban Life* 3: 283–306.

———. 1984. "Lying, Secrecy, and Social Control." In *The Sociology of Deviance,* edited by J. Douglas Newton, 268–79. Newton, MA: Allyn & Bacon.

Marcus, A. 2002. "When Janie Came Marching Home: Women Fought in the Civil War." *New York Times* (March 23). http://www.nytimes.com/2002/03/23/arts/23WOME.html?todaysheadlines.

Marks, G., N. Miller, and G. Maruyama. 1981. "Effects of Targets' Physical Attractiveness on Assumptions of Similarity." *Journal of Personality and Social Psychology* 41: 198–206.

Marks, J., C. Lam, and S. McHale. 2009. "Family Patterns of Gender Role Attitudes." *Sex Roles: A Journal of Research* 61 (3–4): 221–34.

Marlar, M. R., and C. E. Joubert. 2002. "Liking of Personal Names, Self-Esteem, and the Big Five Inventory." *Psychological Reports* 91 (2): 407–10.

Marlowe, C. M., S. L. Schneider, and C. E. Nelson. 1996. "Gender and Attractiveness Biases in Hiring Decisions: Are More Experienced Managers Less Biased?" *Journal of Applied Psychology* 81 (1): 11–21.

Martin, J. B. 2010. "The Development of Ideal Body Image Perceptions in the United States." *Nutrition Today* 45 (3): 98–110.

Martin, M. A., M. L. Frisco, and A. L. May. 2009. "Gender and Face/Ethnic Differences in Inaccurate Weight Perceptions Among U.S. Adolescents." *Women's Health Issues* 19 (5): 292–99.

Martin, P., and E. Midgley. 2006. "Immigration: Shaping and Reshaping America, Revised and Updated 2nd Edition." *Population Bulletin* 61 (4). http://www.prb.org/pdf06/61.4USMigration.pdf.

———. 2010. "Population Bulletin Update: Immigration in America 2010." http://www.prb.org/Publications/PopulationBulletins/2010/immigrationupdate1.aspx.

Martin, P., and G. Zürcher. 2008. "Managing Migration: The Global Challenge." *Population Bulletin* 63 (1).

Martin, R. A., and S. Svebak. 2001. "Stress." In *Motivational Styles in Everyday Life,* edited by M. J. Apter, 130–159. Washington, DC: American Psychological Association.

Martin, T. C., and L. L. Bumpass. 1989. "Recent Trends in Marital Disruption." *Demography* 26: 41.

Marx, G. T. 1999. "What's in a Name? Some Reflections on the Sociology of Anonymity." *Information Society* 15 (2): 99–112.

Massey, D. S. 2008. *Categorically Unequal: The American Stratification System.* New York: Russell Sage.

Mastro, D., and S. Stern. 2003. "Representations of Race in Television Commercials: A Content Analysis of Prime-Time Advertising." *Journal of Broadcasting & Electronic Media* 47 (4): 638–47.

Mather, M. 2007. "The New Generation Gap." Population Reference Bureau. http://www.prb.org/Articles/2007/NewGenerationGap.aspx.

Mathers, J., and M. Manior. 2005. "How It Works: Cap-and-Trade Systems." *Catalyst* 4 (1): 20–1.

Mathisen, J. A. 1989. "A Further Look at 'Common Sense' in Introductory Sociology." *Teaching Sociology* 17 (3): 307–15.

Matousek, M. 2007. "Long-Distance Living." *AARP: The Magazine* (July/August): 36–37.

Mawson, A. R. 2005. "Understanding Mass Panic and Other Collective Responses to Threat and Disaster." *Psychiatry-Interpersonal and Biological Processes* 68 (2): 95–113.

Max Planck Institute for Demographic Research. 2010. "Marriage and Life Expectancy." http://www.demogr.mpg.de/en/press/1813.htm>.

May, E. 1988. *Homeward Bound: American Families in the Cold War Era.* New York: Basic Books.

Mayflower Society. 2007. http://www.themayflowersociety.com/index.htm.

McCabe, M. P., and L. A. Ricciardelli. 2003. "Sociocultural Influences on Body Image and Body Changes Among Adolescent Boys and Girls." *Journal of Social Psychology* 143 (1): 5–26.

McCall, G., M. McCall, G. Suttles, and S. Kurth. 2010. *Friendship as a Social Institution.* New York: Aldine/Transaction.

McConahay, J. B. 1981. "Reducing Racial Prejudice in Desegregated Schools." In *Effective School Desegregation,* edited by W. D. Hawley, 35–48. Beverly Hills, CA: Sage.

McCue, C. 2008. "Tween Avatars: What Do Online Personas Convey about Their Makers?" In *Proceedings of the Society for Information Technology and Teacher Education International Conference 2008,* edited by K. McFerrin et al., 3067–72. Chesapeake, VA: AACE.

McDonald, P. 2001."Low Fertility Not Politically Sustainable." *Population Today* (August/September): 3–8.

McDonald, W. F. 2003. "Immigrant Criminality: In the Eye of the Beholder?" http://www.stranieriinitalia.com/briguglio/immigrazione-e-asilo/2003/maggio/mcdonald-criminalita.'html.

McLanahan. S. 2005. "Life Without Father: What Happens to the Children?" *Contexts* 1 (1): 35–44.

McLean, C. 1998. "Name Your Baby Carefully." *Alberta Report* 25 (23): 34–36.

McLeod, J. D., and T. J. Owens. 2004. "Psychological Well-Being in the Early Life Course: Variations by Socioeconomic Status, Gender, and Race/Ethnicity." *Social Psychology Quarterly* 67 (3): 257–78.

McPherson, M., L. Smith-Lovin, and J. M. Cook. 2001. "Birds of a Feather: Homophily in Social Networks." *Annual Review of Sociology* 27: 415–44.

McSwain, C., and R. Davis. 2007. "College Access for the Working Poor: Overcoming Burdens to Succeed in Higher Education." Institute for Higher Education Policy. http://www.ihep.org/assets/files/publications/a-f/CollegeAccessWorkingPoor.pdf.

Media Matters for America. 2010. "Gender and Ethnic Diversity in Prime-Time Cable News." http://mediamatters.org/reports/diversity_report/.

Mehan, H., L. Hubbard, A. Lintz, and I. Villanueva. 1994. *Tracking Untracking: The Consequences of Placing Low Track Students in High Track Classes* (Research Report 10). Santa Cruz, CA: National Center for Research on Cultural Diversity and Second Language Learning.

Mehrabian, A. 2001. "Characteristics Attributed to Individuals on the Basis of Their First Names." *Genetic, Social and General Psychology Monographs* 127 (1): 59–89.

Meier, B. P., M. D. Robinson, M. S. Carter, and V. B. Hinsz. 2010. "Are Sociable People More Beautiful? A Zero-Acquaintance Analysis of Agreeableness, Extraversion, and Attractiveness." *Journal of Research in Personality* 44 (2): 293–96.

Mercer, K. R. 2009. "The Importance of Funding Postsecondary Correctional Educational Programs." *Community College Review* 37 (2): 153–64.

Merskin, D. 2007. "Three Faces of Eva: Perpetuation of the Hot-Latina Stereotype in *Desperate Housewives.*" *Howard Journal of Communications* 18 (2): 133–51.

Merton, R. K. 1938."Social Structure and Anomie." *American Sociological Review* 3: 672–82.

———. 1957. *Social Theory and Social Structure.* Glencoe, IL: Free Press.

Mesch, G. S., and I. Talmud. 2010. "Internet Connectivity, Community Participation, and Place Attachment: A Longitudinal Study." *American Behavioral Scientist* 53 (8): 1095–1110.

Messinger, P. R., X. Ge, E. Stroulia, K. Lyons, K. Smirnov, and M. Bone. 2008. "On the Relationship Between My Avatar and Myself." *Journal of Virtual Worlds Research* 1 (2): 1–17.

Messner, M. 2007. *Out of Play: Critical Essays on Gender and Sport.* Albany: State University of New York Press.

Meyers, M., and J. Nidiry. 2004. "Our Public Schools Are Still Separate and Unequal." *Society* 41 (5): 12–16.

Meyers, P. M., and S. R. Crull. 1994. "Question Order Effect: A Preliminary Analysis." Paper presented at the Annual Meetings of the Association of Applied Sociology, Detroit, MI.

Miceli, T. J. 1992. "The Welfare Effects of Non-Price Competition Among Real Estate Brokers." *Journal of the American Real Estate and Urban Economics Association* 20 (4): 519–32.

Mignon, S., C. Larson, and W. Holmes. 2002. *Family Abuse: Consequences, Theories, and Responses.* Boston: Allyn & Bacon.

Migration Policy Institute Hub. 2010. "Foreign-Born Population and Foreign Born as Percentage of the Total US Population, 1850 to 2008." http://www.migrationinformation.org/datahub/charts/MPIDataHub-Number-Pct-FB-1850–2008.xls.

Mihailescu, M. 2005. "Dampening the Powder Keg: Understanding Interethnic Cooperation in Post-Communist Romania (1990–96)." *Nationalism & Ethnic Politics* 11 (1): 25–59.

Milkie, M. A. 1999. "Social Comparisons, Reflected Appraisals, and Mass Media: The Impact of Pervasive Beauty Images on Black and White Girls' Self-Concepts." *Social Psychology Quarterly* 62 (2): 190–210.

Miller, E. 2006. "Other Economies Are Possible! Special Collaboration with Grassroots Economic Organizing: Organizing toward an Economy of Cooperation and Solidarity." *Dollars and Sense* 266 (11): 5.

Miller, J. 2004. *Writing About Numbers: Effective Presentation of Quantitative Information.* Chicago: University of Chicago Press.

Miller, M., and A. Summers. 2007. "Gender Differences in Video Game Characters' Roles, Appearances, and Attire as Portrayed in Video Game Magazines." *Sex Roles: A Journal of Research* 57 (9–10): 733–42.

Millman, M. 1980. *Such a Pretty Face: Being Fat in America.* New York: W. W. Norton.

———. 2002. *Seven Stories of Love.* New York: Harper Collins.

Mills, C. 1992. "The War on Drugs: Is It Time to Surrender?" In *Taking Sides: Clashing Views on Controversial Social Issues,* 7th ed., edited by K. Finsterbusch and G. McKenna, 300–306. Guilford, CT: Dushkin Publishing.

Mills, C. W. 1959. *The Sociological Imagination.* London: Oxford University Press.

Minas, J. S., A. Scodel, D. Marlowe, and H. Rawson. 1960. "Some Descriptive Aspects of Two-Person, Zero-Sum Games." *Journal of Conflict Resolution* 4: 193–97.

Mintz, S., and S. Kellogg. 1988. *Domestic Revolutions: A Social History of American Family Life.* New York: Free Press.

Mirowsky, J., and C. E. Ross. 1989. *Social Causes of Psychological Stress.* New York: Aldine de Gruyter.

Mirsky, S. 2000."What's in a Name?" *Scientific American* 283 (3): 112.

Mischel, L., J. Bernstein, and H. Shierholz. 2009. *The State of Working America 2008/2009.* Ithaca, NY: Cornell University Press.

Mitja, D. B., Schmukle, S. C., and Egloff, B. (2008). "Becoming Friends by Chance." *Psychological Science* 5: 19(5) 439–440.

Mitra, A. 2002. "Mathematics Skill and Male-Female Wages." *Journal of Socio-Economics* 31 (5): 443–56.

Mok, T. A. 1998. "Asian Americans and Standards of Attractiveness: What's in the Eye of the Beholder?" *Cultural Diversity and Mental Health* 4 (1): 1–18.

Molnar, S. 1991. *Human Variation: Races, Types and Ethnic Groups,* 3rd ed. Englewood Cliffs, NJ: Prentice Hall.

Mondschein, E., K. Adolph, and C. Tamis-LeMonda. 2000. "Gender Bias in Mothers' Expectations about Infant Crawling." *Journal of Experimental Child Psychology* 77 (4): 304–16.

Monger, R. 2010. "U.S. Legal Permanent Residents: 2009." *Annual Flow Report* (April). http://www.dhs.gov/xlibrary/assets/statistics/publications/lpr_fr_2009.pdf.

Monger, R., and M. Barr. 2010. "Nonimmigrant Admissions to the United States: 2009." *Annual Flow Report* (April). http://www.dhs.gov/xlibrary/assets/statistics/publications/ni_fr_2009.pdf.

Montoya, R. M. 2008. "I'm Hot, So I'd Say You're Not: The Influence of Objective Physical Attractiveness on Mate Selection." *Personality and Social Psychology Bulletin* 34 (10): 1315–31.

Moore, J. F. 1997. *The Death of Competition: Leadership and Strategy in the Age of Business Ecosystems.* New York: Harper Business.

Morgan, M., J. Shanahan, and N. Signorielli. 2009. "Growing Up with Television: Cultivation Processes." In *Media Effects: Advances in Theory and Research,* 3rd ed., edited by J. Bryant and M. B. Oliver, 34–49. New York: Routledge.

Morgan, S. P., and M. Taylor. 2006. "Low Fertility at the Turn of the Twenty-First Century." *Annual Review of Sociology* 32: 375–99.

Morris, M. 2010. "Countering Discrimination and Mortgage Lending in America." NAACP. http://naacp.3cdn.net/7987bf9ee739bf597e_fsm6beky0.pdf.

Morris, S. 2008. "Second Life Affair Leads to Couple's Real-Life Divorce." *The Guardian* (November 14). http://www.guardian.co.uk/technology/2008/nov/14/second-life-virtual-worlds-divorce.

Morrongiello, B., and K. Hogg. 2004. "Mothers' Reactions to Children Misbehaving in Ways That Can Lead to Injury: Implications for Gender Differences in Children's Risk Taking and Injuries." *Sex Roles* 50 (1–2): 103–118.

Mortimer, J. T., and R. G. Simmons. 1978. "Adult Socialization." *Annual Review of Sociology* 4: 421–54.

Moss, P., and C. Tilly. 2001. *Stories Employers Tell: Race, Skill, and Hiring in America.* New York: Russell Sage.

Mruk, C. 2006. *Self-Esteem: Research, Theory, and Practice,* 3rd ed. New York: Springer.
Mughal, S., J. Walsh, and J. Wilding. 1996. "Stress and Work Performance: The Role of Trait Anxiety." *Personality and Individual Differences* 20 (6): 685–91.
Mui, A. C. 2001. "Coping and Depression Among Elderly Korean Immigrants." *Journal of Human Behavior in the Social Environment* 3 (3–4): 281–99.
Mulford, M., J. Orbell, C. Shatto, and J. Stockard. 1998. "Physical Attractiveness, Opportunity, and Success in Everyday Exchange." *American Journal of Sociology* 103 (6): 1565–92.
Mullen, B., and J. M. Smyth. 2004. "Immigrant Suicide Rates as a Function of Ethnophaulisms: Hate Speech Predicts Death." *Psychosomatic Medicine* 66: 343–48.
Murray, S. 2010. "In U.S., 14% Rely on Food Stamps." *The Wall Street Journal* (November 4). http://blogs.wsj.com/economics/2010/11/04/some-14-of-us-uses-food-stamps/.
Mustonen, A., and L. Pulkkinen. 2009. "Television Violence: A Coding Scheme." In *The Content Analysis Reader,* edited by K. Krippendorff and M. A. Bock, 311–21. Thousand Oaks, CA: Sage.
Muthusamy, S. K., J. V. Wheeler, and B. L. Simmons. 2005. "Self-Managing Work Teams: Enhancing Organizational Innovativeness." *Organizational Development Journal* 23 (3): 53–66.
Myers, D. G. 2002. *Social Psychology,* 7th ed. New York: McGraw-Hill.
Myers, P. N., Jr., and F. A. Biocca. 1992. "The Elastic Body Image: The Effect of Television Advertising and Programming on Body Image Distortions in Young Women." *Journal of Communication* 42 (3): 108–34.
Myers, P. N., Jr., F. A. Biocca, G. Wilson, D. Nias, S. B. Kaiser, M. G. Frank, T. Gilovich, F. B. Furlow, and R. K. Aune. 1999. "Appearance and Adornment Cues." In *The Nonverbal Communication Reader: Classic and Contemporary Readings,* 2nd ed., edited by L. K. Guerro, J. A. DeVito, and M. Hecht, 90–91. Prospect Heights, IL: Waveland.
NAACP. 2007. *Still Looking to the Future. Voluntary K–12 School Integration.* The NAACP Legal Defense and Education Fund, The Civil Rights Project/Proyecto Derechos Civiles. http://naacpldf.org/files/case_issue/Still_Looking_to_the_Future_Voluntary_K-12_School_Integration;_A_Manual_for_Parents,_Educators_and_Advocates.pdf.
Nagata, M. L. 1999. "Why Did You Change Your Name? Name Changing Patterns and the Life Course in Early Modern Japan." *History of the Family* 4 (3): 315–38.
Nakao, K., and J. Treas. 1992. "The 1989 Socioeconomic Index of Occupations: Construction From the 1989 Occupational Prestige Scores." *GSS Methodological Report,* No. 74. Chicago: NORC.
Nass, C. I., and S. Brave. 2007. *Wired for Speech: How Voice Activates and Advances the Human-Computer Relationship.* Cambridge, MA: MIT Press.
National Center for Children in Poverty. 2009. "Child Poverty." http://www.nccp.org/topics/childpoverty.html.
National Center for Health Statistics. 2009. *Health, United States, 2009: With Special Feature on Medical Technology.* Hyattsville, MD. 2010. http://www.cdc.gov/nchs/data/hus/hus09.pdf#009.
———. 2010. "Health, United States, 2009: With Special Feature on Medical Technology." Hyattsville, MD. http://www.cdc.gov/nchs/data/hus/hus09.pdf#057.
National Center for Injury Prevention and Control. 2007. "Suicide: Fact Sheet." Centers for Disease Control and Prevention. http://www.cdc.gov/ncipc/factsheets/suifacts.htm.
National Center for Public Policy and Higher Education. 2008. "Measuring Up 2008: The National Report Card on Higher Education." http://measuringup2008.highereducation.org/.
National Center for State Courts. 2005. "An Empirical Overview of Civil Trial Litigation." *Caseload Highlights* 11: 1.
National Coalition for the Homeless. 2009a. "Who Is Homeless?" http://www.nationalhomeless.org/factsheets/Whois.pdf.
———. 2009b. "Why Are People Homeless?" http://www.nationalhomeless.org/factsheets/why.html.
National Conference of State Legislatures. 2010. "2010 Immigration-Related Bills and Resolutions in the States: (January–March 2010)." http://www.ncsl.org/default.aspx?tabid=20244.
National Fair Housing Alliance. 2007. *The Crisis of Housing Segregation.* http://www.nationalfairhousing.org/LinkClick.aspx?fileticket=F1USduPfzQ0%3d&tabid=3917&mid=5321.
———. 2008. "Residential Segregation and Housing Discrimination in the United States: Violations of the International Convention on the Elimination of All Forms of Racial Discrimination." U.S. Housing Scholars

and Research and Advocacy Organizations. http://www.prrac.org/pdf/FinalCERDHousingDiscrimination Report.pdf.

———. 2010. "A Step in the Right Direction: 2010 Fair Housing Trends Report." http://www.nationalfairhousing .org/LinkClick.aspx?fileticket=APout1nxpwg%3d&tabid=3917&mid=5321.

National Highway Traffic Safety Administration. 2009. "2008 Traffic Safety Annual Assessment—Highlights." *Traffic Safety Facts* (June). http://www-nrd.nhtsa.dot.gov/Pubs/811172.pdf.

———. 2010. "Early Estimate of Motor Vehicle Traffic Fatalities in 2009." *Traffic Safety Facts* (March). http:// www-nrd.nhtsa.dot.gov/Pubs/811291.PDF.

National Institute of Mental Health. 2009. "Suicide in the U.S.: Statistics and Prevention." http://www.nimh.nih .gov/health/publications/suicide-in-the-us-statistics-and-prevention/index.shtml#adults.

———. 2010. "The Numbers Count: Mental Disorders in America." http://www.nimh.nih.gov/health/publications/ the-numbers-count-mental-disorders-in-america/index.shtml#Intro.

National Marriage Project. 2009. "The State of Our Unions: Marriage in America 2009." http://www.virginia.edu/ marriageproject/pdfs/Union_11_25_09.pdf.

National Park Service. 1998. "Peak Immigration Years." Ellis Island Exhibit, New York.

National Partnership for Women & Families. 2010. "Fact Sheet: Working Women Need Paid Sick Days." http:// paidsickdays.nationalpartnership.org/site/DocServer/PSD_FactSheet_WorkingWomen_080926.pdf?doc ID=4188.

National Science Board. 2008. "Science and Engineering Indicators 2008." http://www.nsf.gov/statistics/seind08/ start.htm.

National Television Violence Study. 1996–1998. Vols. 1–3, edited by Center for Communication and Social Policy, University of California at Santa Barbara. Thousand Oaks, CA: Sage.

National Women's Law Center. 2010. "Pay Equity Fact Sheet: Congress Must Act to Close the Wage Gap for Women: Facts on Women's Wages and Pending Legislation." http://www.nwlc.org/pdf/PayEquityFactSheetFinal .pdf.

NationMaster. 2007. Crime Statistics. http://www.nationmaster.com/cat/cri-crime.

Nelson, D. L., and B. L. Simmons. 2005. "Eustress and Attitudes at Work: A Positive Approach." In *Research Companion to Organizational Health Psychology,* edited by A. G. Antoniou and C. L. Cooper, 102–10. Northhampton, MA: Edward Elgar Publishing.

Nelson, T. 2001. "Tracking, Parental Education and Child Literacy Development: How Ability Grouping Perpetuates Poor Education Attainment Within Minority Communities." *Georgetown Journal on Poverty Law and Policy* VIII (2): 363–375.

———. 2002. *Ageism: Stereotyping and Prejudice Against Older Persons.* Cambridge, MA: MIT Press.

Nes, L. S., and S. C. Segerstrom. 2006. "Dispositional Optimism and Coping: A Meta-Analytic Review." *Personality and Social Psychology Review* 10 (3): 235–51.

Networkforgood.org. 2010. "About Network for Good." http://www1.networkforgood.org/about-us.

Neumark-Sztainer, D., N. Faulkner, M. Story, C. Perry, P. J. Hannan, and S. Mulert. 2002. "Weight-Teasing Among Adolescents: Correlations with Weight Status and Disordered Eating Behaviors." *International Journal of Obesity and Related Metabolic Disorders* 26 (1): 123–31.

Nevins, M. E. 2008. "They Live in Lonesome Dove: Media and Contemporary Western Apache Place-Naming Practices." *Language and Society* 37 (2): 191–215.

New York Times. 2006. "A New Name for Sale? No Way for a Marine." (December 13): 27.

New Zealand Herald. 2004. "South Korean Teens Buy Dreams of Beauty in Plastic Alley." (August 28). http://www .nzherald.co.nz/world/news/article.cfm?c_id=2&objectid=3587669.

Newport, F. 2005. "Who Are the Evangelicals?" Gallup (June 24). http://www.gallup.com/poll/17041/Who-Evangelicals.aspx.

———. 2007. "Blacks Convinced Discrimination Still Exists in College Admission Process." Gallup (August 24). http:// www.gallup.com/poll/28507/Blacks-Convinced-Discrimination-Still-Exists-College-Admission-Process.aspx.

———. 2009. "Little 'Obama Effect' on Views About Race Relations" Gallup (October 29). http://www.gallup .com/poll/123944/Little-Obama-Effect-Views-Race-Relations.aspx.

Newport, F., and J. Carroll. 2005. "Another Look at Evangelicals in America Today." Gallup (December 2). http://www.gallup.com/poll/20242/Another-Look-Evangelicals-America-Today.aspx.

NewsHour. 2006. "Melding Denominations." Broadcast on WNET, New York City (June 19).

Nie, N., and L. Erbring. 2002. "Internet and Society." http://www.eesc.usp.br/nomads/tics_arq_urb/internet_society%20report.pdf.

Nippert-Eng, C. 1996. *Home and Work.* Chicago: University of Chicago Press.

Nock, S. 1993. *The Costs of Privacy: Surveillance and Reputation in America.* New York: Aldine de Gruyter.

Nock, S., and C. Einolf. 2008. "The One Hundred Billion Dollar Man: The Annual Public Costs of Father Absence." The National Fatherhood Initiative. http://www.fatherhood.org/Page.aspx?pid=729

NPR. 2010. "The Bush Tax Cuts and What They Mean for You." *All Things Considered* (November 15).

Oates, G. 2009. "An Empirical Test of Five Prominent Explanations for the Black-White Academic Performance Gap." *Social Psychology of Education* 12 (4): 415–41.

O'Brien, L., and J. Murnane. 2009. "An Investigation Into How Avatar Appearance Can Affect Interactions in the Virtual World." *International Journal of Social and Humanistic Computing* 1 (2): 192–202.

Office of Management and Budget. 2010. "Analytical Perspectives, Budget of the United States Government, Fiscal Year 2011." http://www.whitehouse.gov/omb/budget/Analytical_Perspectives/.

Office of Research on Women's Health. 2007. "FY 2007 NIH Research Priorities for Women's Health." http://orwh.od.nih.gov/research/FY07ResearchPriorities.html.

———. 2009. "Moving Into the Future: New Dimensions and Strategies for Women's Health Research at the National Institutes of Health." http://orwh.od.nih.gov/moveFuture.html.

Ofovwe, C. E., and A. Awaritefe. 2009. "An Exploratory Study of the Meaning and Perception of Names Among Students in a Nigerian University." *African Journal for the Psychological Study of Social Issues* 12 (1–2): 560–81.

Ogletree, C. 2007. "The Demise of *Brown vs. the Board of Education?* Creating a Blueprint to Achieving Racial Justice in the 21st Century." NAACP Special Edition (January/February). http://findarticles.com/p/articles/mi_qa4081/is_200701/ai_n18621978/.

Oliker, S. 2000. "Examining Care at Welfare's End." In *Care Work: Gender, Class and the Welfare State,* edited by M. Meyer 167–185. New York: Routledge.

Omotani, B., and L. Omotani. 1996. "Expect the Best: How Your Teachers Can Help All Children Learn." *The Executive Educator* 18 (8): 27–31.

Online Newshour. 2007. "Pulitzer-Winning Book Examines Media and Civil Rights Movement." http://www.pbs.org/newshour/bb/social_issues/jan-june07/racebeat_05–18.html.

Ono, H. 2006. "Homogamy Among the Divorced and the Never Married on Marital History in Recent Decades: Evidence from Vital Statistics." *Social Science Research* 35 (2): 356–83.

Orfield, G., and S. Eaton. 2003."Back to Segregation." *Nation* 276 (8): 5–8.

Orfield, G. and C. Lee. 2007. "Historic Reversals, Accelerating Resegregation, and the Need for New Integration Strategies." The Civil Rights Project. http://civilrightsproject.ucla.edu/research/k-12-education/integration-and-diversity/historic-reversals-accelerating-resegregation-and-the-need-for-new-integration-strategies-1/orfield-historic-reversals-accelerating.pdf.

Ottati, V. C., and M. Deiger. 2002. "Visual Cues and the Candidate Evaluation Process." In *The Social Psychology of Politics: Social Psychological Applications to Social Issues,* edited by V. C. Ottati and R. S. Tindale, 75–87. New York: Kluwer Academic/Plenum.

Owen-Blakemore, J. 2003. "Children's Beliefs About Violating Gender Norms: Boys Shouldn't Look Like Girls, and Girls Shouldn't Act Like Boys." *Sex Roles: A Journal of Research* 48 (9–10): 411–19.

Owens, L. 2008. "Network News: The Role of Race in Source Selection and Story Topic." *The Howard Journal of Communications* 19 (4): 355–70.

Owsley, H. 1977. "The Marriage of Rachel Donelson." *Tennessee Historical Quarterly* 36 (Winter): 479–92.

Padavic, I., and B. Reskin. 2002. *Women and Men at Work,* 2nd ed. Thousand Oaks, CA: Pine Forge.

Pager, D., and H. Shepherd. 2008. "The Sociology of Discrimination: Racial Discrimination in Employment, Housing, Credit and Consumer Markets." *Annual Review of Sociology* 34: 181–209.

Pahl, R., and D. J. Pevalin. 2005. "Between Family and Friends: A Longitudinal Study of Friendship Choice." *British Journal of Sociology* 56 (3): 433–50.

Park, S. 2006. "The Influence of Presumed Media Influence on Women's Desire to Be Thin." *Communication Research* 32 (5): 594–614.

Parks, F. R., and J. H. Kennedy. 2007. "The Impact of Race, Physical Attractiveness and Gender on Education Majors' and Teachers' Perceptions of Student Competence." *Journal of Black Studies* 37 (6): 936–43.

Parks, J. 2009. "Executive PayWatch: CEO Perks Rise as Workers' Wages, Jobs Wilt." AFL-CIO Now Blog (April 14). http://blog.aflcio.org/2009/04/14/executive-paywatch-ceo-perks-rise-as-worker-wages-jobs-wilt/.

Parrott, S. 2005. "House Reconciliation Bill Targets Key Low-Income Programs in Ways and Means Committee's Jurisdiction." Center on Budget and Policy Priorities. http://www.cbpp.org/cms/?fa=view&id=769.

———. 2008. "Recession Could Cause Large Increases in Poverty and Push Millions Into Deep Poverty: Stimulus Package Should Include Policies to Ameliorate Harshest Effects of Downturn." Center on Budget and Policy Priorities. http://www.cbpp.org/cms/index.cfm?fa=view&id=1290.

Parsons, T. (1951) 1964. *The Social System.* Glencoe, IL: Free Press.

Passel, J. S., and D. Cohn. 2009. "A Portrait of Unauthorized Immigrants in the United States." Pew Hispanic Center, Pew Research Center. http://pewhispanic.org/files/reports/107.pdf.

Passel, J. S., W. Wang, and P. Taylor. 2010. "One-in-Seven New U.S. Marriages Is Interracial or Interethnic." http://pewsocialtrends.org/pubs/755/trends-attitudes-interracial-interethnic-marriage.

Patzer, G. L. 2006. *The Power and Paradox of Physical Attractiveness.* Boca Raton, FL: Brown/Walker Press.

———. 2008. *Looks: Why They Matter More Than You Ever Imagined.* New York: AMACOM.

Pearlin, L. I. 1989. "The Sociological Study of Stress." *Journal of Health and Social Behavior* 30 (3): 242–56.

Pearlin, L. I., S. Schieman, E. M. Fazio, and S. C. Meersman. 2005. "Stress, Health, and the Life Course: Some Conceptual Perspectives." *Journal of Health and Social Behavior* 46 (2): 205–19.

Pearson, D. E. 1993. "Post Mass Culture." *Society* 30 (5): 17–22.

Pearson, J., S. Crissey, and C. Riegle-Crumb. 2009. "Gendered Fields: Sports and Advanced Course Taking in High School." *Sex Roles: A Journal of Research* 61 (7–8): 519–35.

Perlini, A. H., A. Marcello, S. D. Hansen, and W. Pudney. 2001. "The Effects of Male Age and Physical Appearance on Evaluations of Attractiveness, Social Desirability, and Resourcefulness." *Social Behavior and Personality* 29 (3): 277–87.

Pescosolido, B., E. Grauerholz, and M. Milkie. 1997. "Culture and Conflict: The Portrayal of Blacks in U.S. Children's Picture Books Through the Mid and Late Twentieth Century." *American Sociological Review* 62 (3): 443–64.

Peterson, J. 2002. "Feminist Perspectives on TANF Reauthorization: An Introduction to Key Issues for the Future of Welfare Reform." Institute for Women's Policy Research. http://bhrp.sowo.unc.edu/susanparish/files/Peterson_feminist_social_security.pdf.

Peterson, R. 1996. "A Re-Evaluation of the Economic Consequences of Divorce." *American Sociological Review* 61: 528–36.

Peterson, T. J. 2007. "Another Level: Friendships Transcending Geography and Race." *Journal of Men's Studies* 15 (1): 71–82.

Pettit, K., and K. Rueben. 2010. "Investor-Owners in the Boom and Bust." Urban Institute. http://metrotrends.org/mortgagelending.html.

Pew Forum on Religion and Public Life. 2008. "U.S. Religious Landscape Survey." http://religions.pewforum.org/reports.

———. 2009a. "Mapping the Global Muslim Population." http://pewforum.org/Muslim/Mapping-the-Global-Muslim-Population(2).aspx.

———. 2009b. "Faith in Flux: Changes in Religious Affiliation in the U.S." http://pewforum.org/Faith-in-Flux.aspx.

———. 2010a. "U.S. Religious Knowledge Survey." http://pewforum.org/other-beliefs-and-practices/u-s-religious-knowledge-survey.aspx.

———. 2010b. "Religion among the Millennials." http://pewforum.org/Age/Religion-Among-the-Millennials.aspx.

Pew Foundation. 2007. "Internet and American Life Project: Demographics of Internet Users." http://www.pewinternet.org/Static-Pages/Trend-Data/Whos-Online.aspx.

Pew Hispanic Center. 2010. "Statistical Portrait of the Foreign-Born Population in the United States, 2008." Pew Research Center. http://pewhispanic.org/files/factsheets/foreignborn2008/Table%2040.pdf.

Pew Research Center. 2007. "As Marriage and Parenthood Drift Apart, Public Is Concerned About Social Impact." http://pewresearch.org/pubs/526/marriage-parenthood.

———. 2009. "Growing Old in America: Expectations vs. Reality." http://pewsocialtrends.org/pubs/736/getting-old-in-america.

———. 2010a. "The Return of the Multi-Generational Family Household." http://pewsocialtrends.org/pubs/752/the-return-of-the-multi-generational-family-household.

———. 2010b. "Distrust, Discontent, Anger, and Partisan Rancor: The People and Their Government." http://pewresearch.org/pubs/1569/trust-in-government-distrust-discontent-anger-partisan-rancor.

———. 2010c. "Demographics of Internet Users." http://www.pewinternet.org/Static-Pages/Trend-Data/Whos-Online.aspx.

Pfohl, S. 1977. "The Discovery of Child Abuse." *Social Problems* 24 (3): 310–23.

Pham, L. B., S. E. Taylor, and T. E. Seeman. 2001."Effects of Environmental Predictability and Personal Mastery on Self-Regulatory and Physiological Processes." *Personality and Social Psychology Bulletin* 27: 611–20.

Philipson, I. 2002. *Married to the Job: Why We Live to Work and What We Can Do About It.* New York: Free Press.

Phillips, R. G., and A. J. Hill. 1998. "Fat, Plain, but Not Friendless: Self-Esteem and Peer Acceptance of Obese Adolescent Girls." *International Journal of Obesity* 22 (4): 287–95.

Philogene, G. 1999. *From Black to African American: A New Social Representation.* Westport, CT: Praeger.

Pierce, J. W., and J. Wardle. 1997. "Cause and Effect Beliefs and Self-Esteem of Overweight Children." *Journal of Child Psychology and Psychiatry* 38 (6): 645–50.

Pindus, N., B. Theodos, and G. T. Kingsley. 2007. "Place Matters: Employers, Low-Income Workers, and Regional Economic Development." The Urban Institute. http://www.urban.org/UploadedPDF/411534_place_matters.pdf.

PIRG. 2010. Affordable Higher Education. http://www.uspirg.org/higher-education.

Pitman, S. 2010. "China Cosmetic Sales Growth Powers Ahead, but Rate Slows." Cosmeticsdesign-Europe.com (January 21). http://www.cosmeticsdesign-europe.com/Financial/China-cosmetic-sales-growth-powers-ahead-but-rate-slows.

Planty, M., W. Hussar, T. Snyder, G. Kena, A. Kewal-Ramani, J. Kemp, K. Bianco, and R. Dinkes. 2009. *The Condition of Education 2009* (NCES 2009–081). Washington, DC: National Center for Education Statistics, Institute of Education Sciences, U.S. Department of Education. http://nces.ed.gov/pubs2009/2009081.pdf.

Plunkett, S., A. Behnke, T. Sands, and B. Choi. 2009. "Adolescents' Reports of Parental Engagement and Academic Achievement in Immigrant Families." *Journal of Youth and Adolescence* 38 (2): 257–68.

Poblete, P. 2000. "The Price to Pay for an 'American' Nose and Eyes Is More Than $2,500." *San Francisco Chronicle* (February 24): E1.

———. 2001. "Beauty Ideals Still in the Dark Ages." *San Francisco Chronicle* (June 24): B4.

Policy Alert. 2005. "Income of U.S. Workforce Projected to Decline if Education Doesn't Improve." National Center for Public Policy and Education. http://www.highereducation.org/reports/pa_decline/index.shtml.

Pollack, A. 2000. "Is Everything for Sale? Patenting a Human Gene as If It Were an Invention." *New York Times* (June 28): C2.

PollingReport.com. 2009. "Race and Ethnicity." CNN/Opinion Research Corporation Poll (July 31–August 3). http://pollingreport.com/race.htm.

Popenoe, D. 2002. "Debunking 10 Divorce Myths." Discovery Health. http://health.discovery.com/centers/loverelationships/articles/divorce.html.

Portes, A. 2002."English-Only Triumphs, but the Costs Are High." *Contexts* 1 (1): 10–15.

Portes, A., and R. Rumbaut. 1996. *Immigrant America: A Portrait.* Berkeley: University of California Press.

Poulin-Dubois, D., L. Serbin, J. Eichstedt, M. Sen, and C. Beissel, C. 2002. "Men Don't Put on Make-Up: Toddlers' Knowledge of the Gender Stereotyping of Household Activities." *Social Development* 11 (2): 166–81.

Powell, B., C. Bolzendahl, C. Geist, and L. Steelman. 2010. *Counted Out: Same-Sex Relations and Americans' Definitions of Family.* New York: Russell Sage Foundation.

Power, A. 2010. "The Online Public or Cybercitizen." *SCRIPTed—A Journal of Law, Technology and Society* 7 (1): 185–95.

Prabhakaran, V. 1999. "A Sociolinguistic Analysis of South African Teluga Surnames." *South African Journal of Linguistics* 17 (2/3): 149–61.

President's Commission on Law Enforcement and Administration of Justice. 1968. *The Challenge of Crime in a Free Society.* Washington, DC: U.S. Government Printing Office.

President's Council on Bioethics. 2005. *Taking Care: Ethical Caregiving in our Aging Society.* http://bioethics.georgetown.edu/pcbe/reports/taking_care/

Press, K. 2007. "Divide and Conquer? The Role of Governance for the Adaptability of Industrial Districts." *Advances in Complex Systems* 10 (1): 3–92.

Preston, C., and N. Wallace. 2010. "Donations to Help Haiti Exceed $560-Million as of January 29." *The Chronicle of Philanthropy* (January 29). http://philanthropy.com/article/Donations-to-Help-Haiti-Exceed/63824/.

Preston, J. 2006. "Public Misled on Air Quality After 9/11 Attack, Judge Says." *New York Times* (February 3): B4.

Proctor, C. P., D. August, M. Carlo, and C. Barr. 2010. "Language Maintenance Versus Language of Instruction: Spanish Reading Development among Latino and Latina Bilingual Learners." *Journal of Social Issues* 66 (1): 79–94.

Provasnik, S., P. Gonzales, and D. Miller. 2009. *U.S. Performance Across International Assessments of Student Achievement: Special Supplement to the Condition of Education 2009* (NCES 2009–083). Washington, DC: National Center for Education Statistics, Institute of Education Sciences, U.S. Department of Education. http://nces.ed.gov/pubs2009/2009083.pdf.

Provasnik, S., and M. Planty. 2008. *Community Colleges: Special Supplement to The Condition of Education 2008* (NCES 2008–033). Washington, DC: National Center for Education Statistics, Institute of Education Sciences, U.S. Department of Education.

Public Citizen. 2010. "Medical Malpractice Payments Continue to Fall, Public Citizen Analysis Shows." http://www.citizen.org/pressroom/pressroomredirect.cfm?ID=3060.

Purcell, K. 2010. "Teens and the Internet: The Future of Digital Diversity." Pew Internet & American Life Project. http://www.pewinternet.org/Presentations/2010/Mar/Fred-Forward.aspx.

Putnam, R. 1995. "Bowling Alone: America's Declining Social Capital." *Journal of Democracy* (January): 65–78.

———. 2000. *Bowling Alone: The Collapse and Revival of American Community.* New York: Simon and Schuster.

Puurtinen, M., and T. Mappes. 2009. "Between-Group Competition and Human Cooperation." *Proceedings of the Royal Society—Biological Sciences* 276 (1655): 355–60.

Quarantelli, E. L. 2008. "Conventional Beliefs and Counterintuitive Realities." *Social Research* 75 (3): 873–904.

Quick, J. C., C. L. Cooper, D. L. Nelson, J. D. Quick, and J. H. Gavin. 2003. "Stress, Health, and Well-Being at Work." In *Organizational Behavior: The State of the Science,* 2nd ed., edited by J. Greenberg, 53–89. Mahwah, NJ: Lawrence Erlbaum Associates.

Quiggin, J. 2006. "Blogs, Wikis and Creative Innovation." *International Journal of Cultural Studies* 9 (4): 481–96.

Quinnipiac Poll. 2009. U.S. Voters Disagree 3–1 With Sotomayor on Key Case, Quinnipiac University National Poll Finds; Most Say Abolish Affirmative Action. http://www.quinnipiac.edu/x1295.xml?ReleaseID=1307.

Rabois, D., and D. A. F. Haaga. 2002. "Facilitating Police-Minority Youth Attitude Change: The Effects of Cooperation Within a Competitive Context and Exposure to Typical Exemplars." *Journal of Community Psychology* 30 (2): 189–95.

"Race: The Power of an Illusion." 2003. PBS. http://www.pbs.org/race/000_General/000_00-Home.htm.

Rada, J., and K. T. Wulfemeyer. 2005. "Color Codes: Racial Descriptors in Television Coverage of Intercollegiate Sports." *Journal of Broadcasting & Electronic Media* 49 (1): 65–85.

RAINN. 2010. "Statistics." http://www.rainn.org/statistics.

Raley, S., S. Bianchi, K. Cook, and D. Massey. 2006. "Sons, Daughters, and Family Processes: Does Gender of Children Matter?" *Annual Review of Sociology* 32 (1): 401–21.

Ram, R. 2006. "Further Examination of the Cross-Country Association between Income Inequality and Population Health." *Social Science & Medicine* 62 (3): 779–91.

Ramey, A. 2010. "Analysis of Education and Training Data." U.S. Bureau of Labor Statistics. http://www.bls.gov/emp/ep_education_training.htm.

Ramsey, J. L., and J. H. Langlois. 2002. "Effects of the 'Beauty Is Good' Stereotype on Children's Information Processing." *Journal of Experimental Child Psychology* 81 (3): 320–40.

Ramsey, J. L., J. H. Langlois, R. A. Hoss, A. J. Rubenstein, and A. M. Griffin. 2004. "Origins of a Stereotype: Categorization of Facial Attractiveness by 6 Month Old Infants." *Developmental Science* 7 (2): 201–11.

RAND Institute for Civil Justice. 2007. "Anatomy of an Insurance Class Action." Research Brief. http://www.rand.org/pubs/research_briefs/2007/RAND_RB9249.pdf.

Rapoport, A. 1960. *Fights, Games, and Debates.* Ann Arbor: University of Michigan Press.

Rasmussen Reports. 2010a. "Louisiana Voters Strongly Support Offshore Drilling, Deepwater Drilling." (June 28). http://www.rasmussenreports.com/public_content/politics/general_state_surveys/louisiana/louisiana_voters_strongly_support_offshore_drilling_deepwater_drilling.

———. 2010b. "87% Say English Should Be U.S. Official Language." http://www.rasmussenreports.com/public_content/politics/general_politics/may_2010/87_say_english_should_be_u_s_official_language.

Rawlins, W. 1992. *Friendship Matters: Communication, Dialectics, and the Life Course.* New York: Aldine de Gruyter.

Reeves, B., and Nass, C. 1996. *The Media Equation: How People Treat Computers, Television, and New Media Like Real People and Places.* New York: Cambridge University Press.

Regan, P. C. 2008. *The Mating Game: A Primer on Love, Sex, and Marriage,* 2nd ed. Thousand Oaks, CA: Sage.

Reiman, J., and P. Leighton. 2009. *The Rich Get Richer and the Poor Get Prison,* 9th ed. Prentice Hall.

Reis, H. T., J. Nezlek, and L. Wheeler. 1980. "Physical Attractiveness in Social Interaction." *Journal of Personality and Social Psychology* 38: 604–17.

Rensberger, B. 1981."Racial Odyssey." *Science Digest* (January/February).

Renzetti, C. M., and D. J. Curran. 1989. *Women, Men and Society: The Sociology of Gender.* Boston: Allyn & Bacon.

Reuters. 2007a. "Is It a Bird? Is It a Plane? No, It's a Baby." (August 8). http://www.reuters.com/article/oddlyEnoughNews/idUSWEL7556320070808?feedType=RSS.

———. 2007b. "Chinese Couple to Name Baby '@.'" (August 16). http://uk.reuters.com/article/2007/08/16/oukoe-uk-china-language-idUKPEK36827920070816.

Rhodes, G., K. Geddes, L. Jeffrey, S. Dziurawiec, and A. Clark. 2002. "Are Average and Symmetric Faces Attractive to Infants' Discrimination and Looking Preferences?" *Perception* 31 (3): 315–21.

Rich, F. 2009. "Small Beer, Big Hangover" Opinion. *New York Times* (August 1). http://www.nytimes.com/2009/08/02/opinion/02rich.html?_r=2.

Richardson, L. W. 1988. *The Dynamics of Sex and Gender: A Sociological Perspective.* New York: Harper and Row.

Richter-Menge, J., and J. E. Overland, eds., 2010. "Arctic Report Card 2010." http://www.arctic.noaa.gov/reportcard.

Riddle, K. 2010. "Always on My Mind: Exploring How Frequent, Recent, and Vivid Television Portrayals Are Used in the Formation of Social Reality Judgements." *Media Psychology* 13 (2): 155–79.

Riegle-Crumb, C., G. Farkas, and C. Muller. 2006. "The Role of Gender and Friendship in Advanced Course Taking." *Sociology of Education* 79 (3): 206–28.

Riley, D. 1988. *Am I That Name?* Minneapolis: University of Minnesota Press.

Riley, G. (1991) 1997. *Divorce: An American Tradition.* Lincoln: University of Nebraska Press.

Riniolo, T. C., K. C. Johnson, T. R. Sherman, and J. A. Misso. 2006. "Hot or Not: Do Professors Perceived as Physically Attractive Receive Higher Student Evaluations?" *Journal of General Psychology* 133 (1): 19–35.

Ritzer, G. 1995. *Expressing America: A Critique of the Global Credit Card Society.* Thousand Oaks, CA: Pine Forge.

———. 2001. *Explorations in the Sociology of Consumption: Fast Food, Credit Cards, and Casinos.* Thousand Oaks, CA: Sage.

Robert, S. A., and J. S. House. 2000. "Socioeconomic Inequalities in Health: An Enduring Sociological Problem." In *Handbook of Medical Sociology,* 5th ed., edited by C. Bird, P. Conrad, and A. Fremont, 79–97. Upper Saddle River, NJ: Prentice Hall.

Robert Wood Johnson Foundation and the State Health Access Data Assistance Center. 2010. "Barely Hanging On: Middle-Class and Uninsured. A State-by-State Analysis." National Coalition on Health Care. http://nchc.org/facts-resources/study-illustrates-continuing-decline-insurance-coverage.

Roberts, G., and H. Klibanoff. 2006. *The Race Beat: The Press, the Civil Rights Struggle and the Awakening of a Nation.* New York: Knopf.

Robinson, J. 2003. *Work to Live.* New York: Perogee.

Robison, K. K., and E. M. Crenshaw. 2010. "Reevaluating the Global Digital Divide: Socio-Demographic and Conflict Barriers to the Internet Revolution." *Sociological Inquiry* 80 (1): 34–62.

Roediger, D. 2005. *Working Toward Whiteness: How America's Immigrants Became White.* New York: Basic Books.

Rofé, Y. 2006. "Affiliation Tendencies on the Eve of the Iraqi War: A Utility Theory Perspective." *Journal of Applied Social Psychology* 36 (7): 1781–89.

Rose-Redwood, R. S. 2008. "From Number to Name: Symbolic Capital, Places of Memory, and the Politics of Street Renaming in New York City." *Social and Cultural Geography* 9 (4): 431–52.

Rosenfeld, M. 2007. *The Age of Independence: Interracial Unions, Same-Sex Unions and the Changing American Family.* Cambridge: Harvard University Press.

Rosenhan, D. L. 1973."On Being Sane in Insane Places." *Science* 179: 250–58.

Rosenkrantz, L., and P. R. Satran. 2004. *Beyond Jennifer and Jason, Madison and Montana: What to Name Your Baby Now.* New York: St. Martin's Griffin.

Rosenthal, R., and L. Jacobson. 1968. *Pygmalion in the Classroom.* New York: Holt, Rinehart and Winston.

Ross, C. E., and B. A. Broh. 2000. "The Roles of Self-Esteem and the Sense of Personal Control in the Academic Achievement Process." *Sociology of Education* 73 (4): 270–84.

Ross, C. E., and J. Huber. 1985. "Hardship and Depression." *Journal of Health and Social Behavior* 26 (4): 312–27.

Ross, C., and J. Mirowsky, J. 2002."Family Relationships, Social Support and Subjective Life Expectancy." *Journal of Health and Social Behavior* 43 (4): 469–89.

Ross, H., and H. Taylor. 1989. "Do Boys Prefer Daddy or His Physical Style of Play?" *Sex Roles* 20 (January): 23–33.

Rossi, A. S., and P. Rossi. 1990. *Of Human Bonding: Parent–Child Relations Across the Life Course.* New York: Aldine de Gruyter.

Rostosky, S. S., M. D. Otis, E. D. B. Riggle, S. Kelly, and C. Brodnick. 2008. "An Exploratory Study of Religiosity and Same-Sex Couple Relationships." *Journal of GLBT Family Studies* 4 (1): 17–36.

Roussi, P. K., H. Vagia, and I. Koutri. 2007. "Patterns of Coping and Psychological Distress in Women Diagnosed With Breast Cancer." *Cognitive Therapy and Research* 31 (1): 97–109.

Rowe, G., and M. Murphy. 2009. "Welfare Rules Databook: State TANF Policies as of July 2008." The Urban Institute. http://anfdata.urban.org/wrd/databook.cfm.

Royo, S. 2006. "Beyond Confrontation." *Comparative Political Studies* 38 (8): 969–95.

Ruane, J., K. Cerulo, and J. Gerson. 1994. "Professional Deceit: Normal Lying in an Occupational Setting." *Sociological Focus* 27 (2): 91–109.

Ruane, J. 2005. *Essentials of Research Methods.* Malden, MA: Blackwell.

Rubenstein, A. J., L. Kalakanis, and J. H. Langlois. 1999. "Infant Preferences for Attractive Faces: A Cognitive Explanation." *Developmental Psychology* 35 (3): 848–55.

Rubin, B. 1996. *Shifts in the Social Contract: Understanding Change in American Society.* Thousand Oaks, CA: Pine Forge.

Rubin, J. Z., F. J. Provenzano, and Z. Luria. 1974. "The Eye of the Beholder: Parents' Views on Sex of Newborns." *American Journal of Orthopsychiatry* 44: 512–19.

Rubin, L. B. 1993. *Just Friends: The Role of Friendship in Our Lives.* New York: Harper and Row.

Rubin, N., C. Shmilovitz, and M. Weiss. 1993. "From Fat to Thin: Informal Rites Affirming Identity Change." *Symbolic Interaction* 16 (1): 1–17.

———. 1994. "The Obese and the Slim: Personal Definition Rites of Identity Change in a Group of Obese People Who Became Slim After Gastric Reduction Surgery." *Megamot* 36 (1): 5–19.

Rubinstein, S., and B. Caballero. 2000. "Is Miss America an Undernourished Role Model?" *Journal of the American Medical Association* 283 (2): 1569.

Rudd, N. A., and S. J. Lennon. 1999. "Social Power and Appearance Management Among Women." In *Appearance and Power: Dress, Body, Culture,* edited by K. K. P. Johnson and S. J. Lennon, 153–72. New York: Oxford University Press.

Rudman, L., J. Feinberg, and K. Fairchild. 2002. "Minority Members' Implicit Attitudes: Automatic Ingroup Bias as a Function of Group Status." *Social Cognition* 20 (4): 294–320.

Rumbaut, R. 2009. "A Language Graveyard? The Evolution of Language Competencies, Preferences, and Use Among Young Adult Children of Immigrants." In *The Education of Language Minority Immigrants in the United States,* edited by T. Wiley, J. Lee, and R. Rumberger, 35–71. Tonawanda, NY: Multilingual Matters.

Rust, J., S. Golombok, M. Himes, K. Johnston, and J. Golding. 2000. "The Role of Brothers and Sisters in the Gender Development of Preschool Children." *Journal of Experimental Child Psychology* 77 (4): 292–303.

Ryan, A. 1996. "Professional Liars." *Social Research* 63 (Fall): 619–41.

Saad, L. 2007. "Black–White Education Opportunity Widely Seen as Equal." Gallup (July 2). http://www.gallup.com/poll/28021/blackwhite-educational-opportunities-widely-seen-equal.aspx.

———. 2010. "Nearly 4 in 10 Americans Still Fear Walking Alone at Night." Gallup (November 5). http://www.gallup.com/poll/144272/Nearly-Americans-Fear-Walking-Alone-Night.aspx.

Sacks, H. 1975. "Everyone Has to Lie." In *Sociocultural Dimensions of Language Use,* edited by M. Sanches and B. Blount, 57–79. New York: Academic Press.

Saguy, A. C., and K. Gruys. 2010. "Morality and Health: News Media Constructions of Overweight and Eating Disorders." *Social Problems* 57 (2): 231–50.

SalaryList.com. 2010. "Jobs Salary by Job." http://www.salarylist.com/jobs-salary-by-jobs.htm.

Salmon, M. 1986. *Women and the Law of Property in Early America.* Chapel Hill: University of North Carolina Press.

Sanabria, H. 2001. "Exploring Kinship in Anthropology and History: Surnames and Social Transformation in the Bolivian Andes." *Latin American Research Review* 36 (2): 137–55.

Sandefur, R. 2008. "Access to Civil Justice and Race, Class, and Gender Inequality." *Annual Review of Sociology* 34: 339–358.

Sanderson, C. A., J. M. Darley, and C. S. Messinger. 2002. "I'm Not as Thin as You Think I Am: The Development and Consequences of Feeling Discrepant From the Thinness Norm." *Personality and Social Psychology Bulletin* 28 (2): 172–83.

Sands, E. R., and J. Wardle. 2003. "Internalization of Ideal Body Shapes in 9–12 Year-Old Girls." *International Journal of Eating Disorders* 33 (2): 193–204.

Sangmpam, S. N. 1999. "American Civilization, Name Change, and African-American Politics." *National Political Science Review* 7: 221–48.

Sapkidis, O. 1998. "To Whom Do You Belong?: Catholic and Orthodox Names at Syros (Greece)." In *Naming and Social Structure: Example From Southeast Europe,* edited by P. H. Stahl, 73–88. New York: Columbia University Press.

Satran, O. R., and L. Rosenkrantz. 2008. *Cool Names for Babies.* New York: St. Martin's Griffin.

Sauer, M., and M. Chuchmach. 2009. "Scandal-Ridden Blackwater Changes Name." ABC News Blotter (February 13). http://abcnews.go.com/Blotter/Blackwater/story?id=6873331&page=1.

Savacool, J. 2009. *The World Has Curves: The Global Quest for the Perfect Body.* New York: Rodale Books.

Scassa, T. 1996."National Identity, Ethnic Surnames and the State." *Canadian Journal of Law and Society* 11 (2): 167–91.

Schachter, S. 1959. *The Psychology of Affiliation.* Stanford, CA: Stanford University Press.

Schaffner, B., and M. Gadson. 2004. "Reinforcing Stereotypes? Race and Local Television News Coverage of Congress." *Social Science Quarterly* 85 (3): 604–21.

Scharrer, E., D. Kim, K. Lin, and Z. Liu. 2006. "Working Hard or Hardly Working? Gender, Humor, and the Performance of Domestic Chores in Television Commercials." *Mass Communication & Society* 9 (2): 215–38.

Schatz, A. 2007. "Long Races Force Ad Ingenuity." *Wall Street Journal* (June 19). http://online.wsj.com/article/SB118221211658539826.html?mod=hpp_us_ editors_picks.

Schein, E. H. 2004. "Learning When and How to Lie: A Neglected Aspect of Organizational and Occupational Socialization." *Human Relations* 57 (3): 260–73.

Schmid, R. E. 2007. "Raising Prices Enhances Wine Sales." *Yahoo News.* http://neurocritic.blogspot.com/2008/01/delineation-of-two-buck-chuck.html.

Schmidley, D. 2003. "The Foreign-Born Population in the U.S.: March 2002." *Current Population Reports,* P20–539. Washington, DC: U.S. Census Bureau.

Schmidt, P. 2008. "Researchers Accuse Selective Colleges of Giving Admissions Tests Too Much Weight." *Chronicle of Higher Education* 54 (35): A20–A21.

Schneiderman, R. 1967. *All for One.* New York: P.S. Eriksson.

Schopler, J., C. A. Insko, S. Drigotas, and K. A. Graetz. 1993. "Individual/Group Discontinuity: Further Evidence for Mediation by Fear and Greed." *Personality and Social Psychology Bulletin* 19 (4): 419–31.

Schott, L. 2009. "Policy Basics: An Introduction to TANF." Center on Budget and Policy Priorities. http://www.cbpp.org/cms/index.cfm?fa=view&id=936.

Schrock, D., and M. Schwalbe. 2009. "Men, Masculinity, and Manhood Acts." *Annual Review of Sociology* 35: 277–95.

Schroeder, K., and J. Ledger. 1998. *Life and Death on the Internet.* Menasha, WI: Supple.

Schul, Y., and A. Vinokur. 2000. "Projection in Person Perception Among Spouses as a Function of the Similarity in Their Shared Experiences." *Personality and Social Psychology Bulletin* 26: 987–1001.

Schulman, D. 2004. "Labeling Theory." In *Encyclopedia of Social Theory,* edited by G. Ritzer, 428. Thousand Oaks, CA: Sage.

Schultz, J. W., and D. G. Pruitt. 1978. "The Effect of Mutual Concern on Joint Welfare." *Journal of Experimental Social Psychology* 14: 480–92.

Schuman, D., and R. Olufs. 1995. *Diversity on Campus.* Boston: Allyn & Bacon.

Schur, E. 1984. *Labeling Women Deviant.* Philadelphia: Temple University Press.

Schwartz, M. A., and B. Scott. 2007. *Marriages and Families: Diversity and Change,* 5th ed. Upper Saddle River, NJ: Prentice Hall.

Scott, E., K. Edin, A. London, and R. Kissane. 2004. "Unstable Work, Unstable Income: Implications for Family Well-Being in the Era of Time-Limited Welfare." *Journal of Poverty* 8 (1): 61–88.

Seccombe, K., H. Hartley, J. Newsom, K. Hoffman, G. Marchand, C. Albo, C. Gordon, T. Zaback, R. Lockwood, and C. Pope. 2007. "The Aftermath of Welfare Reform: Health, Health Insurance, and Access to Care Among Families Leaving TANF in Oregon." *Journal of Family Issues* 28 (2): 151–81.

Segrin, C., and R. Nabi. 2002. "Does Television Viewing Cultivate Unrealistic Expectations About Marriage?" *Journal of Communication* 52 (2): 247–63.

Seiter, J., J. Bruschke, and C. Bai. 2002. "The Acceptability of Deception as a Function of Perceivers' Culture, Deceiver's Intention, and Deceiver-Deceived Relationship." *Western Journal of Communication* 66 (2): 158–80.

Sennett, R. 1977. *The Fall of Public Man.* New York: Knopf.

Sennitt, A. 2010. "Governor Wants to Change Name of Guam to Guahan." *Radio Netherlands Worldwide* (February 15). http://blogs.rnw.nl/medianetwork/governor-wants-to-change-name-of-guam-to-guahan.

Serbin, L., D. Poulin-Dubois, and J. Eichstedt. 2002. "Infants' Responses to Gender-Inconsistent Events." *Infancy* 3 (4): 531–42.

Seshamani, M. 2009. "Roadblocks to Health Care: Why the Current Health Care System Does Not Work for Women." Department of Health and Human Services. http://www.healthreform.gov/reports/women/index.html.

Shaffer, D. R., N. Crepaz, and C. Sun. 2000. "Physical Attractiveness Stereotyping in Cross-Cultural Perspective: Similarities and Differences Between Americans and Taiwanese." *Journal of Cross-Cultural Psychology* 31 (5): 557–82.

Shalala, D. 2003. "Older Americans: Living Longer, Living Better." *World Almanac and Book of Facts 2003.* New York: World Almanac Books.

Shapiro, A., and R. Leone. 1999. *The Control Revolution: How the Internet Is Putting Individuals in Charge and Changing the World We Know.* New York: Public Affairs/Century Foundation.

Shaw, A. 2006. "The Arranged Transnational Cousin Marriages of British Pakistanis: Critique, Dissent and Cultural Continuity." *Contemporary South Asia* 15 (2): 209–20.

Shaw, B. A., J. Liang, and N. Krause. 2010. "Age and Race Differences in the Trajectories of Self-Esteem." *Psychological Aging* 25 (1): 84–94.

Sherif, M. 1966. *In Common Predicament: Social Psychology of Intergroup Conflict and Cooperation.* Boston: Houghton Mifflin.

Sherif, M., O. J. Harvey, B. J. White, W. R. Hood, and C. W. Sherif. 1961. *Intergroup Conflict and Cooperation: The Robbers' Cave Experiment.* Norman, OK: University Book Exchange.

Sherman, A., and C. Stone. 2010. "Income Gaps Between Very Rich and Everyone Else More Than Tripled in Last Three Decades, New Data Show." (June 25). Center on Budget and Policy Priorities. http://www.cbpp.org/cms/index.cfm?fa=view&id=3220.

Shinn, L., and M. O'Brien. 2008. "Parent-Child Conversational Styles in Middle Childhood: Gender and Social Class Differences." *Sex Roles: A Journal of Research* 59 (1–2): 61–67.

Shipler, D. 2005. *The Working Poor.* New York: Vintage Books.

Shipman, P. 1994. *The Evolution of Racism: Human Differences and the Use and Abuse of Science.* New York: Simon and Schuster.

Sigelman, L., C. K. Sigelman, and C. Fowler. 1987. "A Bird of a Different Feather? An Investigation of Physical Attractiveness and the Electability of Female Candidates." *Social Psychology Quarterly* 50 (1): 32–43.

Signorielli, N. 2009. "Race and Sex in Prime Time: A Look at Occupations and Occupational Prestige." *Mass Communication & Society* 12 (3): 332–52.

Signorielli, N., and M. Vasan, M. 2005. *Violence in the Media: A Reference Handbook.* Santa Barbara, CA: ABC-CLIO Publishers.

Silverman, R. M. 2005. "Redlining in a Majority Black City?: Mortgage Lending and the Racial Composition of Detroit Neighborhoods." *Western Journal of Black Studies* 29 (1): 531–41.

Silverstein, M. J., and N. Fiske. 2003. "Luxury for the Masses." *Harvard Business Review* (April): 48–54.

Simmel, G. 1950a. "The Lie." In *The Sociology of Georg Simmel,* edited by K. Wolff, 312–16. New York: Free Press.

———. 1950b. "The Stranger." In *The Sociology of Georg Simmel,* edited by K. Wolff, 402–08. New York: Free Press.

Simmons, J. L. 1966."Public Stereotypes of Deviants." *Social Problems* 13: 223–32.

Simon, R. W. 1997. "The Meanings Individuals Attach to Role Identities and Their Implications for Mental Health." *Journal of Health and Social Behavior* 38 (3): 256–74.

Simon, R., and K. Marcussen. 1999. "Marital Transitions, Marital Beliefs, and Mental Health." *Journal of Health and Social Behavior* 40 (2): 111–25.

Simpson, J., B. Campbell, and E. Berscheid. 1986. "The Association Between Romantic Love and Marriage: Kephart (1967) Twice Revisited." *Personality and Social Psychology Bulletin* 12: 363–72.

Simpson, R. 2004. "Masculinity at Work: The Experiences of Men in Female Dominated Occupations." *Work, Employment and Society* 18 (2): 349–68.

Sinha, R. 2008. "Chronic Stress, Drug Use, and Vulnerability to Addiction." *Annals of the New York Academy of Sciences* 1141 (October): 105–30.

Skolnick, A. 1991. *Embattled Paradise: The American Family in an Age of Uncertainty.* New York: Basic Books.

Slater, A., G. Bremner, S. P. Johnson, P. Sherwood, R. Hayes, and E. Brown, E. 2000. "Newborn Infants' Preference for Attractive Faces: The Role of Internal and External Facial Features." *Infancy* 1 (2): 265–74.

Slater, P. 1970. *The Pursuit of Loneliness: American Culture at the Breaking Point.* Boston: Beacon.

Slavin, R. E., and N. A. Madden. 1979. "School Practices That Improve Race Relations." *American Educational Research Journal* 16: 169–80.

Smedley, A., and B. Smedley. 2005. "Race as Biology Is Fiction; Racism as a Social Problem Is Real." *American Psychologist* 60 (1): 16–26.

Smith, A. 1991. *National Identity.* Reno: University of Nevada Press.

———. 2009. "The Internet's Role in Campaign 2008." Pew Internet & American Life Project. (April 15). http://www.pewinternet.org/Reports/2009/6—The-Internets-Role-in-Campaign-2008.aspx.

———. 2010. "Neighbors Online." Pew Internet & American Life Project. (June 9). http://www.pewinternet.org/Reports/2010/Neighbors-Online.aspx.

Smith, A., K. L. Schlozman, S. Verba, and H. Brady. 2009. "The Internet and Civic Engagement." Pew Internet & American Life Project (September 1). http://www.pewinternet.org/Reports/2009/15—The-Internet-and-Civic-Engagement.aspx.

Smith, B. 2010. "Tea Parties Stir Evangelicals' Fears." *Politico* (March 12). http://www.politico.com/news/stories/0310/34291.html.

Smith, F., F. Hardman, and S. Higgins. 2007. "Gender Inequality in the Primary Classroom: Will Interactive Whiteboards Help?" *Gender and Education* 19 (4): 455–469.

Smith, G. 2008. "Palin V.P. Nomination Puts Pentecostalism in the Spotlight." Pew Forum on Religion and Public Life. http://pewforum.org/Palin-VP-Nomination-Puts-Pentecostalism-in-the-Spotlight.aspx.

Smith, J. C. 2007. *Supreme Conflict: The Inside Story of the Struggle for Control of the United States Supreme Court.* New York: Penguin.

Smith, K. 2005. "Prebirth Gender Talk: A Case Study in Prenatal Socialization." *Women and Language* 28 (1): 49–53.

Smith, S. 2006. "Counter Culture." *MediaWeek* 16 (27): 24–25.

Snyder, M. 2001. "Self-Fulfilling Stereotypes." In *Self and Society,* edited by A. Branaman, 30–35. Malden, MA: Blackwell.

Sobal, J., and D. Maurer. (Eds.). 1999. *Interpreting Weight: The Social Management of Fatness and Thinness.* New York: Aldine de Gruyter.

Solman, P. 2006a. "Global Warming Presents New Business Opportunities." Online NewsHour (June 5). http://www.pbs.org/newshour/bb/environment/jan-june06/globalwarming_06–05.html.

———. 2006b. "Biofuels as Oil Alternative." Online NewsHour (April 13). http://www.pbs.org/newshour/bb/economy/jan-june06/biofuels_4–13.html.

Solomon, D., V. Battistich, and A. Hom. 1996. "Teacher Beliefs and Practices in Schools Serving Communities that Differ in Socioeconomic Level." *Journal of Experimental Education* 64 (4): 327–347.

Sonnert, G. 2009. "Parents Who Influence Their Children to Become Scientists: Effects of Gender and Parental Education." *Social Studies of Science* 39 (6): 927–41.

South, S. 2001. "Time-Dependent Effects of Wives' Employment on Marital Dissolution." *American Sociological Review* 66 (2): 226–45.

South, S., and K. Lloyd. 1995. "Spousal Alternatives and Marital Dissolution." *American Sociological Review* 60 (1): 21–35.

Spain, D. 1999. "America's Diversity: On the Edge of Two Centuries." *PBR Reports on America* 1 (2).

Spamlaws.com. 2009. "Spam Statistics and Facts." http://www.spamlaws.com/spam-stats.html.

Spencer, L., and R. Pahl. 2006. *Rethinking Friendship: Hidden Solidarities Today.* Princeton, NJ: Princeton University Press.

Spencer, R., M. Porche, and D. Tolman. 2003. "We've Come a Long Way—Maybe: New Challenges for Gender Equity in Education." *Teachers College Record* 105 (9): 1774–1807.

Sprecher, S., and P. C. Regan. 2002. "Liking Some Things (in Some People) More Than Others: Partner Preferences in Romantic Relationships and Friendships." *Journal of Social and Personal Relationships* 19 (4): 463–81.

Squires, G., and J. Chadwick. 2006. "Linguistic Profiling." *Urban Affairs Review* 41 (3): 400–15.

Stack, S., and J. Eshleman. 1998. "Marital Status and Happiness: A 17-Nation Study." *Journal of Marriage & Family* 60 (2): 527–37.

Stafford, L., and A. Merolla. 2007. "Idealization, Reunions, and Stability in Long-Distance Dating Relationships." *Journal of Social & Personal Relationships* 24 (1): 37–54.

Stake, J., and S. Nickens. 2005. "Adolescent Girls' and Boys' Science Peer Relationships and Perceptions of the Possible Self as Scientist." *Sex Roles* 52 (1–2): 1–11.

Stankiewicz, J., and F. Rosselli. 2008. "Women as Sex Objects and Victims in Print Advertisements." *Sex Roles: A Journal of Research* 58 (7–8): 579–89.

Stanley, R., V. Sullivan, and J. Wardle. 2009. "Self-Perceived Weight in Adolescents: Overestimation or Underestimation?" *Body Image* 6 (1): 56–59.

Stapel, D. A., and W. Koomen. 2005. "Competition, Cooperation, and the Effects of Others on Me." *Journal of Personality and Social Psychology* 88 (6): 1029–38.

Staples, B. 2010. "Cutting and Pasting: A Senior Thesis by (Insert Name)." *New York Times* (July 12). http://www.nytimes.com/2010/07/13/opinion/13tue4.html?_r=1.

Stenson, J. 2003. "Extra Stress Stresses Immune System, Too." preventdisease.com/news/articles/extra_stress_stresses_immune_system.shtml.

———. 2009. "'Phantom Fat' Can Linger After Weight Loss." http://www.msnbc.msn.com/id/31489881/ns/health-womens_health/.

Stevens, J. 1999. *Reproducing the State.* Princeton, NJ: Princeton University Press.

Stice, E., D. Spangler, and W. S. Agras. 2001. "Exposure to Media-Portrayed Thin-Ideal Images Adversely Affects Vulnerable Girls: A Longitudinal Experiment." *Journal of Social and Clinical Psychology* 20 (3): 270–88.

Stinchcombe, A. 1963. "Some Empirical Consequences of the Davis-Moore Theory of Stratification." *American Sociological Review* 28 (5): 805–08.

———. 1997. "On the Virtues of the Old Institutionalism." *Annual Review of Sociology* 23: 1–18.

Stock, P. 1978. *Better Than Rubies: A History of Women's Education.* New York: G. P. Putnam.

Stolberg, S. 2003. "Senate Becomes OK Corral for a Surgeon and a Lawyer." *New York Times* (July 11).http://query.nytimes.com/gst/fullpage.html?res=9B00E2DD113DF932A25754C0A9659C8B63.

Stone, L. 1989. "The Road to Polygamy." *New York Review of Books* (March): 13.

Stout, C., and R. Kline. 2008. "Ashamed Not to Vote for an African-American; Ashamed to Vote for a Woman: An Analysis of the Bradley Effect from 1982–2006." Center for the Study of Democracy, UC Irvine. http://www.escholarship.org/uc/item/3q6437f6.

Strand, K., and E. Mayfield. 2000. "The Effects of 'Female Friendly' Teaching Strategies on College Women's Persistence in Mathematics." Presentation at Southern Sociological Society.

Straughan, R., and M. Lynn. 2002. "The Effects of Salesperson Compensation on Perceptions of Salesperson Honesty." *Journal of Applied Social Psychology* 32 (4): 719–31.

Straus, M., with D. Donnelly. 2001. *Beating the Devil out of Them: Corporal Punishment in American Families and Its Effects on Children.* New Brunswick, NJ: Transaction.

Straus, M., R. Gelles, and S. Steinmetz. 2006. *Behind Closed Doors: Violence in American Families.* New Brunswick, NJ: Transaction.

Streitfeld, D. 2010. "No Help in Sight, More Homeowners Walk Away." *New York Times* (February 2). http://www.nytimes.com/2010/02/03/business/03walk.html?pagewanted=1&_r=1.

Suarez, E. 1997. "A Woman's Freedom to Choose Her Surname: Is It Really a Matter of Choice?" *Women's Right Law Reporter* 18 (2): 233–42.

Suarez-Orozco, C., J. Rhodes, and M. Milburn. 2009. "Unraveling the Immigrant Paradox: Academic Engagement and Disengagement Among Recently Arrived Immigrant Youth." *Youth & Society* 41 (2): 151–85.

Suarez-Orozco, C., M. Susarez-Orozco, and I. Todorova. 2008. *Learning a New Land: Immigrant Students in American Society.* Cambridge, MA: Belknap Press.

Sue, C., and E. Telles. 2007. "Assimilation and Gender in Naming." *American Journal of Sociology* 12 (5): 1383–1415.

Sullins, P. 2000. "The Stained Glass Ceiling: Career Attainment for Women Clergy." *Sociology of Religion* 61 (3): 243–66.

Sullivan, B. A., J. L. Sullivan, and M. Snyder. 2008. *Cooperation: The Political Psychology of Effective Human Interaction.* Malden, MA: John Wiley and Sons.

Sumner, W. G. 1963. "Sociology." In *Social Darwinism: Selected Essays of William Graham Sumner,* 9–29. Englewood Cliffs, NJ: Prentice Hall.

Swartz, D. 2003. "From Correspondence to Contradiction and Change: Schooling in Capitalist America Revisited." *Sociological Forum* 18 (1): 167–86.

Swartz, T. 2009. "Intergenerational Family Relations in Adulthood: Patterns, Variations and Implications in the Contemporary United States." *Annual Review of Sociology* 35: 191–212.

Sweeney, J., and M. R. Bradbard. 1988. "Mothers' and Fathers' Changing Perceptions of Their Male and Female Infants Over the Course of Pregnancy." *Journal of Genetic Psychology* 149: 393–404.

Swidler, A. 2003. *Talk of Love: How Culture Matters.* Chicago: University of Chicago Press.

Sykes, G., and D. Matza. 1957. "Techniques of Neutralization: A Theory of Delinquency." *American Sociological Review* 22: 664–70.

Tach, L., and G. Farkas. 2006. "Learning-Related Behaviors, Cognitive Skills, and Ability Grouping When Schooling Begins." *Social Science Research* 35 (4): 1048–79.

Tait, C. 2006. "Namesakes and Nicknames: Naming Practices in Early Modern Ireland." *Continuity and Change* 21 (2): 313–40.

Tajfel, H. 1982. "Social Psychology of Intergroup Relations." *Annual Review of Psychology* 33: 1–39.

Takeuchi, S. A. 2006. "On the Matching Phenomenon in Courtship: A Probability Matching Theory of Mate Selection." *Marriage & Family Review* 40 (1): 25–51.

Taleporos, G., and M. P. McCabe. 2002. "The Impact of Sexual Esteem, Body Esteem and Sexual Satisfaction on Psychological Well-Being in People With Physical Disability." *Sexuality and Disability* 20 (3): 177–183.

Tang, L. 2010. "Development of Online Friendship in Different Social Spaces: A Case Study." *Information, Communication and Society* 13 (4): 615–33.

Tanur, J. 1992. *Questions About Questions*. New York: Russell Sage Foundation.

Taylor, T. L. 2002. "Living Digitally: Embodiments in Virtual Worlds." In *The Social Life of Avatars: Presence and Interactions in Shared Virtual Environments,* edited by R. Schroeder, 40–62. London: Springer-Verlag.

Teachout, T. 2002. "Is Tony Soprano Today's Ward Cleaver?" *New York Times* (September 15).http://www.nytimes .com/2002/09/15/weekinreview/ideas-trends-is-tony-soprano-today-s-ward-cleaver.html.

Teichner, G., E. Ames, and P. Kerig. 1997. "The Relation of Infant Crying and the Sex of the Infant to Parents' Perceptions of the Infant and Themselves." *Psychology—A Quarterly Journal of Human Behavior* 34 (3–4): 59–60.

Teigen, K. H. 1986. "Old Truths or Fresh Insights? A Study of Students' Evaluations of Proverbs." *Journal of British Social Psychology* 25 (1): 43–50.

TenBensel, R., M. Rheinberger, and S. Radbill. 1997. "Children in a World of Violence: The Roots of Child Maltreatment." In *The Battered Child,* edited by M. Helfer, R. Kempe, and R. Krugman, 3–28. Chicago: University of Chicago Press.

Tenenbaum, H., and M. Ruck. 2007. "Are Teachers' Expectations Different for Racial Minority Than for European American Students? A Meta-Analysis." *Journal of Educational Psychology* 99 (2): 253–73.

Tennen, H., and Affleck, G. 1999. "Finding Benefits in Adversity." In *Coping: The Psychology of What Works,* edited by C. R. Snyder, 279–304. New York: Oxford University Press.

Terhune, C., and K. Epstein. 2009. "The Health Insurers Have Already Won." *Bloomberg BusinessWeek.* http:// www.businessweek.com/magazine/content/09_33/b4143034820260.htm?campaign_id=magazine_related.

Tews, M. J., K. Stafford, and J. Zhu. 2009. "Beauty Revisited: The Impact of Attractiveness, Ability, and Personality in the Assessment of Employment Suitability." *International Journal of Selection and Assessment* 17 (1): 92–100.

Thelwall, M. 2009. "Homopholy in MySpace." *Journal of the American Society for Information Science* 60 (2): 219–31.

Thernstrom, M. 2005. "The New Arranged Marriage." *New York Times Magazine* (December 13). http://www .nytimes.com/2005/02/13/magazine/13MATCHMAKING.html.

Thio, A. 2009. *Deviant Behavior,* 10th ed. Boston: Allyn & Bacon.

Thoits, P. 1983. "Dimensions of Life Events That Influence Psychological Distress: An Evaluation and Synthesis of the Literature." In *Psychosocial Stress: Trends in Theory and Research,* edited by H. Kaplan, 33–103. New York: Academic Press.

———. 1994. "Stressors and Problem-Solving: The Individual as Psychological Activist." *Journal of Health and Social Behavior* 35 (1): 143–60.

———. 1995. "Stress, Coping, and Social Support Processes: Where Are We? What Next?" *Journal of Health and Social Behavior* 36 (extra issue): 53–79.

———. 2006. "Personal Agency in the Stress Process." *Journal of Health and Social Behavior* 47 (4): 309–23.

Thompson, J. K., and E. Stice. 2001. "Thin-Ideal Internalization: Mounting Evidence for a New Risk Factor for Body-Image Disturbance and Eating Pathology." *Current Directions in Psychological Science* 10 (5): 181–83.

Thompson, R. 2006. "Bilingual, Bicultural, and Binominal Identities: Personal Name Investment and the Imagination in the Lives of Korean-Americans." *Journal of Language, Identity and Education* 5 (3): 179–208.

Thompson, S. H., R. G. Sargent, and K. A. Kemper. 1996. "Black and White Adolescent Males' Perceptions of Ideal Body Type." *Sex Roles* 34 (March): 391–406.

Thomsen, S. R., M. M. Weber, and L. Beth-Brown. 2002. "The Relationship Between Reading Beauty and Fashion Magazines and the Use of Pathogenic Dieting Methods Among Adolescent Females." *Adolescence* 37 (145): 1–18.

Thorne, B. 1995. "Girls and Boys Together . . . But Mostly Apart: Gender Arrangements in Elementary School." In *Sociology: Exploring the Architecture of Everyday Life*, edited by D. M. Newman, 93–102. Thousand Oaks, CA: Pine Forge.

Thumma, S., and W. Bird. 2009. "Not Who You Think They Are: A Profile of the People Who Attend America's Megachurches." http://hirr.hartsem.edu/megachurch/megachurch_attender_report.htm.

Tierney, K., C. Bevc, and E. Kuligowsky. 2006. "Metaphors Matter: Disaster Myths, Media Frames, and Their Consequences in Hurricane Katrina." *Annals of the American Academy of Political and Social Science* 604: 57–81.

Time-Life Books. 1988. *This Fabulous Century 1920–1930.* New York: Time-Life Books.

Tisdell, C., C. Wilson, and H. S. Nantha. 2005. "Association of Public Support for Survival of Wildlife Species With Their Likeability." *Anthrozoos: A Multidisciplinary Journal of The Interactions of People & Animals* 18 (2): 160–74.

Tjosvold, D. 2008. "Is the Way You Resolve Conflicts Related to Your Psychological Health? An Empirical Investigation." *Peace and Conflict: Journal of Peace Psychology,* 14 (4): 395–428.

Torff, B. 2008. "Using the Critical Thinking Belief Appraisal to Assess the Rigor Gap." *Learning Inquiry* 2 (1): 29–52.

Tracy, B. 2010. "Housing Crisis Getting Uglier in 2010: Nearly 6 Million Foreclosures in Past 3 Years—3 Million More Expected in 2010." CBSNews.com (February 2). http://www.cbsnews.com/stories/2010/02/02/eveningnews/main6167610.shtml.

Treas, J., and E. de Ruijter. 2008. "Earnings and Expenditures on Household Services in Married and Cohabiting Unions." *Journal of Marriage and Family* 70 (3): 796–805.

Trebay, G. 2003. "From Woof to Warp." *New York Times* (April 6): Section 9:1.

Treharne, G. J., A. C. Lyons, and R. E. Tupling. 2001. "The Effects of Optimism, Pessimism, Social Support, and Mood on Lagged Relationship Between Daily Stress and Symptoms." *Current Research in Social Psychology* 7 (5): 60–81.

Tropp, L. 2003. "The Psychological Impact of Prejudice: Implications for Intergroup Contact." *Group Processes & Intergroup Relations* 6 (2): 131–49.

Tuggle, J., and M. Holmes. 1997. "Blowing Smoke Status Politics and the Smoking Ban." *Deviant Behavior* 18: 1.

Tumin, M. 1967. *Social Stratification: The Forms and Functions of Inequality.* Englewood Cliffs, NJ: Prentice Hall.

Turkel, G. 2002. "Sudden Solidarity and the Rush to Normalization: Toward an Alternative Approach." *Sociological Focus* 35 (1): 73–79.

Turkle, S. 1996. *Life on the Screen.* New York: Simon and Schuster.

———. 1997. "Multiple Subjectivity and Virtual Community in the End of the Freudian Century." *Sociological Inquiry* 67 (1): 72–84.

———. 2007. *Evocative Objects: Things We Think With.* Cambridge, MA: MIT Press.

Turner, M. A., and L. Rawlings. 2009. "Promoting Neighborhood Diversity: Benefits, Barriers, and Strategies." The Urban Institute. http://www.urban.org/UploadedPDF/411955promotingneighborhooddiversity.pdf.

Twenge, J. M., and M. Manis. 1998. "First-Name Desirability and Adjustment: Self-Satisfaction, Others' Ratings, and Family Background." *Journal of Applied Social Psychology* 28 (1): 41–51.

Umberson, D. 1996. "Relations Between Adult Children and Their Parents: Psychological Well-Being." *Journal of Marriage and Family* 51: 999–1012.

UnderGodProCon.org. 2006. http://www.undergodprocon.org/pop/PledgeHistory.htm.

UNESCO. 2003. "Education in a Multilingual World." UNESCO Education Position Paper. http://unesdoc.unesco.org/images/0012/001297/129728e.pdf.

UNICEF. 2009. "The State of the World's Children: Special Edition. Table 3. Health." http://www.unicef.org/rightsite/sowc/pdfs/statistics/SOWC_Spec_Ed_CRC_TABLE%203.%20HEALTH_EN_111309.pdf.

———. 2010. "Child Info: Monitoring the Situation of Women and Children." http://www.childinfo.org/low_birthweight_table.php.

Union of Concerned Scientists. 2009a. "Global Climate Change Impacts in the United States." http://www.ucsusa.org/global_warming/science_and_impacts/impacts/global-climate-change.html.

———. 2009b. "What Do Tropical Forests Have to Do With Global Warming?" http://www.ucsusa.org/global_warming/solutions/forest_solutions/tropical-deforestation-and.html.

———. 2010a. "Global Thermometer Still Climbing." http://www.ucsusa.org/global_warming/science_and_impacts/science/global-thermometer-still-climbing.html.

———. 2010b. "How Certain Is Climate Science?" http://www.ucsusa.org/assets/documents/global_warming/how-certain-is-climate-science.pdf.

United Nations. 2005. "Fostering Economic Policy Coordination in Latin America: The REDIMA Approach to Escaping the Prisoner's Dilemma." New York: United Nations.

———. 2007. *Demographic Year Book.* http://unstats.un.org/unsd/demographic/products/dyb/dyb2007/Table21.xls.

Urban Institute. 1999. "Do We Need a National Report Card on Discrimination?" http://www.urban.org/url.cfm?ID=900310.

U.S. Bureau of Labor Statistics. 2009. "May 2009 National Occupational Employment and Wage Estimates United States." http://www.bls.gov/oes/current/oes_nat.htm#00–0000.

U.S. Census Bureau. 2002. Table F-1. :Race and Hispanic Origin, for the United States and Historical Sections and Subsections of the United States: 1790 to 1990—Con." http://www.census.gov/population/www/documentation/twps0056/appF.xls.

———. 2007. "Current Population Survey: Annual Social and Economic (ASEC) Supplement." http://pubdb3.census.gov/macro/032007/perinc/new02_000.htm.

———. 2008. "American Fact Finder. United States Selected Social Characteristics in the United States: 2008." http://factfinder.census.gov/servlet/ADPTable?_bm=y&-geo_id=01000US&-ds_name=ACS_2008_1YR_G00_&-_lang=en&-_caller=geoselect&-format.

———. 2009a. "America's Families and Living Arrangements: 2009." http://www.census.gov/population/www/socdemo/hh-fam/cps2009.html.

———. 2009b. "Current Population Survey, Annual Social and Economic (ASEC) Supplement: 2008 Poverty Table of Contents." http://www.census.gov/hhes/www/cpstables/032009/pov/toc.htm.

———. 2009c. Current Population Survey. 2009 Annual Social and Economic (ASEC) Supplement. Person Income Table of Contents. http://www.census.gov/hhes/www/cpstables/032009/perinc/toc.htm

———. 2009d. "Current Population Survey, Annual Social and Economic (ASEC) Supplement: Health Insurance Coverage Table of Contents." http://www.census.gov/hhes/www/cpstables/032009/health/toc.htm.

———. 2009e. "Current Population Survey, Table H1: Households by Type and Tenure of Householder for Selected Characteristics: 2009." http://www.census.gov/population/socdemo/hh-fam/cps2009/tabH1-all.xls.

———. 2010a. "State & County QuickFacts." http://quickfacts.census.gov/qfd/states/23000.html.

———. 2010b. "Census Bureau Reports Families with Children Increasingly Face Unemployment." http://www.census.gov/newsroom/releases/archives/families_households/cb10–08.html.

———. 2010c. "Poverty: Poverty Thresholds 2009." http://www.census.gov/hhes/www/poverty/data/threshld/thresh09.html.

———. 2010d. "Poverty: Overview/Highlights." http://www.census.gov/hhes/www/poverty/about/overview/index.html.

———. 2010e. "Facts for Features. Hispanic Heritage Month 2010: September 15–October 15." http://www.census.gov/newsroom/releases/archives/facts_for_features_special_editions/cb10-ff17.html.

———. 2011. *The 2011 Statistical Abstracts: The National Data Book.* http://www.census.gov/compendia/statab/2011/tables/11s0689.pdf

U.S. Conference of Catholic Bishops. 2009. "Poverty by Age." http://www.usccb.org/cchd/povertyusa/povfacts_age.shtml.

U.S. Courts. 2010. "Bankruptcy Statistics: Filings." http://www.uscourts.gov/Statistics/BankruptcyStatistics.aspx.

U.S. Department of Education. National Center for Education Statistics,1997. *The Condition of Education 1997* (NCES 97–388). T. Smith, B. Aronstamm, B. Young, Y. Bae, S. Choy, and N. Alsalam. Washington, DC: U.S. Government Printing Office.

———. 2005. *Teacher Follow-up Survey* (TFS). National Center for Education Statistics. http://nces.ed.gov/surveys/sass/tables_list.asp#2005.

———. 2006. *The Condition of Education 2006.* National Center for Educational Statistics (NCES 2006–071). Washington, DC: U.S. Government Printing Office.

———. 2007. *The Condition of Education 2007.* National Center for Educational Statistics. Washington, DC: U.S. Government Printing Office. http://nces.ed.gov/programs/coe/list/index.asp.

———. 2008. *Digest of Education Statistics: Table 110.* Institute of Education Sciences. http://nces.ed.gov/programs/digest/d08/tables/dt08_110.asp.

———. 2009. *Digest of Education Statistics: 2009.* http://nces.ed.gov/programs/digest/d09/index.asp.

U.S. Department of Health and Human Services. 2008. "Employment, Economic Security and Father Involvement." Responsible Fatherhood Spotlight. http://library.fatherhood.gov/cwig/ws/library/docs/FATHERHD/Blob/65036.pdf?w=NATIVE%28%27TI+ph+is+%27%27Employment%27%27%2C%27%27Economic+Security+and+Father+Involvement%27%27+AND+YEAR+%3D+2008%27%29&upp=0&rpp=25&order=native%28%27year%2FDescend%27%29&r=1&m=1.

———. 2009a. "Health, United States 2009." http://www.cdc.gov/nchs/data/hus/hus09.pdf.

———. 2009b. " Indicators of Welfare Dependence: Annual Report to Congress 2008." http://aspe.hhs.gov/hsp/indicators08/index.shtml.

———. 2010. Child Maltreatment 2008. http://www.acf.hhs.gov/programs/cb/pubs/cm08/cm08.pdf.

U.S. Department of Homeland Security. 2008. *Yearbook of Immigration Statistics.* Washington, DC: Office of Immigration Statistics. See also http://www.dhs.gov/xlibrary/assets/statistics/yearbook/2008/ois_yb_2008.pdf.

———. 2010. *Yearbook of Immigration Statistics: 2009.* Washington, DC: Office of Immigration Statistics. http://www.dhs.gov/xlibrary/assets/statistics/yearbook/2009/ois_yb_2009.pdf.

U.S. Department of Housing and Urban Development. 2005. "Homeownership Gaps among Low-Income and Minority Borrowers and Neighborhoods." Washington, DC: Office of Policy Development and Research. http://www.huduser.org/publications/pdf/HomeownershipGapsAmongLow-IncomeAndMinority.pdf#xml.

U.S. Department of Justice. 2008a. "Violent Crime: Crime in the United States." http://www2.fbi.gov/ucr/cius2008/offenses/violent_crime/index.html.

———. 2008b. "Criminal Victimization, 2008." http://bjs.ojp.usdoj.gov/content/pub/pdf/cv08.pdf.

———. 2008c. "Violent Crime Rate Remained Unchanged While Theft Rate Decline in 2008." http://bjs.ojp.usdoj.gov/content/pub/press/cv08pr.cfm#.

———. 2009a. "2008 Crime in the United States." http://www2.fbi.gov/ucr/cius2008/index.html.

———. 2009b. "Criminal Victimization, 2008." http://pacelawlibrary.blogspot.com/2009/09/criminal-victimization-2008.html.

———. 2010. "Victim Characteristics." http://bjs.ojp.usdoj.gov/index.cfm?ty=tp&tid=92.

U.S. Department of Labor. 2006. *Women in the Labor Force: A Databook.* http://www.bls.gov/cps/wlf-databook2006.htm.

———. 2009a. "American Time Use Survey, 2008." *http://www.bls.gov/news.release/archives/atus_06242009.pdf.*

———. 2009b. "Quick Facts on Nontraditional Occupations for Women." http://www.dol.gov/wb/factsheets/nontra2008.htm.

———. 2009c. *Women in the Labor Force: A Databook (2009 edition).* http://www.bls.gov/cps/wlftable16.htm.

———. 2010a. "Current Population Survey, Characteristics of the Employed—Table 11." http://www.bls.gov/cps/tables.htm#charemp.

———. 2010b. "Labor Force Statistics From the Current Population Survey." http://www.bls.gov/cps/demographics.htm.

———2010c. *Occupational Outlook Handbook, 2010–11 Edition.* http://www.bls.gov/oco/ocos114.htm#outlook.

———. 2010d. "Annual Report: A Profile of the Working Poor, 2008." Bureau of Labor Statistics. http://www.bls.gov/cps/earnings.htm#workpoor.

———. 2010e. "Current Population Survey, Characteristics of the Employed—Table 10." http://www.bls.gov/cps/cpsaat10.pdf.

———. 2010f. "Overview of the 2008–18 Projections." *Occupational Outlook Handbook, 2010–11.* http://www.bls.gov/oco/oco2003.htm.

———. 2011a. "The Employment Situation—February 2011" Bureau of Labor Statistics. http://www.bls.gov/news.release/empsit.nr0.htm.

———. 2011b. "International Comparisons of Hourly Compensation Costs in Manufacturing, 200–9." http://www.bls.gov/news.release/pdf/ichcc.pdf.

U.S. Equal Employment Opportunity Commission. 2009. "Race-Based Charges FY 1997–FY 2010." http://www.eeoc.gov/eeoc/statistics/enforcement/race.cfm.

U.S. Global Change Research Program. 2009. *Global Climate Change Impacts in the United States,* edited by T. R. Karl, J. M. Melillo, and T. C. Peterson. New York: Cambridge University Press. http://www.globalchange.gov/publications/reports/scientific-assessments/us-impacts.

U.S. Social Security Administration. 2009. "Fast Facts & Figures about Social Security, 2009." http://www.ssa.gov/policy/docs/chartbooks/fast_facts/2009/fast_facts09.html#highlights.

Valdesolo, P., J. Ouyang, and D. DeSteno. 2010. "The Rhythm of Joint Action: Synchrony Promotes Cooperative Ability." *Journal of Experimental Social Psychology* 46 (4): 693–95.

Vallas, R. 2008. "Of (Public Interest): UVA and the Need for Legal Aid Lawyers." *Virginia Law Weekly* 62 (3). http://www.lawweekly.org/?module=displaystory&story_id=2148&edition_id=96&format=html.

Vallas, S., W. Finley, and A. S. Wharton. 2009. *Sociology of Work: Structures and Inequalities.* New York: Oxford University Press.

Van Avermaet, E., H. Buelens, N. Vanbeselaere, and G. Van Vaerenbergh. 1999. "Intragroup Social Influence Processes in Intergroup Behavior." *European Journal of Social Psychology* 29 (5–6): 815–23.

Van den Broeck, A., N. De Cuyoer, H. De Witte, and M. Vansteenkiste. 2010. "Not All Job Demands Are Equal: Differentiating Job Hindrances and Job Challenges in the Job Demands-Resources Model." *European Journal of Work and Organizational Psychology* 20 (1): 1–25.

Van den Buick, J. 2000. "Is Television Bad for Your Health? Behavior and Body Image of the Adolescent 'Couch Potato.'" *Journal of Youth and Adolescence* 29 (3): 273–88.

VanderHart, P. 2006. "Why Do Some Schools Group by Ability?" *The American Journal of Economics and Sociology* 65 (2): 435–62.

Vanman, E. J., B. Y. Paul, and T. A. Ito. 1997. "The Modern Face of Prejudice and the Structural Features That Moderate the Effect of Cooperation on Affect." *Journal of Personality and Social Psychology* 73 (November): 941–59.

Van Overwalle, F. 1997. "Dispositional Attributions Require the Joint Applications of the Methods of Difference and Agreement." *Personality and Social Psychology Bulletin* 23: 974–80.

Vargas, J. A. 2008. "Obama Raised Half a Billion Online." *The Washington Post* (November 20). http://voices.washingtonpost.com/44/2008/11/obama-raised-half-a-billion-on.html.

Vartanian, L. R., C. L. Giant, and R. M. Passino. 2001. "Ally McBeal vs. Arnold Schwarzeneggar: Comparing Mass Media, Interpersonal Feedback and Gender as Predictors of Personal Satisfaction with Body Thinness and Muscularity." *Social Behavior and Personality* 29 (7): 2001.

Verhaeghe, R., P. Vlerick, P. Gemmel, G. Van Maele, and G. De Backer. 2006. "Impact of Recurrent Changes in the Work Environment on Nurses' Psychological Well-Being and Sickness Absence." *Journal of Advanced Nursing* 56 (6): 646–56.

Veroff, J., E. Douvan, and R. Kulka. 1981. *The Inner American: A Self-Portrait From 1957 to 1976.* New York: Basic Books.

Vinorskis, M. 1992. "Schooling and Poor Children in 19th Century America." *American Behavioral Scientist* 35 (3): 313–31.

Visher, E., J. Visher, and K. Pasley. 2003. "Remarriage Families and Stepparenting." In *Normal Family Processes: Growing Diversity and Complexity,* 3rd ed., edited by F. Walsh, 153–75. New York: Guilford.

Vogel, C. 2006. "A Pollock Is Sold, Possibly for a Record Price." *New York Times* (November 2). http://www.nytimes.com/2006/11/02/arts/design/02drip.html?ei=5090 &en=53ef078e6646b854&ex=1320123600&adxnnl=1&partner=rssuserland& emc=rss&adxnnlx=1162883074-hD7+iTcDZc6SWlkvwJtuPQ.

Volunteermatch.org. 2010. "VolunteerMatch by the Numbers." http://www.volunteermatch.org/volunteers/resources/newsletter.jsp?id=178.

Von Ah, D., D. Kang, and J. S. Carpenter. 2007. "Stress, Optimism, and Social Support: Impact on Immune Responses in Breast Cancer." *Research in Nursing and Health* 30 (1): 72–83.

Voss, K., D. Markiewicz, and A. B. Doyle. 1999. "Friendship, Marriage, and Self-Esteem." *Journal of Social and Personal Relationships* 16 (1): 103–22.

Vrij, A., and H. R. Firmin. 2002. "Beautiful Thus Innocent? The Impact of Defendants' and Victims' Physical Attractiveness and Participants' Rape Beliefs on Impression Formation in Alleged Rape Cases." *International Review of Victimology* 8 (3): 245–55.

Walker, K. 1995. "Always There for Me: Friendship Patterns and Expectations Among Middle and Working Class Men and Women." *Sociological Forum* 10 (2): 273–96.

Wang, F., T. C. Wild, W. Kipp, S. Kuhle, and P. J. Veugelers. 2009. "The Influence of Childhood Obesity on the Development of Self Esteem," *Health Reports* 20 (2): 21–27.

Wang, H., and B. Wellman. 2010. "Social Connectivity in America: Changes in Adult Friendship Network Size From 2002 to 2007." American Behavioral Scientist 53 (8): 1148–69.

Wang, M. K. 2002. "The Ancient Foundations of Modern Nation-Building in China: The Case of the Offspring of Yan and Yellow Emperors." *Bulletin of the Institute of History and Philology Academia Sinica* 73 (3): 583–624.

Wansink, B., S. Seymour, and N. M. Bradburn. 2004. *Asking Questions: The Definitive Guide to Questionnaire Design—For Market Research, Political Polls, and Social and Health Questionnaires.* San Francisco: Jossey-Bass.

Ward, L. M., E. Hansbrough, and E. Walker. 2005. "Contributions of Music Video Exposure to Black Adolescents' Gender and Sexual Schemas." *Journal of Adolescent Research* 20 (2): 143–66.

Wartik, N. 2002."Hurting More, Helped Less?" *New York Times* (June 23). http://query.nytimes.com/gst/fullpage.html?res=980CE6DD133CF930A15755C0A9649C8B63.

Waskul, D. D., and P. van der Riet. 2002. "The Abject Embodiment of Cancer Patients: Dignity, Selfhood, and the Grotesque Body." *Symbolic Interaction* 25 (4): 487–513.

Waters, M., and T. Jimenez. 2005. "Assessing Immigrant Assimilation: New Empirical and Theoretical Challenges." *Annual Review of Sociology* 31: 105–25.

Waters, M., R. Ueda, and H. Marrow. 2007. *The New Americans: A Guide to Immigration Since 1965.* Cambridge, MA: Harvard University Press.

Watkins, L. M., and L. Johnston. 2000. "Screening Job Applicants: The Impact of Physical Attractiveness and Application Quality." *International Journal of Selection and Assessment* 8 (2): 76–84.

Watson, D., B. Hubbard, and D. Wiese. 2000. "Self-Other Agreement in Personality and Affectivity: The Role of Acquaintanceship, Trait Visibility, and Assumed Similarity." *Journal of Personality and Social Psychology* 78: 546–58.

Watson, W. H., and R. J. Maxwell, eds. 1977. *Human Aging and Dying: A Study in Sociocultural Gerontology.* New York: St. Martin's.

Wattenberg, L. 2005. *The Baby Name Wizard: A Magical Method for Finding the Perfect Name for Your Baby.* New York: Broadway.

Weber, M. (1922) 1968. *Economy and Society.* New York: Bedminster.

Weisberg, K. 1975. "'Under Great Temptations Here': Women and Divorce in Puritan Massachusetts." *Feminist Studies* 2 (2/3): 183–93.

Weiss, M. 1998. "Parents' Rejection of Their Appearance-Impaired Newborns: Some Critical Observations Regarding the Social Myth of Bonding." *Marriage and Family Review* 27 (3–4): 191–209.

Weiss, R. 2005. *The Experience of Retirement.* Ithaca, NY: Cornell University Press.

Wellman, B. 2004. "Connecting Communities: On and Offline." *Contexts* 3 (4): 22–28.

Wells, R. 1982. *Revolutions in Americans' Lives: A Demographic Perspective on the History of Americans, Their Families and Their Society.* Westport, CT: Greenwood.

Wertheimer, B. 1977. *We Were There: The Story of Working Women in America.* New York: Pantheon.

West, A., J. Lewis, and P. Currie. 2009. "Students' Facebook 'Friends': Public and Private Spheres." *Journal of Youth Studies* 12 (6): 615–27.

Wetzel, D. 2007. "Products of the System." Yahoo! Sports (September 11). http://sports.yahoo.com/nfl/news?slug=dw-productsofthesystem091107&prov=yhoo&type=lgns.

Wheaton, B. 1982. "A Comparison of the Moderating Effects of Personal Coping Resources on the Impact of the Exposure to Stress in Two Groups." *Journal of Community Psychology* 10: 293–311.

———. 1983. "Stress, Personal Coping, Resources, and Psychiatric Symptoms: An Investigation of Interactive Models." *Journal of Health and Social Behavior* 24 (3): 208–29.

———. 1990. "Life Transitions, Role Histories, and Mental Health." *American Sociological Review* 55 (2): 209–23.

White, M. A., J. R. Kohlmaier, P. Varnado-Sullivan, and D. A. Williamson. 2003. "Racial/Ethnic Differences in Weight Concerns: Protective and Risk Factors for the Development of Eating Disorders and Obesity Among Adolescent Females." *Eating and Weight Disorders* 8 (1): 20–25.

Whitehead, B. 1993."Dan Quayle Was Right." *Atlantic Monthly* (April): 47–84.

Whitehead, B. D., and D. Popenoe 2008. "Life Without Children: The Social Retreat From Children and How It Is Changing America." The National Marriage Project. http://www.virginia.edu/marriageproject/pdfs/2008LifeWithoutChildren.pdf.

Wichman, H. 1970. "Effects of Isolation and Communication on Cooperation in a Two-Person Game." *Journal of Personality and Social Psychology* 16: 114–20.

Wight, V. R., and M. Chau. 2009. "Basic Facts About Low-Income Children, 2008: Children Under Age 18." National Center for Children in Poverty. http://www.nccp.org/publications/pub_892.html.

Wight, V. R., M. Chau, and Y. Aratani. 2010. "Who Are America's Poor Children? The Official Story." National Center for Children in Poverty. http://www.nccp.org/publications/pub_912.html.

Wilcox, W. B. (ed.). 2009. *The State of Our Unions: 2009.* http://www.virginia.edu/marriageproject/pdfs/Union_11_25_09.pdf.

Wilder, D. A., and P. N. Shapiro. 1984. "Role of Outgroup Cues in Determining Social Identity." *Journal of Personality and Social Psychology* 47: 342–48.

Wilkinson, I., and L. Young. (2002). "On Cooperating: Firms, Relations, and Networks." *Journal of Business Research* 55 (2): 123–32.

Wilkinson, P. J., and P. G. Coleman. 2010. "Strong Beliefs and Coping in Old Age: A Case-Based Comparison of Atheism and Religious Faith." *Ageing and Society* 30 (2): 337–61.

Will, J. A., P. A. Self, and N. Dalton. 1976. "Maternal Behavior and Perceived Sex of Infant." *American Journal of Orthopsychiatry* 49: 135–39.

Williams, D. 2005. "Controversial Issues Coming to the Fore at Bishops' Gathering: Vatican Moves to Limit Information Provided to Public." *Washington Post* (October 6): A18.

Williams, K., and A. Dunne-Bryant. 2006. "Divorce and Adult Psychological Well-Being: Clarifying the Role of Gender and Child Age." *Journal of Marriage & Family* 68 (5): 1178–96.

Williams, R. M., Jr. 1970. *American Society: A Sociological Interpretation,* 3rd ed. New York: Alfred A. Knopf.

Williams, S. 2001. "Sexual Lying Among College Students in Close and Casual Relationships." *Journal of Applied Social Psychology* 31 (November): 2322–2338.

Williamson, D. A., N. L. Zucker, C. K. Martin, and M. A. M. Smeets. 2001. "Etiology and Management of Eating Disorders." In *Comprehensive Handbook of Psychopathology,* edited by P. B. Sutker and H. E. Adams, 187–213. New York: Kluwer Academic/Plenum.

Willis, F. N., L. A. Willis, and J. A. Grier. 1982. "Given Names, Social Class, and Professional Achievement." *Psychological Reports* 54: 543–49.

Wilson, D. M. 2010. "The Casualties of the Twenty-First-Century Community College." *Academe* 96 (3): 12–18.

Wilson, J. 2000. "Volunteering." *Annual Review of Sociology* 26: 215–40.

Winneg, K., K. Kenski, and K. H. Jamieson. 2005. "Detecting the Effects of Deceptive Presidential Advertisements in the Spring of 2004." *American Behavioral Scientist* 49 (1): 114–29.

Winterfield, L., M. Coggeshall, M. Burke-Storer, V. Correa, and S. Tidd. 2009. "The Effects of Postsecondary Correctional Education, Final Report." The Urban Institute. http://www.urban.org/UploadedPDF/411952_PSE_FINAL_5_29_09_webedited.pdf.

Wiseman, C. V., J. J. Gray, J. E. Mosimann, and A. H. Ahrens. 1992. "Cultural Expectations of Thinness in Women: An Update." *International Journal of Eating Disorders* 11 (1): 85–89.

Wisman, J. D. 2000. "Competition, Cooperation, and the Future of Work." *Peace Review* 12 (2): 197–203.

Witkin-Lanoil, G. 1984. *The Female Stress Syndrome: How to Recognize and Live With It.* New York: Newmarket.

Wolf, N. 1991. *The Beauty Myth: How Images of Beauty Are Used Against Women.* New York: W. Morrow.

Wolff, E. N. 2010. "Recent Trends in Household Wealth in the United States: Rising Debt and the Middle-Class Squeeze—An Update to 2007." Working Paper No. 589. Levy Economics Institute of Bard College. http://www.levyinstitute.org/publications/?docid=1235.

Wong, F. Y., D. R. McCreary, C. C. Bowden, and S. M. Jenner. 1991. "The Matching Hypothesis: Factors Influencing Dating Preferences." *Psychology* 28 (3–4): 27–31.

Wong, J. 2002. "What's in a Name?: An Examination of Social Identities." *Journal for the Theory of Social Behavior* 32 (4): 451–64.

Wood, D., R. Kaplan, and V. McLoyd. 2007. "Gender Differences in the Educational Expectations of Urban, Low-Income African American Youth: The Role of Parents and the School." *Journal of Youth and Adolescence* 36 (4): 417–27.

World Almanac and Book of Facts (2010). 2010. New York: World Almanac Books.

World Bank. 2010. "Education and Development." http://web.worldbank.org/WBSITE/EXTERNAL/TOPICS/EXTEDUCATION/0,,contentMDK:20591648~menuPK:1463858~pagePK:148956~piPK:216618~theSitePK:282386,00.html.

World Health Organization. 2010a. "Poverty and Health." http://www.who.int/hdp/poverty/en/.

———. 2010b *World Heath Statistics 2010.* Paris: World Health Organization. http://www.who.int/whosis/whostat/EN_WHS10_Full.pdf.

———. 2010c. "Global Health Observatory: Health Workforce." http://apps.who.int/ghodata/.

Wu, Z., and L. MacNeill. 2002. "Education, Work, and Childbearing After Age 30." *Journal of Comparative Family Studies* 33 (2): 191–213.

Yamamoto, R., D. Ariely, W. Chi, D. D. Langelben, and I. Elman. 2009. "Gender Differences in the Motivational Processing of Babies Are Determined by Their Facial Features." *PLoS One* 4 (6): e6042.

Yao, M. Z., C. Mahood, and D. Linz, D. 2010. "Sexual Priming, Gender Stereotyping, and Likelihood to Sexually Harass: Examining the Cognitive Effects of Playing a Sexually Explicit Video Game." *Sex Roles* 62 (1–2): 77–88.

Yerkes, M. J., and T. F. Pettijohn II. 2008. "Developmental Stability of Perceived Physical Attractiveness From Infancy to Young Adulthood." *Social Behavior and Personality* 36 (5): 691–92.

Young, R. K., A. H. Kennedy, A. Newhouse, P. Browne, and D. Thiessen. 2006. "The Effects of Names on Perception of Intelligence, Popularity, and Competence." *Journal of Applied Social Psychology* 23 (21): 1770–88.

Zaidel, D. W., S. Bava, and V. A. Reis. 2003. "Relationship Between Facial Asymmetry and Judging Trustworthiness in Faces." *Laterality* 8 (3): 225–232.

Zebrowitz, L. A., M. A. Collins, and R. Dutta. 1998. "The Relationship Between Appearance and Personality Across the Life Span." *Personality and Social Psychology Bulletin* 24 (7): 736–49.

Zeitlin, M., K. G. Lutterman, and J. W. Russell, 1977. "Death in Vietnam: Class, Poverty, and the Risks of War." In *American Society Incorporated,* 2nd ed., edited by M. Zeitlin, 143–55. Chicago: Rand McNally.

Zelizer, V. 1985. *Pricing the Priceless Child.* New York: Basic Books.

Zernike, K. 2006. "The Remarrying Kind." *New York Times* (January 29). http://www.nytimes.com/2006/01/29/weekinreview/29zernike.html.

Ziegler A., and H. Stoeger. 2008. "Effects of Role Models from Films on Short-Term Ratings of Intent, Interest, and Self-Assessment of Ability by High School Youth: A Study of Gender-Stereotyped Academic Subjects." *Psychological Reports* 102 (2): 509–31.

Zimmer-Gembeck, M. J., and E. M. Locke. 2007. "The Socialization of Adolescent Coping: Relationships at Home and School." *Journal of Adolescence* 30 (1): 1–16.

Zitner, A. 2003. "Nation's Birthrate Drops to Its Lowest Level Since 1909." *Los Angeles Times* (June 26): 1.

Zuckerman, M., K. Miyake, and C. S. Elkin. 1995. "Effects of Attractiveness and Maturity of Face and Voice on Interpersonal Impressions." *Journal of Research in Personality* 29 (2): 253–72.

Index

About the Authors

Janet M. Ruane (PhD, Rutgers University) is a professor of sociology at Montclair State University. She has served as her department's coordinator of undergraduate advising and as her department's graduate advisor. Professor Ruane's research interests include formal and informal social control mechanisms, media and technology, research methods, and applied sociology. She is the author of *Essentials of Research Methods* (Blackwell) and has contributed chapters to edited volumes and articles to several journals, including *Sociological Forum, Sociological Inquiry, Law and Policy, Communication Research, Sociological Focus, Journal of Applied Sociology,* and *Science as Culture.* Since stepping into her first classroom, Professor Ruane has acquired more than 20 years of valuable teaching experience in both introductory and advanced-level sociology courses. These years have afforded her a wealth of opportunities for generating and fine-tuning many second thoughts.

Karen A. Cerulo (PhD, Princeton University) is a professor of sociology at Rutgers University. Her research interests include culture and cognition, symbolic communication, media and technology, and comparative historical studies. Professor Cerulo's articles appear in a wide variety of journals, including the *American Sociological Review, Contemporary Sociology, Poetics, Social Forces, Sociological Forum, Sociological Inquiry, Communication Research,* and annuals such as the *Annual Review of Sociology* and *Research in Political Sociology.* She is the author of three books: *Identity Designs: The Sights and Sounds of a Nation,* a work that won the ASA Culture Section's award for the Best Book of 1996 (Rose Book Series of the ASA, Rutgers University Press); *Deciphering Violence: The Cognitive Structure of Right and Wrong* (Routledge); and *Never Saw It Coming: Cultural Challenges to Envisioning the Worst* (University of Chicago Press). She has also edited a collection entitled *Culture in Mind: Toward a Sociology of Culture and Cognition* (Routledge). Professor Cerulo's teaching earned her the Rutgers University Award for Distinguished Contributions to Undergraduate Education.